学习 Highcharts 4（影印版）

Joe Kuan 著

南京　东南大学出版社

图书在版编目(CIP)数据

学习 Highcharts 4:英文/(英)宽(Kuan,J.)著.—影印本.—南京:东南大学出版社,2016.1

书名原文:Learning Highcharts 4

ISBN 978-7-5641-6093-7

Ⅰ.①学… Ⅱ.①宽… Ⅲ.①网页制作工具—英文 Ⅳ.①TP393.02

中国版本图书馆 CIP 数据核字(2015)第 256609 号

学习 Highcharts 4(影印版)

出版发行:东南大学出版社
地　　址:南京四牌楼 2 号　　邮编:210096
出 版 人:江建中
网　　址:http://www.seupress.com
电子邮件:press@seupress.com
印　　刷:常州市武进第三印刷有限公司
开　　本:787 毫米×980 毫米　16 开本
印　　张:29.75
字　　数:583 千字
版　　次:2016 年 1 月第 1 版
印　　次:2016 年 1 月第 1 次印刷
书　　号:ISBN 978-7-5641-6093-7
定　　价:88.00 元

本社图书若有印装质量问题,请直接与营销部联系。电话(传真):025-83791830

About the Reviewers

Philip Arad is a web architect and a frontend developer who works mainly on enterprise large scale applications. He has more than 10 years of experience in user interface design and UI infrastructures on the web platform, using mainly Sencha and Highcharts. Nowadays, Philip designs the UI architecture for the SolarEdge monitoring portal and its administration web applications. He is responsible for the integration of technologies, information, and business processes for the Web and the mobile platforms of the company.

You can see his work at `http://monitoring.solaredge.com`.

Zhang Bo is a web developer with 6 years of experience on PHP. He currently works as an R and D Engineer at NSFOCUS (`http://www.nsfocus.com`), leading development work of a new version of the next-generation firewall product. Zhang has a strong background in using Highcharts to visually present network traffic and security. He is also a qualified professional in the web security field. Apart from work, he enjoys spending time on his interests of travel, gourmet, and photography.

Katherine Y. Chuang is a software engineer and data scientist based in New York. She obtained her PhD in interactive design and worked on interactive data visualizations for decision support systems used in clinical care. Now she works for a consulting company that creates software that helps users quickly model large datasets and automated analytics reporting. In her free time, she organizes open source tech events for communities in NYC and scientific conferences.

Acknowledgments

When I wrote my first ever book, it was a draining 9 months to grind through alongside my demanding work and spending little time with my family. A year later, Packt Publishing approached me to write the second edition. I was in two minds about going through the hectic schedule again. Since then, Highcharts has become a more complete and improved product and it would be a closure for me to see the book end the same way. The first edition was tough but the new experience kept me motivated towards the goal. This journey has taught me one thing: what it takes and means to keep up the level of momentum.

I would like to dedicate this book to my eldest sister, Valamie, who is courageously fighting cancer. Although she is small, we call her "big sister." She never complains and always puts the interests of other people, especially our parents, above her own, despite what she is going through. Special thanks to my wife for her support and patience again and my boy, Ivan, for understanding why his dad has been glued to the computer all this time.

I would also like to thank Adrian Raposo from Packt Publishing for putting up with me and injecting new spark into this book.

> *"Be studious in your profession, and you will be learned. Be industrious and frugal, and you will be rich. Be sober and temperate, and you will be healthy. Be in general virtuous, and you will be happy. At least you will, by such conduct, stand the be."*
>
> *– Benjamin Franklin*

About the Author

Joe Kuan was born in Hong Kong and continued his education in the UK from secondary school to university. He studied computer science at the University of Southampton for his BSc and PhD. After finishing his education, he worked with different technologies and industries in the UK. Currently, he is working for iTrinegy—a company specializing in network emulation, profiling, and performance monitoring. He enjoys working on a variety of projects and applying different programming languages. Part of his role is to develop frontend data and present complex network data in stylish and interactive charts. He has adopted Highcharts in his company products since the early version 2. Since then, he has been contributing blogs (`joekuan.wordpress.com`) and software (`joekuan.org`) on Highcharts, Highstocks, and Highmaps. In December 2012, he published his first book, *Learning Highcharts, Packt Publishing*, which is a comprehensive book on Highcharts covering tutorials, examples, tricks, and tips.

Credits

Author
Joe Kuan

Reviewers
Philip Arad
Zhang Bo
Katherine Y. Chuang
Rémi Lepage
Jon Arild Nygård

Commissioning Editor
Kevin Colaco

Acquisition Editor
Kevin Colaco

Content Development Editor
Adrian Raposo

Technical Editor
Tanvi Bhatt

Copy Editors
Safis Editing
Ameesha Green

Project Coordinator
Sanchita Mandal

Proofreaders
Simran Bhogal
Stephen Copestake
Maria Gould
Ameesha Green
Paul Hindle

Indexer
Tejal Soni

Graphics
Abhinash Sahu

Production Coordinator
Aparna Bhagat

Cover Work
Aparna Bhagat

Rémi Lepage is a freelance software engineer and a computer science technical trainer.

He holds a master's degree in computer science from Supinfo and the ZCE (Zend Certified Engineer) PHP 5.3 certification.

Specializing in web application development and database administration, he has created several management intranets. He currently creates enterprise applications.

Jon Arild Nygård is a web developer whose prime focus and interest is frontend development. He especially appreciates all the endless possibilities the JavaScript libraries offer. Currently, he is employed at Highsoft, where he thrives and enjoys the support of his great colleagues. In his spare time, he likes to exercise, work on his all too many personal projects, and spend time with friends and family.

You can follow him at his Twitter profile: `@JonANygaard`.

www.PacktPub.com

Support files, eBooks, discount offers, and more

For support files and downloads related to your book, please visit www.PacktPub.com.

Did you know that Packt offers eBook versions of every book published, with PDF and ePub files available? You can upgrade to the eBook version at www.PacktPub.com and as a print book customer, you are entitled to a discount on the eBook copy. Get in touch with us at service@packtpub.com for more details.

At www.PacktPub.com, you can also read a collection of free technical articles, sign up for a range of free newsletters and receive exclusive discounts and offers on Packt books and eBooks.

https://www2.packtpub.com/books/subscription/packtlib

Do you need instant solutions to your IT questions? PacktLib is Packt's online digital book library. Here, you can search, access, and read Packt's entire library of books.

Why subscribe?

- Fully searchable across every book published by Packt
- Copy and paste, print, and bookmark content
- On demand and accessible via a web browser

Free access for Packt account holders

If you have an account with Packt at www.PacktPub.com, you can use this to access PacktLib today and view 9 entirely free books. Simply use your login credentials for immediate access.

Table of Contents

Preface

Learning Highcharts 4 aims to be the missing manual for Highcharts from every angle. It is written for web developers who would like to learn about Highcharts. This book has the following features:

- It is a step-by-step guide on building presentable charts from basic looking ones
- There are plenty of examples with real data covering all the Highcharts series types—line/spline, column, pie, donut, scatter, bubble, area range, column range, gauge, solid gauge, pyramid, box plot, spider, waterfall, funnel, error bar, heatmaps, and 3D charts
- Subject areas are included that haven't yet been covered in online reference manuals and demos, such as chart layout structure, color shading, and in-depth explanations of some specific options such as sizeBy in the bubble series, groupZPadding in the column series, and how to modify or create plugins
- Applications demonstrating how to create dynamic and interactive charts using Highcharts APIs and events handling are covered
- Applications demonstrating how to integrate Highcharts with a mobile framework such as jQuery Mobile and a Rich Internet Application framework such as Ext JS are also covered
- Applications demonstrating how to run Highcharts on the server side for automating charts generation and exporting their graphical outputs are also covered
- Using the latest online service Highcharts Cloud, you'll learn to embed the graphs into documents
- You'll also learn the structure of Highcharts plugins and how to create plugins

This book is not a reference manual as the Highcharts team has already done an excellent job in providing a comprehensive online reference, and each configuration is coupled with jsFiddle demos. This book also does not aim to be a chart design guide nor a tutorial on programming design with Highcharts.

In short, this book shows you what you can do with Highcharts.

What this book covers

The second edition includes four new chapters, a rewritten chapter, and new sections in some of the existing chapters. All the contents from the previous edition have been technically revised. As a result, the new edition consists of about 50 percent new material.

As this book contains a myriad of examples, it would be impractical to include all the source code of each example. For step-by-step tutorials, the code is listed incrementally. If you want to experiment with the sample code, you are strongly recommended to download the code from the Packt Publishing website or visit `http://joekuan.org/Learning_Highcharts` for online demos.

Chapter 1, Web Charts, describes how web charts have been done since the birth of HTML to the latest HTML5 standard with SVG and canvas technologies. This chapter also provides a short survey of charting software on the market using the HTML5 standard and discusses why Highcharts is a better product than others.

Chapter 2, Highcharts Configurations, covers the common configuration options in chart components with plenty of examples and explains how the chart layout works.

Chapter 3, Line, Area, and Scatter Charts, demonstrates plotting a simple line, area charts, and scatter charts to plotting a poster-like chart including all three series types.

Chapter 4, Bar and Column Charts, demonstrates bar and column charts as well as various derived charts such as a stacked chart, percentage chart, mirror chart, group chart, overlap chart, mirror stacked chart, and horizontal gauge chart.

Chapter 5, Pie Charts, demonstrates how to build various charts, from a simple pie chart to a multiseries chart, such as multiple pies in a chart and a concentric rings pie chart, that is, a donut chart. The chapter also explores how to create an open donut chart with specific options.

Chapter 6, Gauge, Polar, and Range Charts, is a step-by-step guide on constructing a twin dial speedometer and the creation of a simple solid gauge chart. It also demonstrates the polar chart's characteristics and its similarity to a Cartesian chart. It also illustrates the use of range data on area and column range charts.

Chapter 7, Bubble, Box Plot, and Error Bar Charts, explains the characteristics of bubble charts and their specific options. The chapter establishes a gradual approach on creating a bubble chart similar to a real life sport chart, applies the same exercise to a box plot with environmental data, and provides a tutorial on error bar charts using racing data.

Chapter 8, Waterfall, Funnel, Pyramid, and Heatmap Charts, illustrates how to configure waterfall and funnel charts and uses the drilldown feature to link both charts. Then there is a tutorial on pyramid charts and reconstructing them from a financial brochure. Then, heatmap charts are introduced and different outputs are shown by experiencing a number of series options.

Chapter 9, 3D Charts, discusses what 3D charts in Highcharts really means and demonstrates the concept of 3D charts, such as column, pie, and scatter series. It illustrates specific 3D options with experiments and reconstructs a 3D chart from infographics. The chapter also covers how to create interactive 3D orientations with mouse actions.

Chapter 10, Highcharts APIs, explains the usage of Highcharts APIs and illustrates this by using a stock market demo to draw dynamic charts. The chapter discusses the use of different methods to update the series and analyzes the performance of each method on various browsers, as well as the scalability of Highcharts.

Chapter 11, Highcharts Events, explains Highcharts events and demonstrates them through various user interactions with charts from the portfolio application demos.

Chapter 12, Highcharts and jQuery Mobile, is a short tutorial on the jQuery Mobile framework and demonstrates how to integrate it with Highcharts by creating a mobile web application browsing an Olympic medals table. The chapter also covers the use of touch-based and rotate events with Highcharts.

Chapter 13, Highcharts and Ext JS, is a short introduction to Sencha's Ext JS and describes the components likely to be used in an application with Highcharts. It also shows how to use a module and a Highcharts extension in order to plot Highcharts graphs within an Ext JS application.

Chapter 14, Server-side Highcharts, discusses how Highcharts' server-side solution has evolved using PhantomJS. A quick introduction of PhantomJS is given and a step-by-step experiment is conducted to create a server-side solution using Highcharts. The chapter also demonstrates how to use the Highcharts official server-side script.

Chapter 15, Highcharts Online Services and Plugins, is a quick introduction to the export server service and discusses a recent significant cloud development: Highcharts Cloud. The chapter gives you a tour of what it offers and how to use it, from a basic level without any prior knowledge of Highcharts and JavaScript programming to an advanced user level. The chapter also demonstrates the structure of plugins and shows you how to combine multiple plugins to create a new user experience.

What you need for this book

Readers are expected to have basic knowledge of web development in the following areas:

- The structure of a HTML document and its syntax
- Ajax

As this book is all about Highcharts, which is developed in JavaScript, readers should be comfortable with the language at an intermediate level. Some of the examples use jQuery as a quick way to access document elements and bind methods to events. Hence, a basic knowledge of jQuery should be sufficient. Some knowledge of jQuery UI would also be an advantage, as it is lightly used in *Chapter 10, Highcharts APIs* and *Chapter 11, Highcharts Events*.

Who this book is for

This book is written for web developers who:

- Would like to learn how to incorporate graphical charts into their web applications
- Would like to migrate their Adobe Flash charts for an HTML5 JavaScript solution
- Want to learn more about Highcharts through examples

Conventions

In this book, you will find a number of text styles that distinguish between different kinds of information. Here are some examples of these styles and an explanation of their meaning.

Code words in text, database table names, folder names, filenames, file extensions, pathnames, dummy URLs, user input, and Twitter handles are shown as follows: "The `updatePie` function is called at several places in this demo, such as to remove a series, when the legend checkbox is checked, and more."

A block of code is set as follows:

```
// Bring up the modify dialog box
click: function(evt) {
    // Store the clicked pie slice
    // detail into the dialog box
    $('#updateName').text(evt.point.name);
    $('#percentage').val(evt.point.y);
    $('#dialog-form').dialog("option",
        "pieSlice", evt.point);
```

When we wish to draw your attention to a particular part of a code block, the relevant lines or items are set in bold:

```
<div data-role="content">
    <ul data-role="listview" data-inset="true">
        <li><a href="./gold.html"
           data-ajax="false" >Top 10 countries by gold</a></li>
        <li><a href="./medals.html"
          data-ajax="false" >Top 10 countries by medals</a></li>
        <li><a href="#">A-Z countries</a></li>
        <li><a href="#">A-Z olympians</a></li>
    </ul>
```

Any command-line input or output is written as follows:

```
java -jar batik-rasterizer.jar /tmp/chart.svg
```

New terms and **important words** are shown in bold. Words that you see on the screen, for example, in menus or dialog boxes, appear in the text like this: "The order is such that the values of the subcategories for the **Nintendo** category are before the subcategory data for **Electronic Arts,** and so on."

Warnings or important notes appear in a box like this.

Tips and tricks appear like this.

Reader feedback

Feedback from our readers is always welcome. Let us know what you think about this book—what you liked or disliked. Reader feedback is important for us as it helps us develop titles that you will really get the most out of.

To send us general feedback, simply e-mail `feedback@packtpub.com`, and mention the book's title in the subject of your message.

If there is a topic that you have expertise in and you are interested in either writing or contributing to a book, see our author guide at `www.packtpub.com/authors`.

Customer support

Now that you are the proud owner of a Packt book, we have a number of things to help you to get the most from your purchase.

Downloading the example code

You can download the example code files from your account at `http://www.packtpub.com` for all the Packt Publishing books you have purchased. If you purchased this book elsewhere, you can visit `http://www.packtpub.com/support` and register to have the files e-mailed directly to you.

Errata

Although we have taken every care to ensure the accuracy of our content, mistakes do happen. If you find a mistake in one of our books—maybe a mistake in the text or the code—we would be grateful if you could report this to us. By doing so, you can save other readers from frustration and help us improve subsequent versions of this book. If you find any errata, please report them by visiting `http://www.packtpub.com/submit-errata`, selecting your book, clicking on the **Errata Submission Form** link, and entering the details of your errata. Once your errata are verified, your submission will be accepted and the errata will be uploaded to our website or added to any list of existing errata under the Errata section of that title.

To view the previously submitted errata, go to `https://www.packtpub.com/books/content/support` and enter the name of the book in the search field. The required information will appear under the **Errata** section.

Piracy

Piracy of copyrighted material on the Internet is an ongoing problem across all media. At Packt, we take the protection of our copyright and licenses very seriously. If you come across any illegal copies of our works in any form on the Internet, please provide us with the location address or website name immediately so that we can pursue a remedy.

Please contact us at `copyright@packtpub.com` with a link to the suspected pirated material.

We appreciate your help in protecting our authors and our ability to bring you valuable content.

Questions

If you have a problem with any aspect of this book, you can contact us at `questions@packtpub.com`, and we will do our best to address the problem.

1
Web Charts

In this chapter, you will learn the general background of web charts. This includes a short history of how web charts used to be made before Ajax and HTML5 became the new standard. The recent advances in JavaScript programming will be briefly discussed. Then, SVG support and the new HTML5 feature canvas, the main drive behind JavaScript charts, are introduced and demonstrated. This is followed by a quick guide to the other JavaScript graphing packages that are available on the market. Finally, we are introduced to Highcharts and will explain the advantages of using Highcharts over other products. In this chapter, we will cover the following topics:

- A short history of web charting
- The rise of JavaScript and HTML5
- JavaScript charts on the market
- Why Highcharts?

A short history of web charting

Before diving into Highcharts, it is worth mentioning how web charts evolved from pure HTML with server-side technology to the current client side.

HTML image map (server-side technology)

This technique has been used since the early days of HTML, when server-side operations were the main drive. Charts were only HTML images generated from the web server. Before there were any server-side scripting languages such as PHP, one of the most common approaches was to use **Common Gateway Interface** (**CGI**), which executes plotting programs (such as gnuplot) to output images. Later, when PHP became popular, the GD graphic module was used for plotting. One product that uses this technique is **JpGraph**.

The following is an example of how to include a chart image in an HTML page:

```
<img src="pie_chart.php" border=0 align="left">
```

The chart script file `pie_chart.php` is embedded in an HTML `img` tag. When the page is loaded, the browser sees the `img src` attribute and sends an HTTP request for `pie_chart.php`. As far as the web browser is concerned, it has no knowledge of whether the `.php` file is an image file or not. When the web server (with PHP support) receives the request, it recognizes the `.php` extension and executes the PHP scripts. The following is a cut down JpGraph example; the script outputs the image content and streams it back as an HTTP response, in the same way as normal image content would be sent back:

```
// Create new graph
$graph = new Graph(350, 250);
// Add data points in array of x-axis and y-axis values
$p1 = new LinePlot($datay,$datax);
$graph->Add($p1);
// Output line chart in image format back to the client
$graph->Stroke();
```

Furthermore, this technology combines with an HTML `map` tag for chart navigation, so that, when users click on a certain area of a graph, for example a slice in a pie chart, it can load a new page with another graph.

This technology has the following advantages:

- Ideal for automation tasks, for example scheduled reports or e-mail alerts with the graph attached.
- Doesn't require JavaScript. It is robust, pure HTML, and is light on the client side.

It has the following disadvantages:

- More workload on the server side
- Pure HTML and a limited technology—few interactions can be put on the graphs and none on the animations

Java applet (client-side) and servlet (server-side)

A Java applet enables the web browser to execute multiplatform Java bytecode to achieve what HTML cannot do, such as graphics display, animations, and advanced user interactions. This was the first technology to extend traditional server-based work to the client side. To include a Java applet in an HTML page, HTML's `applet` (deprecated) or `object` tags are used and require a Java plugin to be installed for the browser.

The following is an example of including a Java applet inside an `object` tag. As Java does not run in the same environment in Internet Explorer as in other browsers, the conditional comments for IE were used:

```
<!--[if !IE]> Non Internet Explorer way of loading applet -->
<object classid="Java:chart.class" type="application/x-java-applet"
 height="300" width="550" >
<!--<![endif] Internet way of loading applet -->
  <object classid="clsid:8AD9C840..." codebase="/classes/">
  <param name="code" value="chart.class" />
  </object>
<!--[if !IE]> -->
</object>
<!--<![endif]-->
```

Generally, the Java 2D chart products are built from the `java.awt.Graphics2D` class and the `java.awt.geom` package from Java **Abstract Window Toolkit** (**AWT**). Then, the main chart library allows the users to utilize it in a browser, extending from the `Applet` class, or to run it on the server side, extending from the `Servlet` class.

An example of a Java product is **JFreeChart**. It comes with 2D and 3D solutions and is free for nonprofit use. JFreeChart can be run as an applet, servlet, or standalone application. The following shows part of the code used to plot data points within an applet:

```
public class AppletGraph extends JApplet {
  // Create X and Y axis plot dataset and populate
  // with data.
  XYPlot xyPlot = new XYPlot();
  xyPlot.setDataset(defaultXYDataset);
  CombinedDomainXYPlot combinedDomainXYPlot =
    new CombinedDomainXYPlot();
  combinedDomainXYPlot.add(xyPlot);
  // Create a jFreeChart object with the dataset
```

```
    JFreeChart jFreeChart = new JFreeChart(combinedDomainXYPlot);
    // Put the jFreeChart in a chartPanel
    ChartPanel chartPanel = new ChartPanel(jFreeChart);
    chartPanel.setPreferredSize(new Dimension(900,600));
    // Add the chart panel into the display
    getContentPane().add(chartPanel);
  }
```

To run a chart application on the server side, a servlet container is needed—for example Apache Tomcat. The standard `web.xml` file is defined to bind a URL to a servlet:

```
<?xml version="1.0" encoding="UTF-8"?>
<web-app id="server_charts" version="2.4" xmlns="..." xmlns:xsi="..."
  xsi:schemaLocation="...">
  <servlet>
    <servlet-name>PieChartServlet</servlet-name>
    <servlet-class>charts.PieChartServlet</servlet-class>
  </servlet>
  <servlet-mapping>
    <servlet-name>PieChartServlet</servlet-name>
    <url-pattern>/servlets/piechart</url-pattern>
  </servlet-mapping>
</web-app>
```

When the servlet container, such as Tomcat, receives an HTTP request with the URL `http://localhost/servlets/piechart`, it resolves the request into a servlet application. The web server then executes the chart servlet, formats the output into an image, and returns the image content as an HTTP response.

This technology has the following advantages:

- Advanced graphics, animations, and user interfaces
- Reusable core code for different deployment options: client-side, server-side, or standalone applications

It has the following disadvantages:

- Applet security issues
- If the plugin crashes, it can hang or crash the browser
- Very CPU intensive
- Requires a Java plugin
- Long startup time
- Standardization problems

Adobe Shockwave Flash (client-side)

Flash is widely used because it offers audio, graphics, animation, and video capabilities on web browsers. Browsers are required to have the Adobe Flash Player plugin installed. As for plotting graphs, this technique was a common choice (because there weren't many other options) before the HTML5 standard became popular.

Graphing software adopting this technology ships with its own exported **Shockwave Flash** (**SWF**) files. These SWF files contain compressed vector-based graphics and compiled ActionScript instructions to create a chart. In order for the Flash Player to display the graphs, the SWF file has to be loaded from an HTML page. To do that, an HTML `object` tag is needed. The tag is internally created and injected into the document's DOM by the software's own JavaScript routines.

Inside this `object` tag is dimension and SWF path information for plotting the graph. The graph variable data is also passed inside this tag. So, as soon as the browser sees an `object` tag with specific parameters, it calls the installed Flash Player to process both the SWF file and the parameters. To pass the graph's plot data from the server side to the client-side Flash Player, `flashVars` is embedded inside a `param` tag with the data type. The following is an example from Yahoo YUI 2:

```
<object id="yuiswf1" type="..." data="charts.swf" width="100%"
height="100%">
    <param name="allowscriptaccess" value="always">
    <param name="flashVars" value="param1=value1&param2=value2">
</object>
```

This technology has the following advantages:

- Pretty graphics and animations with rich user interactions

It has the following disadvantages:

- If the plugin crashes, it can hang or crash the browser
- Very CPU-intensive
- Long startup time
- Standardization problems

The rise of JavaScript and HTML5

The role of JavaScript has shifted significantly from a few simple client routines to a dominant language to create and manage web user interfaces. The programming technique has moved towards object-oriented with the introduction of function object, prototype, and closure. This was driven by a group of pioneers such as Douglas Crockford, who was responsible for transforming the language to educate and make JavaScript a better language with his book *JavaScript: The Good Parts, O'Reilly Media / Yahoo Press* Others include Sam Stephenson, creator of the Prototype JavaScript library (`http://www.prototypejs.org`), and John Resig, creator of the JQuery library (`http://jquery.com`), who brought JavaScript into a framework for building more complicated frontend web software.

It is beyond the scope of this book to give an introduction to this new programming style. Readers are expected to know the basics of jQuery and CSS selector syntax, which are used in some of the chapters. Readers should also be familiar with the advanced JavaScript scripting described in the book *JavaScript: The Good Parts, O'Reilly Media / Yahoo Press*, such as prototypes, closure, inheritance, and function objects.

HTML5 (SVG and Canvas)

In this section, two HTML5 technologies, SVG and Canvas, are covered, along with examples.

SVG

HTML5 is the biggest advance so far in the HTML standard. The adoption of the standard is growing fast (also fuelled by Apple mobile devices, which stopped supporting Adobe Flash). Again, it is beyond the scope of this book to cover them. However, the most relevant part to web charting is **Scalable Vector Graphics** (**SVG**). SVG is an XML format for describing vector-based graphics that is composed of components such as paths, text, shapes, color, and so on. The technology is similar to PostScript, except that PostScript is a stack-based language. As implied by its name, one of the major advantages of SVG is that it is a lossless technology (the same as PostScript): it doesn't suffer from any pixelation effects by enlarging the image. A reduced image size will not suffer from loss of original content.

Furthermore, SVG can be scripted with timing animation **Synchronized Multimedia Integration Language** (**SMIL**) and event handling. Apart from IE, this SVG technology is supported by all mainstream browsers, `http://caniuse.com/#feat=svg-smil`.

The following is a simple example of SVG code showing a single curved line between two points:

```
<svg xmlns="http://www.w3.org/2000/svg" version="1.1">
  <path id="curveAB" d="M 100 350 q 150 -300 300 0" stroke="blue"
stroke-width="5" fill="none" />
  <!-- Mark relevant points -->
  <g stroke="black" stroke-width="3" fill="black">
    <circle id="pointA" cx="100" cy="350" r="3" />
    <circle id="pointB" cx="400" cy="350" r="3" />
  </g>
  <!-- Label the points -->
  <g font-size="30" font="sans-serif" fill="black" stroke="none" text-
anchor="middle">
    <text x="100" y="350" dx="-30">A</text>
    <text x="400" y="350" dx="30">B</text>
  </g>
</svg>
```

The preceding SVG code is executed in the following steps:

1. Draw a path with `id="curveAB"` with data (`d`). First, move `M` to an absolute coordinate (`100, 350`), then draw a Bézier quadratic curve from the current position to (`150, -300`) and finish at (`300, 0`).
2. Group (`g`) the two circle elements—`"pointA"` and `"pointB"`—with the center coordinates (`100, 350`) and (`400, 350`) respectively with a radius of `3` pixels. Then fill both circles in black.
3. Group the two text elements `A` and `B`, started at (`100, 350`) and (`400, 350`), which display with the sans-serif font in black, and then shift along the x-axis (dx) `30` pixels left and right, respectively.

The following is the final graph from the SVG script:

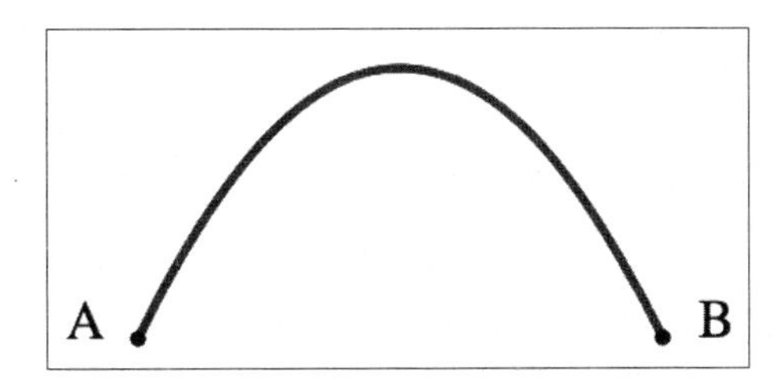

Canvas

Canvas is another new HTML5 standard that is used by some JavaScript chart software packages. The purpose of Canvas is as its name implies; you declare a drawing area on the `canvas` tag, then use the new JavaScript APIs to draw lines and shapes in pixels. This is the distinct difference between the two techniques: Canvas is pixel-based, whereas SVG is vector-based. Canvas has no built-in animation routine, so API calls in timed sequences are used to simulate an animation. Also, there is no event-handling support, so developers need to manually attach event handlers to certain regions on the canvas. Fancy chart animation may prove more complicated to implement.

The following is an example of Canvas code that achieves the same effect as the preceding SVG curve:

```
<canvas id="myCanvas" width="500" height="300" style="border:1px solid
#d3d3d3;">Canvas tag not supported</canvas>
<script type="text/javascript">
    var c=document.getElementById("myCanvas");
   var ctx=c.getContext("2d");
  // Draw the quadratic curve from Point A to B
   ctx.beginPath();
   ctx.moveTo(100, 250);
   ctx.quadraticCurveTo(250, 0, 400, 250);
   ctx.strokeStyle="blue";
   ctx.lineWidth=5;
   ctx.stroke();
  // Draw a black circle attached to the start of the curve
   ctx.fillStyle="black";
   ctx.strokeStyle="black";
   ctx.lineWidth=3;
   ctx.beginPath();
   ctx.arc(100,250,3, 0, 2* Math.PI);
   ctx.stroke();
   ctx.fill();
  // Draw a black circle attached to the end of the curve
   ctx.beginPath();
   ctx.arc(400,250,3, 0, 2* Math.PI);
   ctx.stroke();
   ctx.fill();
  // Display 'A' and 'B' text next to the points
   ctx.font="30px 'sans-serif'";
   ctx.textAlign="center";
   ctx.fillText("A", 70, 250);
   ctx.fillText("B", 430, 250);
</script>
```

As you can see, both canvas and SVG can do the same task, but Canvas requires more instructions:

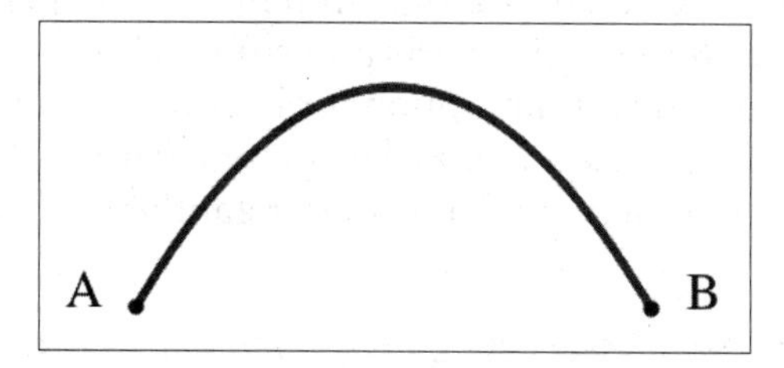

Instead of a continuous path description in SVG, a sequence of JavaScript drawing methods is called. The preceding Canvas code follows these steps to draw the curve:

- In the preceding example, it first calls `beginPath` to start a path in the canvas, then issues the `moveTo` call to move the path starting point to the coordinate (`100, 250`) without creating a line. The `quadraticCurveTo` method creates the curve path from the `moveTo` location to the top of the curve and the end point, which is (`400, 250`). Once the path is set up, we set up the stroke properties before calling the `stroke` method to draw the line along the path.
- We call `beginPath` to reset to a new path to draw and invoke the `arc` method to create a tiny circular path at both the start and end coordinates of the curve path. Once we have set up the stroke and fill styles, we call the `fill` and `stroke` routines to fill it black, to act as a coordinate marker.
- Finally, we set up the font properties and create the text `A` and `B` by calling `fillText` with the position near the coordinate markers.

In a nutshell, instead of a single tag with multiple attributes as in SVG, Canvas adopts multiple attribute-setting routines. SVG is mostly declarative, while Canvas enforces an imperative programming approach.

JavaScript charts on the market

It would be unimaginable to program a chart by hand-coding in SVG or Canvas. Fortunately, there are many different chart libraries on offer on the market. It is impossible to discuss each one of them.

We have omitted some of the products mentioned in the first edition of this book, due to a lack of competitive edge.

The chart libraries are open source, but some of them are short-lived in terms of not having a comprehensive set of basic charts, such as pie, line, and bar charts, and they look rather unfinished. Here, a handful of commercial and open source products are discussed, including all the basic charts and some with extras. Some of them still support the Flash plugin, which is an option for backwards-compatibility because SVG and canvas are not supported in older browsers. Although some of them are not free for commercial development, which is understandable, they are very affordable.

See `http://code.google.com/p/explorercanvas/`.
Many libraries use this add-on to emulate Canvas prior to IE 9.

amCharts

amCharts offers a full set of charts in both 2D and 3D, with other interesting charts such as radar, bubble, candlestick, and polar. All the charts look pretty and support animations. amCharts is free for commercial use, but a credit label will be displayed in the upper-left corner of the charts. The presentation of the charts looks nice and the animations are fully customizable. The default animations, compared to Highcharts, seems slightly overdone.

Ext JS 5 charts

Ext JS is a very popular Ajax application framework developed by Sencha, a pioneering company specializing in web application development. Ext JS 4 comes with a pure JavaScript charts library, unlike its predecessor Ext JS 3 that used the YUI 2 Flash chart library. As the market trend is moving away from Adobe Flash, Sencha responded with a home-brew charting library. Ext JS 4 covers all the basic 2D charts plus gauge and radar charts, and all the charts support animations. The license is free for open source and noncommercial usage, but a developer license is needed for commercial development. One great benefit of Ext JS charts is that integration with a comprehensive set of UI components, for example for a chart with a storage framework, makes displaying/updating both the chart and the table of data very simple to do with editors.

In Ext JS 5, the charting library has been completely restructured. Appearance is a major step forward compared to Ext JS 4: the charts look much more professionally done and on a par with other competitors. Although the Ext JS 5 chart layout, color, and animations may not be as stylish and smooth as Highcharts, it is a close call and still well-received.

Data Driven Documents

Data Driven Documents (**D3**) is the most widely-used charting library. It was created by Bostock et al., much improved from their prior academic research work on Protovis.

[Bostock, Michael; Ogievetsky, Vadim; Heer, Jeffrey (October 2011), *D3: Data-Driven Documents*, IEEE Transactions on Visualization and Computer Graphics, *IEEE Press*.]

The principle of D3 is pretty unique, in that it focuses on transformation in document elements. In other words, it is a decorative framework to manipulate selected elements, known as *selections*, through a myriad of APIs. These APIs allow users to join the selections with chart data and style them like CSS or apply specific effects. This approach enables D3 to create a wide variety of impressive charts with animations that other products cannot produce. It is an ideal tool for plotting specific scientific graphs, or graphs that require complex data visualization presentation, such as Hierarchical Edge Bundling.

The software is free for commercial use and has attracted a large user base and contribution from user communities, especially from the academic sector.

Due to its highly programmable and comparatively low-level approach, it requires a much steeper learning curve that may not appeal to less technical users or developers looking for a chart-ready solution on a production level. As for constructing 3D charts in D3, it can be a challenging task. Although D3 trumps in performance, control, and presentation, it is not for everyone.

FusionCharts

FusionCharts is probably one of the most impressive-looking tools, and has the most comprehensive range of charts on the market. Not only does it come with a full variety of interesting 2D charts (radar, dial, map, and candlestick) available as a separate product, but it also offers fully interactive 3D charts. All the chart animations are very professionally done. FusionCharts can be run in two modes: Flash or JavaScript. Although FusionCharts comes with a higher price tag, it offers the best looking charts bundled with rotatable, animated 3D charts in column, pie, funnel, and pyramid series.

Raphaël

Raphaël is another free graphics library. It supports both SVG and VML for earlier browsers. It is designed like a generic graphics library, handling SVG elements, but it can be used as a charting solution. The software offers APIs to create a Canvas-like object, paper, that users can then create basic shapes or line SVG elements in. The API construct is somewhat similar to D3 in the sense that it can bind data into elements and manipulate them, but with rather more primitive methods, whereas D3 is a data-centric solution. The documentation is basic and it has a smaller user community. Like D3, it requires more effort to program a presentable chart than Highcharts and is not an ideal choice for 3D chart solutions.

Why Highcharts?

Highcharts offers very appealing and professional-looking 2D/3D charts on the market. It is a product that stands out by paying attention to details, not only on the presentation side, but also in other areas that are described later on. It was developed by a Norwegian company called Highsoft AS, created and founded by Torstein Hønsi, and released in late 2009. Highcharts is not their first product, but is by far their best-selling one.

Highcharts and JavaScript frameworks

Although Highcharts is built with the JavaScript framework library, it is implemented in such a way that it doesn't totally rely on one particular framework. Highcharts is packaged with adapters to make its interfaces to frameworks pluggable.

As a result, Highcharts can be incorporated with MooTools, Prototype, or jQuery JavaScript frameworks. Highcharts also has a standalone framework for those who write in pure JavaScript. This empowers users without compromising their already-developed product, or allows them to decide to use the framework that is best suited for their projects. Users who develop their web charting applications in jQuery are only required to load the jQuery library before Highcharts.

To use Highcharts in the MooTools environment, users simply do the following:

```
<script src="http://ajax.googleapis.com/ajax/libs/mootools/1.4.5/
mootools-yui-compressed.js"></script>
<script type="text/javascript"
        src="http://code.highcharts.com/adapters/mootools-adapter.
js"></script>
<script type="text/javascript"
        src="http://code.highcharts.com/highcharts.js"></script>
```

To use Highcharts under Prototype, users need to do the following:

```
<script src="http://ajax.googleapis.com/ajax/libs/prototype/1.7.1.0/
prototype.js"></script>
<script type="text/javascript"
        src="http://code.highcharts.com/adapters/prototype-adapter.
js"></script>
<script type="text/javascript"
        src="http://code.highcharts.com/highcharts.js"></script>
```

Presentation

Highcharts strikes the right balance of look and feel. The charts themselves are visually pleasant and yet the style is simple. The default choices of color are soothing without a sharp contrast and don't conflict with each other, aided by the subtle shadow and white border effects. None of the text nor the colors of the axes are in black or any dark color, which keeps the viewer's attention centered on the colored data presentation. The following is an example of a Highcharts representation:

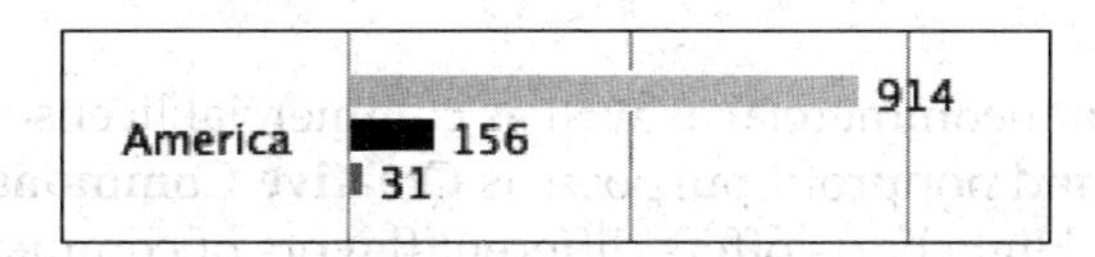

All the animations (initial, update, tool tip) in Highcharts are finely tuned: smooth with a gradual slowdown motion. The initial animation of the donut chart, which is a multi-series pie chart, is the most impressive one. This is the area in which Highcharts is clearly better. The animations in other charts are too mechanical, too much, and sometimes off-putting:

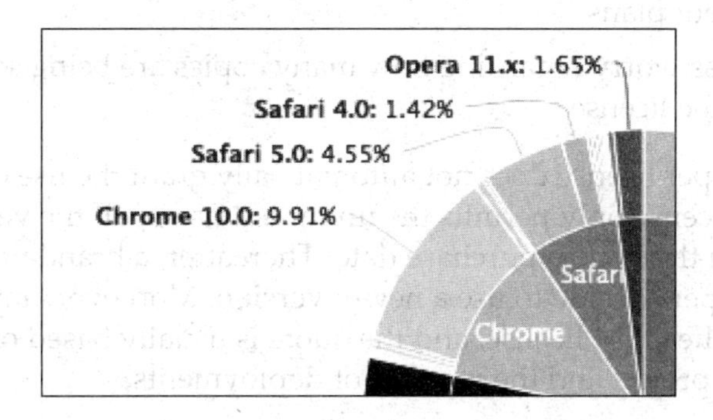

The round corners of tool tips and legends (both inner and outer) with a simple border do not fight for the viewer's attention and nicely blend into the chart. The following is a tool tip sample:

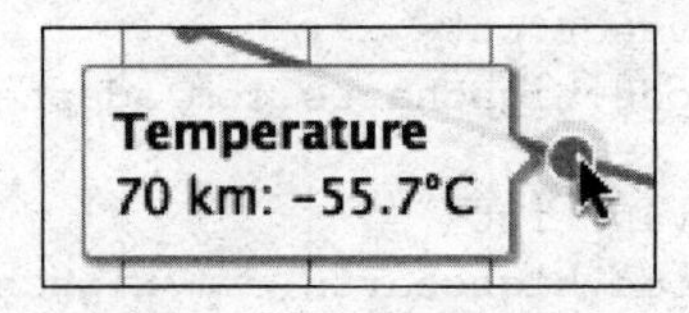

The following is a legend example with two series:

In a nutshell, each element in Highcharts does not compete with others for the viewer's attention, so they share the load equally and work together as a chart.

License

Highcharts has free noncommercial as well as commercial licenses. The free license for personal and nonprofit purposes is **Creative Commons – Attribution Noncommercial 3.0**. Highcharts offers different flavors of commercial license for different purposes. They have a one-off single website license and, thankfully, a developer license. For web development products, a developer license is a better model than charging in units of website use or a very high-priced OEM license because of the following reasons:

- It is easier for software companies to work out the math in their development plans
- There is less worry regarding how many copies are being sold, so as not to violate the license

As usual, a developer license does not automatically grant the use of Highcharts indefinitely. The license only permits the unlimited use of all the versions released within a year from the license purchase date. Thereafter, a brand new license is required if developers decide to use a newer version. Moreover, any condition can be negotiated for the OEM license, and the quote is usually based on the number of developers on the project and the number of deployments.

Simple API model

Highcharts has a very simple API model. To create a chart, the constructor API expects an object specifier with all the necessary settings. To dynamically update an existing chart, Highcharts comes with a small set of APIs. The configuration properties are described in detail in *Chapter 2, Highcharts Configurations*. The API calls are discussed in *Chapter 10, Highcharts APIs*.

Documentations

Highcharts' online documentation is one of the areas that really outshines the others. It is not just a simple documentation page that dumps all the definitions and examples. It's a documentation page built with thought.

The left-hand side of the documentation page is organized in an object structure, as you would pass it to create a chart. You can further expand and collapse the object's attributes as in a JavaScript console. This helps users to become familiar with the product by using it naturally:

```
series : [{
    data            : "",
    dataParser      : ,
    dataURL         : null,
    legendIndex     : undefined,
    name            : "",
    stack           : null,
    type            : "line",
    xAxis           : 0,
    yAxis           : 0
}],
```

The well-thought-out part of the documentation is on the right-hand side with the definitions of attributes. Each definition comes with a description and an online demonstration for each setting, linking to the jsFiddle website:

> **type** : String
> The type of series. Can be one of `area`, `areaspline`, `bar`, `column`, `line`, `pie`, `scatter` or `spline`. Defaults to `"line"`.
> *Try it:* *Line and column in the same chart*

This instant jsFiddle demo invites users to explore different property values and observe the effect on the chart, so the whole documentation-browsing process becomes very effective and fluid.

Openness (feature request with user voice)

One important way that Highcharts decides new features for every major release is via the users' voice (this is not unusual in open source projects, but it is one of the areas in which Highcharts is better than others). Users can submit new feature requests and then vote for them. The company then reviews the feature requests with the most votes and draws up a development plan that includes the new features. The details of the plan are then published on the Highcharts website.

In addition, Highcharts is hosted on GitHub, an online public source control service, which allows JavaScript developers to contribute and clone their own versions:

Highcharts – a quick tutorial

In this section, you will see how to implement your first Highcharts graph. Assuming that we want to run Highcharts from our local web server, first download the latest version from the Highcharts website: `http://www.highcharts.com/download`.

Alternatively, we can load the library via the **Content Delivery Network (CDN)**, as follows:

```
<script type="text/javascript"
        src="http://code.highcharts.com/highcharts.js"></script>
```

For debugging, `highcharts.src.js` is also available from the CDN.

Directory structure

When you unpack the downloaded ZIP file, you should see the following directory structure under the `Highcharts-4.x.x` top-level directory:

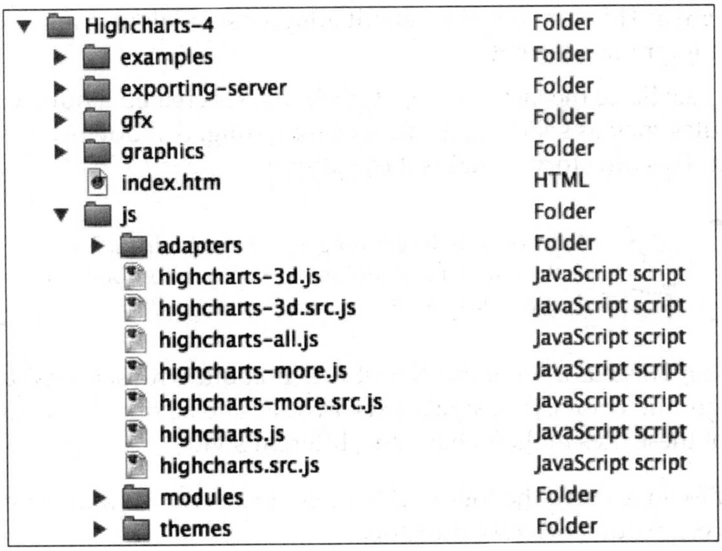

Name	Kind
▼ Highcharts-4	Folder
▶ examples	Folder
▶ exporting-server	Folder
▶ gfx	Folder
▶ graphics	Folder
index.htm	HTML
▼ js	Folder
▶ adapters	Folder
highcharts-3d.js	JavaScript script
highcharts-3d.src.js	JavaScript script
highcharts-all.js	JavaScript script
highcharts-more.js	JavaScript script
highcharts-more.src.js	JavaScript script
highcharts.js	JavaScript script
highcharts.src.js	JavaScript script
▶ modules	Folder
▶ themes	Folder

The following is what each directory contains and is used for:

- `index.html`: This is the demo HTML page, which is the same as the demo page on the Highcharts website, so that you can experiment with Highcharts offline.
- `examples`: This contains all the source files for the examples.
- `graphics`: This contains image files used by the examples.
- `gfx`: This contains an image file required to produce the radial gradient effect for browsers that only support VML.
- `exporting-server`: This directory contains three main components: a Java implementation of the online export server, a PhantomJs toolkit to run Highcharts on the service side, and server script to service the chart export function using Batik. This directory is useful for users who need to set up their own internal export service or run Highcharts as a server process. See *Chapter 14, Server-side Highcharts* and *Chapter 15, Highcharts Online Services and Plugin*.
- `js`: This is the main directory for Highcharts code. Each JavaScript filename has two suffixes: `.src.js`, which contains the source code with comments in it, and `.js`, which is the minified version of the JavaScript source file. The `highcharts.js` file contains the core functionality and the basic chart implementations, `highcharts-3d.js` is the 3D charts extension, and `highcharts.more.js` contains additional series such as polar, gauge, bubble, range, waterfall, errobar, and boxplot. `Highcharts-all.js` lets users provide all the chart series in their application.

- `adapters`: This contains the default adapter standalone-framework in source and compressed format.
- `modules`: Since the last edition, Highcharts has created a number of plugin modules such as solid-gauge, funnel, exporting, drilldown, canvas-tools and so on. This directory contains these plugins.

> A third-party tool, canvg, supports Android 2.x, as the native browser has no SVG support but can display the canvas.

- `themes`: This has a set of JavaScript files prebuilt with settings such as background colors, font styles, axis layouts, and so on. Users can load one of these files in their charts for different styles.

All you need to do is move the top-level `Highcharts-4.x.x/js` directory inside your web server document's root directory.

To use Highcharts, you need to include `Highcharts-4.x.x/js/highcharts.js` library in your HTML file. The following is an example showing the percentage of web browser usage for a public website. The example uses the minimal configuration settings for getting you started quickly. The following is the top half of the example:

```
<!DOCTYPE HTML>
<html>
  <head>
    <meta http-equiv="Content-Type"
          content="text/html; charset=utf-8">
    <title>Highcharts First Example</title>
    <script src="http://ajax.googleapis.com/ajax/libs/jquery/1.8.2 /
jquery.min.js"></script>
    <script type="text/javascript"
        src="Highcharts-4.0.4/js/highcharts.js"></script>
```

We use the Google public library service to load the jQuery library version 1.8.2 before loading the Highcharts library. At the time of writing, the latest jQuery version is 2.1.1 and the Highcharts system requirement for jQuery is 1.8.2.

The second half of the example is the main Highcharts code, as follows:

```
<script type="text/javascript">
var chart;
    $(document).ready(function() {
        Chart = new Highcharts.Chart({
            chart: {
                renderTo: 'container',
```

```
                type: 'spline'
            },
            title: {
                text: 'Web browsers statistics'
            },
            subtitle: {
                text: 'From 2008 to present'
            },
            xAxis: {
                categories: [ 'Jan 2008', 'Feb', .... ],
                tickInterval: 3
            },
            yAxis: {
                title: {
                    text: 'Percentage %'
                },
                min: 0
            },
            plotOptions: {
                series: {
                    lineWidth: 2
                }
            },
            series: [{
                name: 'Internet Explorer',
                data: [54.7, 54.7, 53.9, 54.8, 54.4, ... ]
            }, {
                name: 'FireFox',
                data: [36.4, 36.5, 37.0, 39.1, 39.8, ... ]
            }, {
                // Chrome started until late 2008
                name: 'Chrome',
                data: [ null, null, null, null, null, null,
                        null, null, 3.1, 3.0, 3.1, 3.6, ... ]
            }, {
                name: 'Safari',
                data: [ 1.9, 2.0, 2.1, 2.2, 2.4, 2.6, ... ]
            }, {
                name: 'Opera',
                data: [ 1.4, 1.4, 1.4, 1.4, 1.5, 1.7, ... ]
            }]
        });
    });
</script>
```

```
    </head>
    <body>
    <div>
      <!-- Highcharts rendering takes place inside this DIV -->
      <div id="container"></div>
    </div>
  </body>
  </html>
```

The spline graph is created via an object specifier that contains all the properties and series data required. Once the `chart` object is created, the graph is displayed in the browser. Within this object specifier, there are major components corresponding to the structure of the chart:

```
var chart = new HighCharts.Chart({
    chart: {
        ...
    },
    title: '...'
    ...
});
```

The `renderTo` option instructs Highcharts to display the graph onto the HTML `<div>` element with `'container'` as the ID value, which is defined in the HTML `<body>` section. The `type` option is set to the default presentation type as `'spline'` for any series data, as follows:

```
chart: {
    renderTo: 'container',
    type: 'spline'
}
```

Next is to set `title` and `subtitle`, which appear in the center at the top of the chart:

```
title: {
   text: 'Web browsers ... '
},
subtitle: {
   text: 'From 2008 to present'
},
```

The `categories` option in the `xAxis` property contains an array of x-axis labels for each data point. Since the graph has at least 50 data points, printing each x-axis label will make the text overlap. Rotating the labels still results in the axis looking very packed. The best compromise is to print every third label, (`tickIntervals: 3`), which makes the labels nicely spaced out from each other.

For the sake of simplicity, we use 50 entries in `xAxis.categories` to represent the time. However, we will see a more optimal and logical way to display date and time data in the next chapter:

```
xAxis: {
   categories: [ 'Jan 2008', 'Feb', .... ],
   tickInterval: 3
},
```

The options in `yAxis` are to assign the title of the y-axis and set the minimum possible value to zero; otherwise, Highcharts will display a negative percentage range along the y-axis, which is not wanted for this data set:

```
yAxis: {
     title: {
          text: 'Percentage %'
     },
     min: 0
},
```

The `plotOptions` property is to control how each series is displayed, according to its type (line, pie, bar, and so on). The `plotOptions.series` option is the general configuration applied to all the series types instead of defining each setting inside the `series` array. In this example, the default `linewidth` for every series is set to `2`-pixels wide, as follows:

```
 plotOptions: {
    series: {
        lineWidth: 2
    }
},
```

The `series` property is the heart of the whole configuration object that defines all the series data. It is an array of the `series` object. The `series` object can be specified in multiple ways. For this example, the name is the name of the series that appears in the chart legend and tool tip. The data is an array of y-axis values that have the same length as the `xAxis.categories` array, to form (x,y) data points:

```
series: [{
     name: 'Internet Explorer',
     data: [54.7, 54.7, 53.9, 54.8, 54.4, ... ]
}, {
     name: 'FireFox',
     data: [36.4, 36.5, 37.0, 39.1, 39.8, ... ]
}, {
```

Downloading the example code

You can download the example code files from your account at `http://www.packtpub.com` for all the Packt Publishing books you have purchased. If you purchased this book elsewhere, you can visit `http://www.packtpub.com/support` and register to have the files e-mailed directly to you.

The following screenshot shows how the final Highcharts should look on the Safari browser:

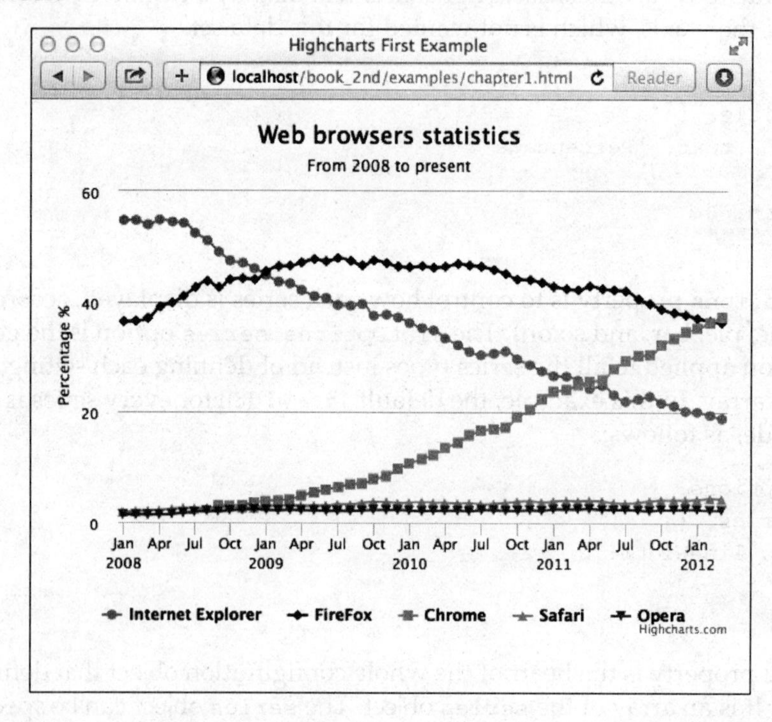

The following screenshot shows how it should look on the Internet Explorer 11 browser:

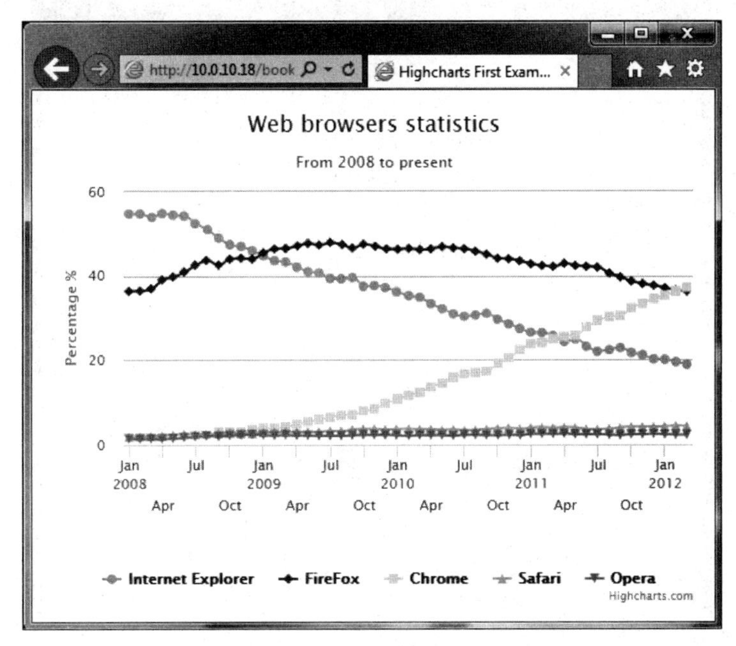

The following screenshot shows how it should look on the Chrome browser:

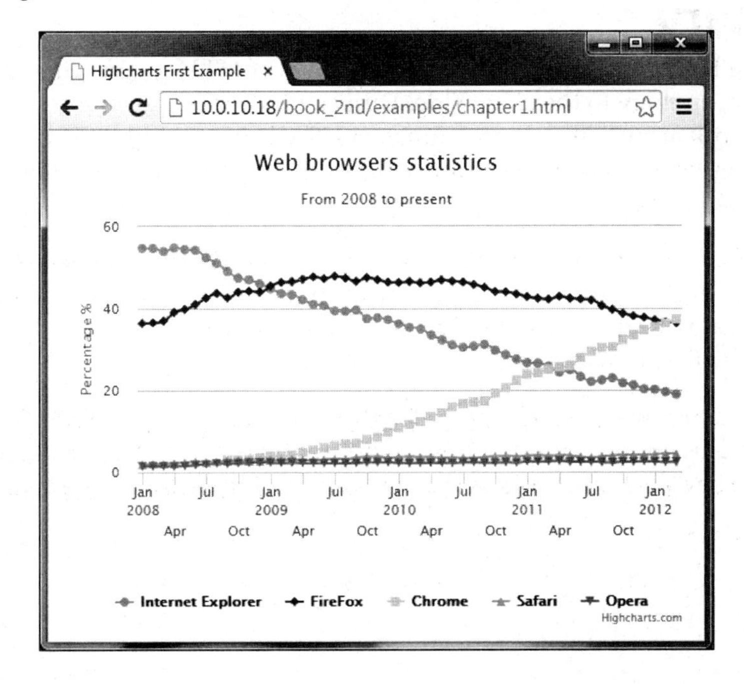

The following screenshot shows how it should look on the Firefox browser:

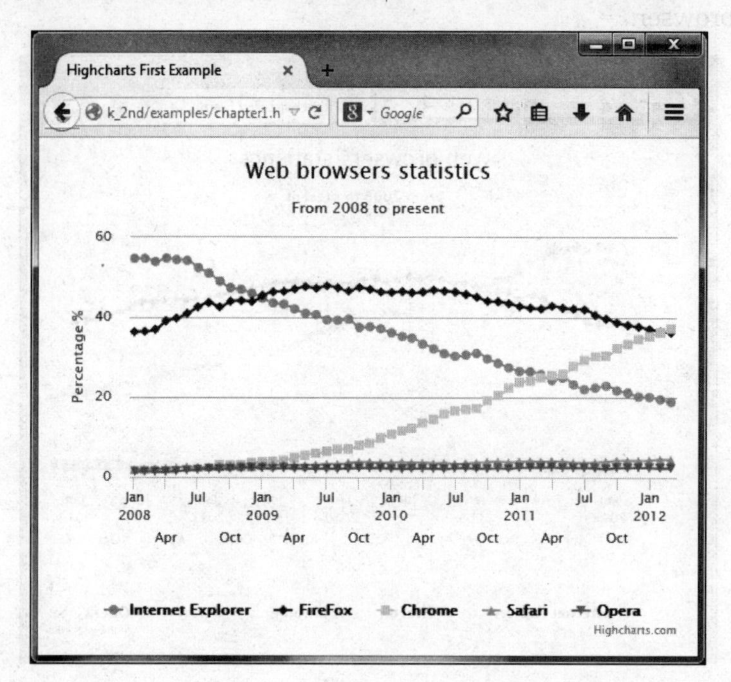

Summary

Web charting has been around since the early days of HTML, emerging from server-side technology to the client side. During this period, several solutions were adopted to work around the shortcomings of HTML. Now, with HTML5, which is rich in features, web charting has come back to HTML and this time it is for good, with the aid of JavaScript.

A number of JavaScript chart products were mentioned in this chapter. Among these, Highcharts emerged with a distinct graphical style and smooth user interactions. We also went through a simple chart example to experience how quick and easy it is to create our very first Highcharts chart.

In the next chapter, we will explore the Highcharts configuration object in greater detail with plenty more examples. The configuration object is the core part of the product that the structure serves as the common prototype for all the charts.

2

Highcharts Configurations

All Highcharts graphs share the same configuration structure and it is crucial for us to become familiar with the core components. However, it is not possible to go through all the configurations within the book. In this chapter, we will explore the functional properties that are most used and demonstrate them with examples. We will learn how Highcharts manages layout, and then explore how to configure axes, specify single series and multiple series data, followed by looking at formatting and styling tool tips in both JavaScript and HTML. After that, we will get to know how to polish our charts with various types of animations and apply color gradients. Finally, we will explore the `drilldown` interactive feature. In this chapter, we will cover the following topics:

- Understanding Highcharts layout
- Framing the chart with axes
- Revisiting the series config
- Styling the tool tips
- Animating charts
- Expanding colors with gradients
- Constructing a chart with a `drilldown` series

Configuration structure

In the Highcharts configuration object, the components at the top level represent the skeleton structure of a chart. The following is a list of the major components that are covered in this chapter:

- `chart`: This has configurations for the top-level chart properties such as layouts, dimensions, events, animations, and user interactions
- `series`: This is an array of series objects (consisting of data and specific options) for single and multiple series, where the series data can be specified in a number of ways

- `xAxis/yAxis/zAxis`: This has configurations for all the axis properties such as labels, styles, range, intervals, plotlines, plot bands, and backgrounds
- `tooltip`: This has the layout and format style configurations for the series data tool tips
- `drilldown`: This has configurations for drilldown series and the ID field associated with the main series
- `title/subtitle`: This has the layout and style configurations for the chart title and subtitle
- `legend`: This has the layout and format style configurations for the chart legend
- `plotOptions`: This contains all the plotting options, such as display, animation, and user interactions, for common series and specific series types
- `exporting`: This has configurations that control the layout and the function of print and export features

For reference information concerning all configurations, go to `http://api.highcharts.com`.

Understanding Highcharts' layout

Before we start to learn how Highcharts layout works, it is imperative that we understand some basic concepts first. To do that, let's first recall the chart example used in *Chapter 1, Web Charts,* and set a couple of borders to be visible. First, set a border around the plot area. To do that we can set the options of `plotBorderWidth` and `plotBorderColor` in the `chart` section, as follows:

```
chart: {
        renderTo: 'container',
        type: 'spline',
        plotBorderWidth: 1,
        plotBorderColor: '#3F4044'
},
```

The second border is set around the Highcharts container. Next, we extend the preceding `chart` section with additional settings:

```
chart: {
        renderTo: 'container',
        ....
        borderColor: '#a1a1a1',
```

```
            borderWidth: 2,
            borderRadius: 3
    },
```

This sets the container border color with a width of 2 pixels and corner radius of 3 pixels.

As we can see, there is a border around the container and this is the boundary that the Highcharts display cannot exceed:

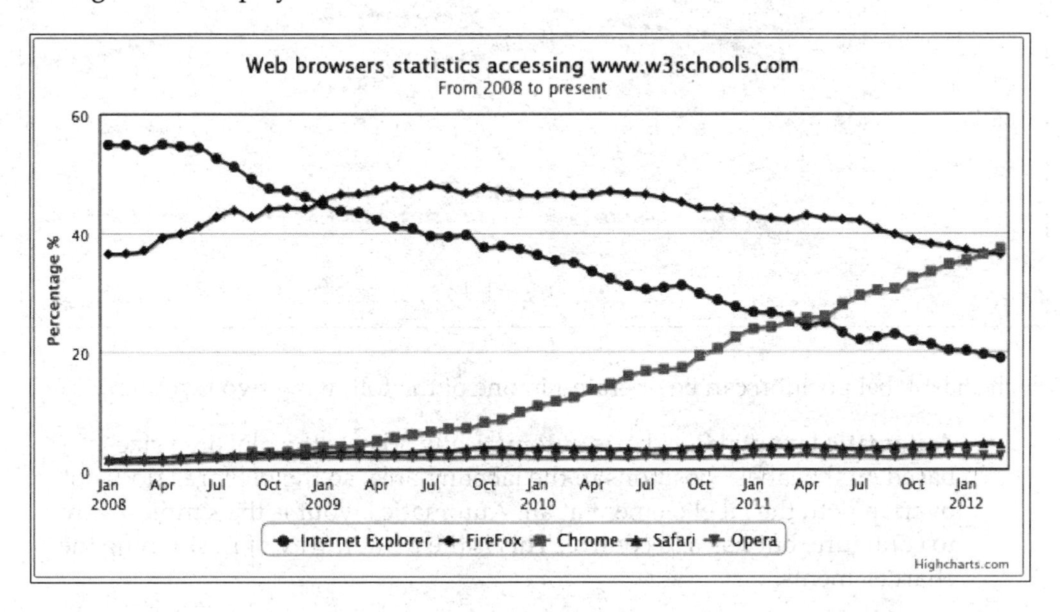

By default, Highcharts displays have three different areas: spacing, labeling, and plot area. The plot area is the area inside the inner rectangle that contains all the plot graphics. The labeling area is the area where labels such as title, subtitle, axis title, legend, and credits go, around the plot area, so that it is between the edge of the plot area and the inner edge of the spacing area. The spacing area is the area between the container border and the outer edge of the labeling area.

The following screenshot shows three different kinds of areas. A gray dotted line is inserted to illustrate the boundary between the spacing and labeling areas.

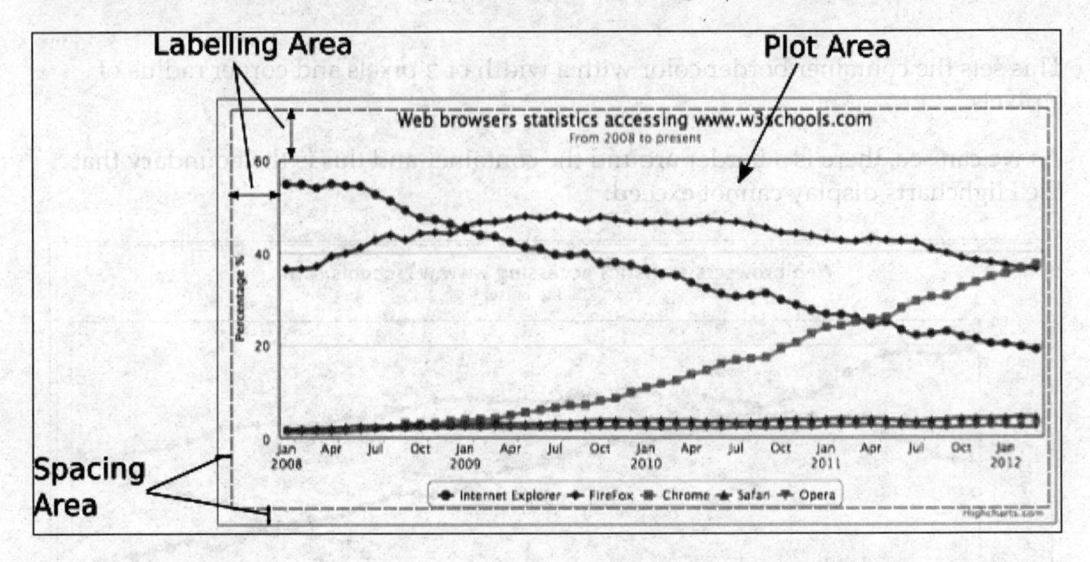

Each chart label position can be operated in one of the following two layouts:

- **Automatic layout**: Highcharts automatically adjusts the plot area size based on the labels' positions in the labeling area, so the plot area does not overlap with the label element at all. Automatic layout is the simplest way to configure, but has less control. This is the default way of positioning the chart elements.
- **Fixed layout**: There is no concept of labeling area. The chart label is specified in a fixed location so that it has a floating effect on the plot area. In other words, the plot area side does not automatically adjust itself to the adjacent label position. This gives the user full control of exactly how to display the chart.

The spacing area controls the offset of the Highcharts display on each side. As long as the chart margins are not defined, increasing or decreasing the spacing area has a global effect on the plot area measurements in both automatic and fixed layouts.

Chart margins and spacing settings

In this section, we will see how chart margins and spacing settings have an effect on the overall layout. Chart margins can be configured with the properties `margin`, `marginTop`, `marginLeft`, `marginRight`, and `marginBottom`, and they are not enabled by default. Setting chart margins has a global effect on the plot area, so that none of the label positions or chart spacing configurations can affect the plot area size. Hence, all the chart elements are in a fixed layout mode with respect to the plot area. The `margin` option is an array of four margin values covered for each direction, the same as in CSS, starting from north and going clockwise. Also, the `margin` option has a lower precedence than any of the directional `margin` options, regardless of their order in the `chart` section.

Spacing configurations are enabled by default with a fixed value on each side. These can be configured in the `chart` section with the property names `spacing`, `spacingTop`, `spacingLeft`, `spacingBottom`, and `spacingRight`.

In this example, we are going to increase or decrease the `margin` or `spacing` property on each side of the chart and observe the effect. The following are the chart settings:

```
chart: {
    renderTo: 'container',
    type: ...
    marginTop: 10,
    marginRight: 0,
    spacingLeft: 30,
    spacingBottom: 0
},
```

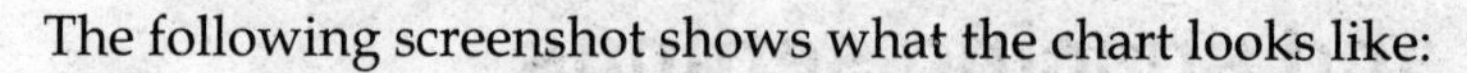

The following screenshot shows what the chart looks like:

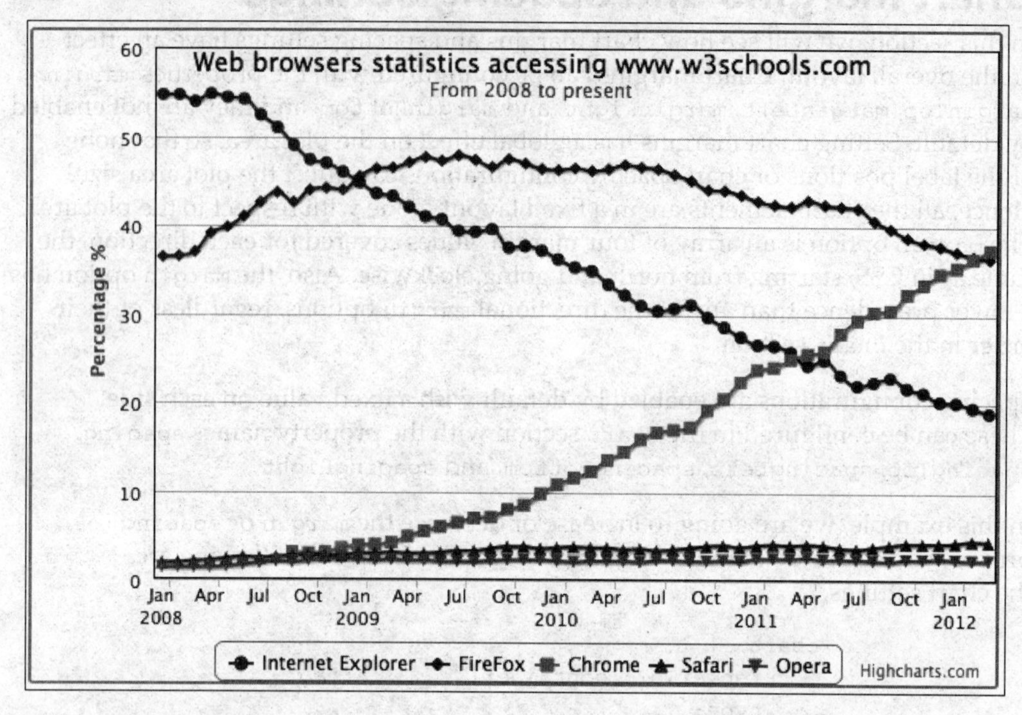

The `marginTop` property fixes the plot area's top border `10` pixels away from the container border. It also changes the top border into fixed layout for any label elements, so the chart title and subtitle float on top of the plot area. The `spacingLeft` property increases the spacing area on the left-hand side, so it pushes the *y* axis title further in. As it is in automatic layout (without declaring `marginLeft`), it also pushes the plot area's west border in. Setting `marginRight` to `0` will override all the default spacing on the chart's right-hand side and change it to fixed layout mode. Finally, setting `spacingBottom` to `0` makes the legend touch the lower bar of the container, so it also stretches the plot area downwards. This is because the bottom edge is still in automatic layout even though `spacingBottom` is set to `0`.

Chart label properties

Chart labels such as `xAxis.title`, `yAxis.title`, `legend`, `title`, `subtitle`, and `credits` share common property names, as follows:

- `align`: This is for the horizontal alignment of the label. Possible keywords are `'left'`, `'center'`, and `'right'`. As for the axis title, it is `'low'`, `'middle'`, and `'high'`.

- `floating`: This is to give the label position a floating effect on the plot area. Setting this to `true` will cause the label position to have no effect on the adjacent plot area's boundary.
- `margin`: This is the margin setting between the label and the side of the plot area adjacent to it. Only certain label types have this setting.
- `verticalAlign`: This is for the vertical alignment of the label. The keywords are `'top'`, `'middle'`, and `'bottom'`.
- `x`: This is for horizontal positioning in relation to alignment.
- `y`: This is for vertical positioning in relation to alignment.

As for the labels' `x` and `y` positioning, they are not used for absolute positioning within the chart. They are designed for fine adjustment with the label alignment. The following diagram shows the coordinate directions, where the center represents the label location:

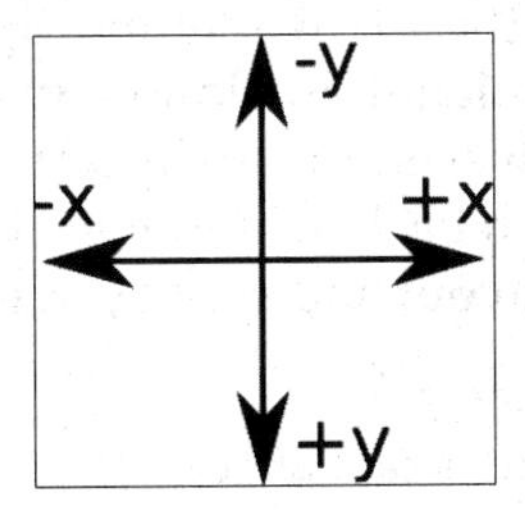

We can experiment with these properties with a simple example of the `align` and `y` position settings, by placing both title and subtitle next to each other. The title is shifted to the left with `align` set to `'left'`, whereas the subtitle alignment is set to `'right'`. In order to make both titles appear on the same line, we change the subtitle's `y` position to `15`, which is the same as the title's default `y` value:

```
title: {
    text: 'Web browsers ...',
    align: 'left'
},
subtitle: {
    text: 'From 2008 to present',
    align: 'right',
    y: 15
},
```

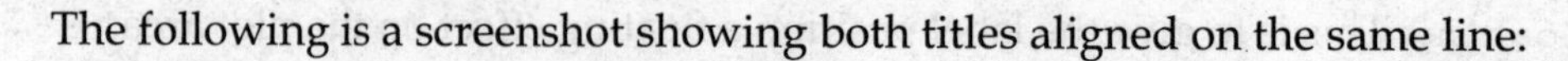

The following is a screenshot showing both titles aligned on the same line:

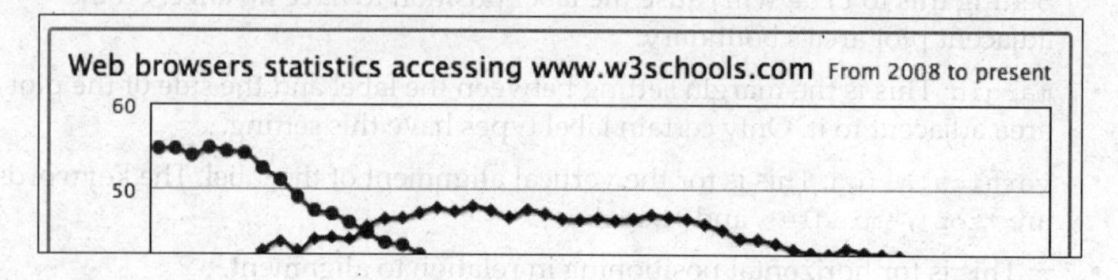

In the following subsections, we will experiment with how changes in alignment for each label element affect the layout behavior of the plot area.

Title and subtitle alignments

Title and subtitle have the same layout properties, and the only differences are that the default values and title have the `margin` setting. Specifying `verticalAlign` for any value changes from the default automatic layout to fixed layout (it internally switches `floating` to `true`). However, manually setting the subtitle's `floating` property to `false` does not switch back to automatic layout. The following is an example of `title` in automatic layout and `subtitle` in fixed layout:

```
title: {
   text: 'Web browsers statistics'
},
subtitle: {
   text: 'From 2008 to present',
   verticalAlign: 'top',
   y: 60
},
```

The `verticalAlign` property for the subtitle is set to `'top'`, which switches the layout into fixed layout, and the `y` offset is increased to `60`. The `y` offset pushes the subtitle's position further down. Due to the fact that the plot area is not in an automatic layout relationship to the subtitle anymore, the top border of the plot area goes above the subtitle. However, the plot area is still in automatic layout towards the title, so the title is still above the plot area:

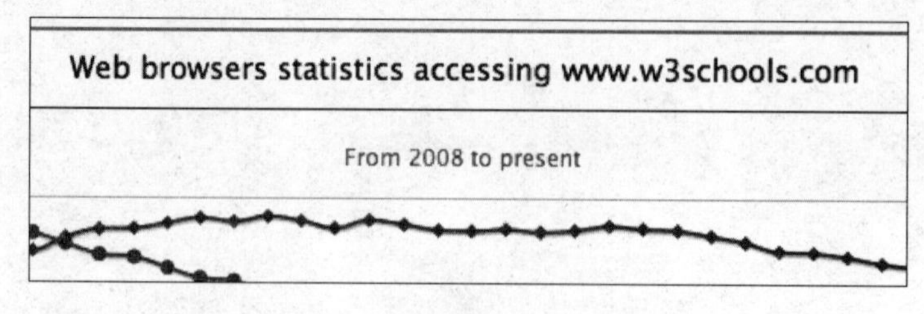

Legend alignment

Legends show different behavior for the `verticalAlign` and `align` properties. Apart from setting the alignment to `'center'`, all other settings in `verticalAlign` and `align` remain in automatic positioning. The following is an example of a legend located on the right-hand side of the chart. The `verticalAlign` property is switched to the middle of the chart, where the horizontal `align` is set to `'right'`:

```
legend: {
        align: 'right',
        verticalAlign: 'middle',
        layout: 'vertical'
},
```

The `layout` property is assigned to `'vertical'` so that it causes the items inside the legend box to be displayed in a vertical manner. As we can see, the plot area is automatically resized for the legend box:

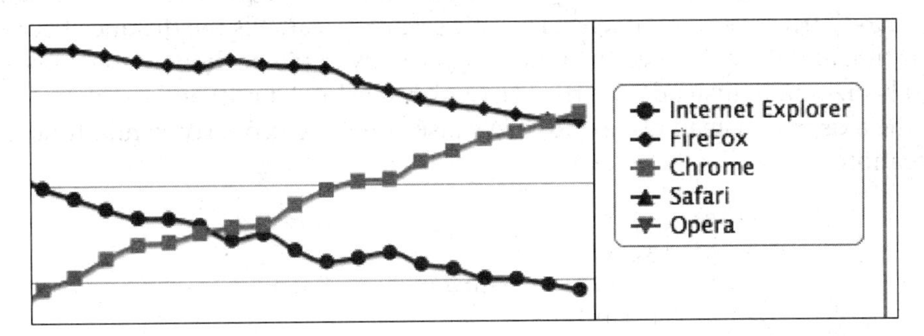

Note that the border decoration around the legend box is disabled in the newer version. To display a round border around the legend box, we can add the `borderWidth` and `borderRadius` options using the following:

```
legend: {
        align: 'right',
        verticalAlign: 'middle',
        layout: 'vertical',
        borderWidth: 1,
        borderRadius: 3
},
```

Here is the legend box with a round corner border:

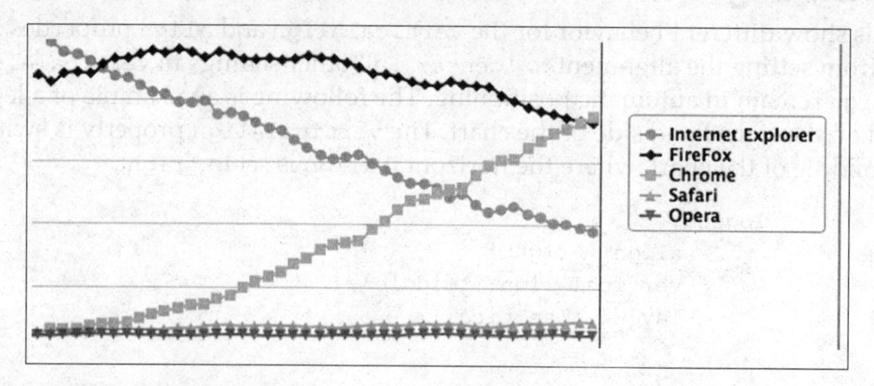

Axis title alignment

Axis titles do not use `verticalAlign`. Instead, they use the `align` setting, which is either `'low'`, `'middle'`, or `'high'`. The title's `margin` value is the distance between the axis title and the axis line. The following is an example of showing the y-axis title rotated horizontally instead of vertically (which it is by default) and displayed on the top of the axis line instead of next to it. We also use the `y` property to fine-tune the title location:

```
yAxis: {
    title: {
        text: 'Percentage %',
        rotation: 0,
        y: -15,
        margin: -70,
        align: 'high'
    },
    min: 0
},
```

The following is a screenshot of the upper-left corner of the chart showing that the title is aligned horizontally at the top of the *y* axis. Alternatively, we can use the `offset` option instead of `margin` to achieve the same result.

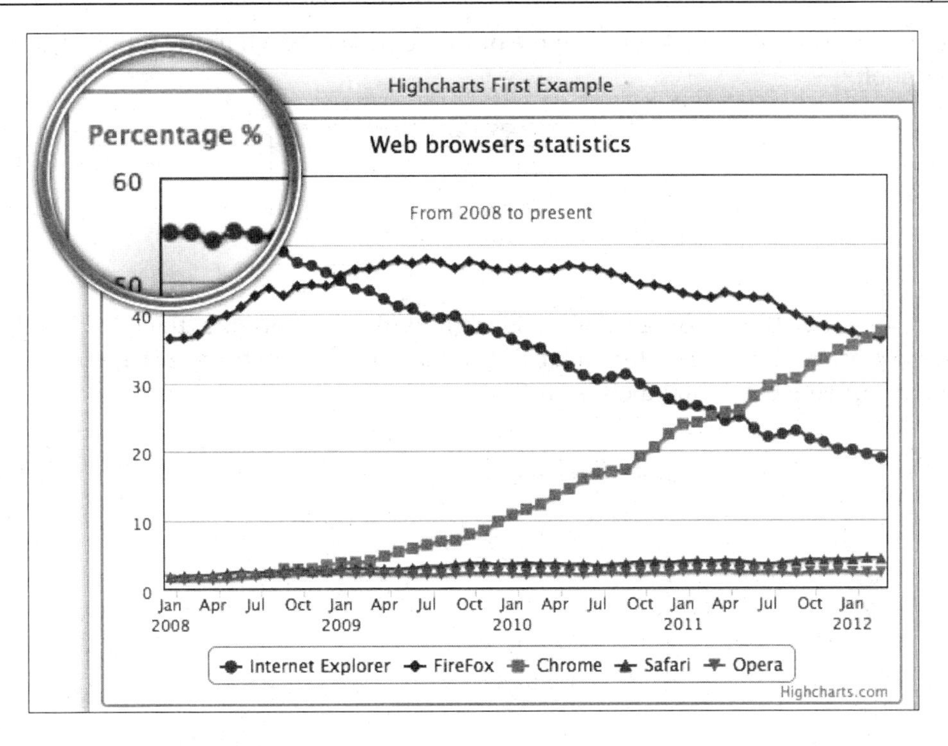

Credits alignment

Credits is a bit different from other label elements. It only supports the `align`, `verticalAlign`, `x`, and `y` properties in the `credits.position` property (shorthand for `credits: { position: … }`), and is also not affected by any spacing setting. Suppose we have a graph without a legend and we have to move the credits to the lower-left area of the chart, the following code snippet shows how to do it:

```
legend: {
    enabled: false
},
credits: {
    position: {
       align: 'left'
    },
    text: 'Joe Kuan',
    href: 'http://joekuan.wordpress.com'
},
```

However, the credits text is off the edge of the chart, as shown in the following screenshot:

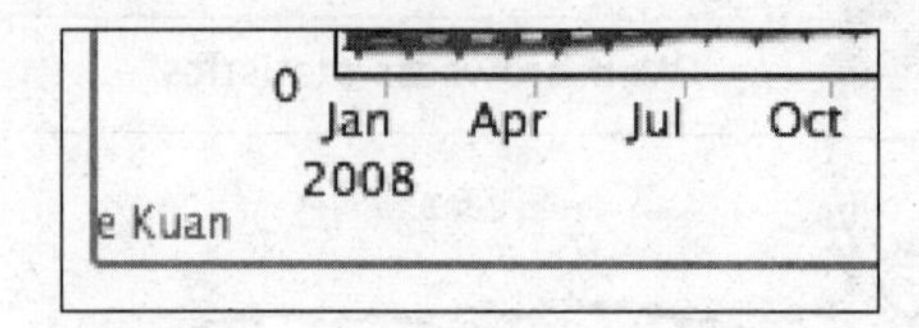

Even if we move the `credits` label to the right with `x` positioning, the label is still a bit too close to the *x* axis interval label. We can introduce extra `spacingBottom` to put a gap between both labels, as follows:

```
chart: {
        spacingBottom: 30,
         ....
},
credits: {
     position: {
        align: 'left',
        x: 20,
        y: -7
     },
},
....
```

The following is a screenshot of the credits with the final adjustments:

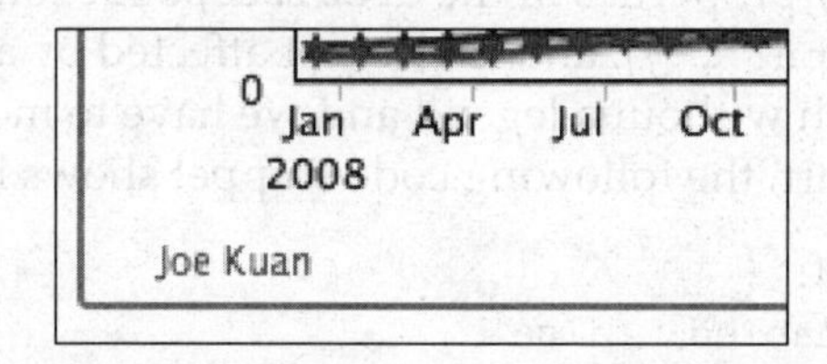

Experimenting with an automatic layout

In this section, we will examine the automatic layout feature in more detail. For the sake of simplifying the example, we will start with only the chart title and without any chart spacing settings:

```
chart: {
    renderTo: 'container',
    // border and plotBorder settings
    borderWidth: 2,
    .....
```

```
    },
    title: {
            text: 'Web browsers statistics,
    },
```

From the preceding example, the chart title should appear as expected between the container and the plot area's borders:

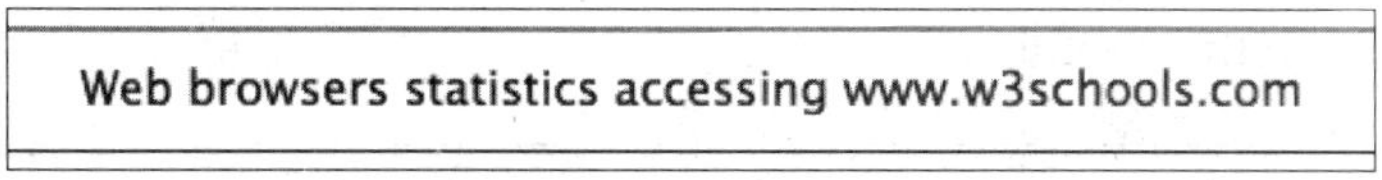

Web browsers statistics accessing www.w3schools.com

The space between the title and the top border of the container has the default setting `spacingTop` for the spacing area (a default value of 10-pixels high). The gap between the title and the top border of the plot area is the default setting for `title.margin`, which is 15-pixels high.

By setting `spacingTop` in the `chart` section to `0`, the chart title moves up next to the container top border. Hence the size of the plot area is automatically expanded upwards, as follows:

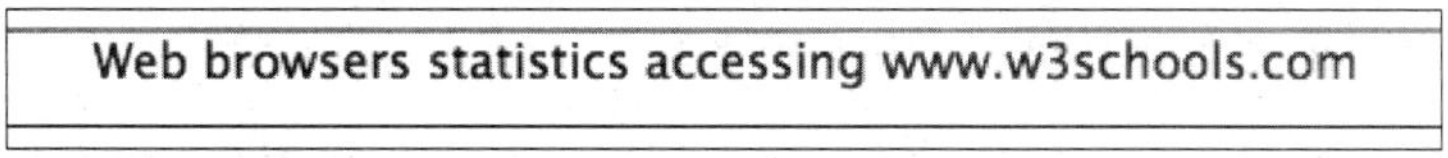

Web browsers statistics accessing www.w3schools.com

Then, we set `title.margin` to `0`; the plot area border moves further up, hence the height of the plot area increases further, as follows:

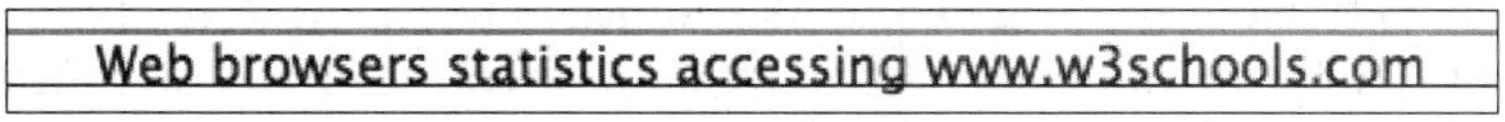

Web browsers statistics accessing www.w3schools.com

As you may notice, there is still a gap of a few pixels between the top border and the chart title. This is actually due to the default value of the title's `y` position setting, which is 15 pixels, large enough for the default title font size.

The following is the chart configuration for setting all the spaces between the container and the plot area to `0`:

```
chart: {
    renderTo: 'container',
    // border and plotBorder settings
    .....
    spacingTop: 0
},
title: {
```

```
        text: null,
        margin: 0,
        y: 0
    }
```

If we set `title.y` to `0`, all the gap between the top edge of the plot area and the top container edge closes up. The following is the final screenshot of the upper-left corner of the chart, to show the effect. The chart title is not visible anymore as it has been shifted above the container:

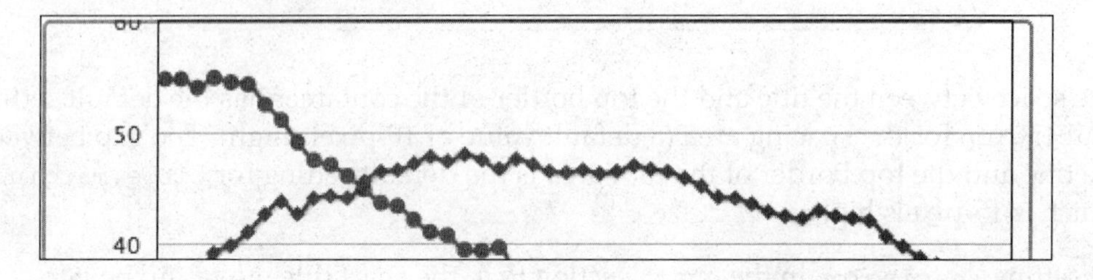

Interestingly, if we work backwards to the first example, the default distance between the top of the plot area and the top of the container is calculated as:

spacingTop + title.margin + title.y = 10 + 15 + 15 = 40

Therefore, changing any of these three variables will automatically adjust the plot area from the top container bar. Each of these offset variables actually has its own purpose in the automatic layout. Spacing is for the gap between the container and the chart content; thus, if we want to display a chart nicely spaced with other elements on a web page, spacing elements should be used. Equally, if we want to use a specific font size for the label elements, we should consider adjusting the `y` offset. Hence, the labels are still maintained at a distance and do not interfere with other components in the chart.

Experimenting with a fixed layout

In the preceding section, we have learned how the plot area dynamically adjusted itself. In this section, we will see how we can manually position the chart labels. First, we will start with the example code from the beginning of the *Experimenting with automatic layout* section and set the chart title's `verticalAlign` to `'bottom'`, as follows:

```
chart: {
    renderTo: 'container',
    // border and plotBorder settings    ....
},
```

```
title: {
    text: 'Web browsers statistics',
    verticalAlign: 'bottom'
},
```

The chart title is moved to the bottom of the chart, next to the lower border of the container. Notice that this setting has changed the title into floating mode; more importantly, the legend still remains in the default automatic layout of the plot area:

Be aware that we haven't specified `spacingBottom`, which has a default value of 15 pixels in height when applied to the chart. This means that there should be a gap between the title and the container bottom border, but none is shown. This is because the `title.y` position has a default value of 15 pixels in relation to spacing. According to the diagram in the *Chart label properties* section, this positive `y` value pushes the title towards the bottom border; this compensates for the space created by `spacingBottom`.

Let's make a bigger change to the `y` offset position this time to show that `verticalAlign` is floating on top of the plot area:

```
title: {
    text: 'Web browsers statistics',
    verticalAlign: 'bottom',
    y: -90
},
```

The negative `y` value moves the title up, as shown here:

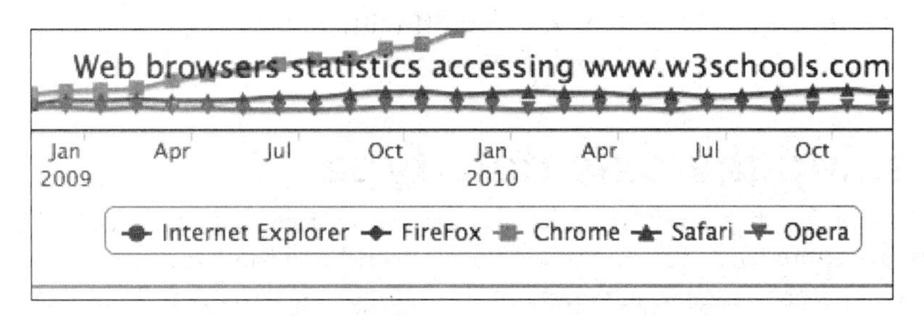

Now the title is overlapping the plot area. To demonstrate that the legend is still in automatic layout with regard to the plot area, here we change the legend's `y` position and the `margin` settings, which is the distance from the axis label:

```
            legend: {
                margin: 70,
                y: -10
            },
```

This has pushed up the bottom side of the plot area. However, the chart title still remains in fixed layout and its position within the chart hasn't been changed at all after applying the new legend setting, as shown in the following screenshot:

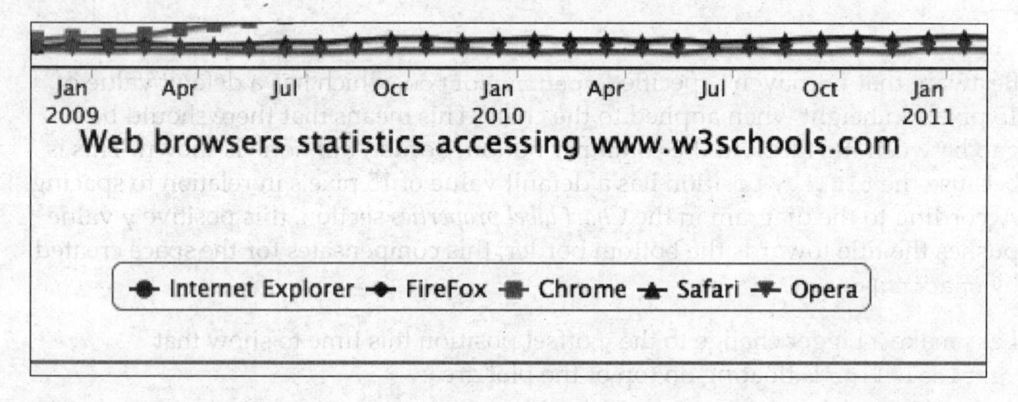

By now, we should have a better understanding of how to position label elements, and their layout policy relating to the plot area.

Framing the chart with axes

In this section, we are going to look into the configuration of axes in Highcharts in terms of their functional area. We will start off with a plain line graph and gradually apply more options to the chart to demonstrate the effects.

Accessing the axis data type

There are two ways to specify data for a chart: categories and series data. For displaying intervals with specific names, we should use the `categories` field that expects an array of strings. Each entry in the categories array is then associated with the series data array. Alternatively, the axis interval values are embedded inside the series data array. Then, Highcharts extracts the series data for both axes, interprets the data type, and formats and labels the values appropriately.

The following is a straightforward example showing the use of categories:

```
chart: {
    renderTo: 'container',
    height: 250,
    spacingRight: 20
},
title: {
    text: 'Market Data: Nasdaq 100'
},
subtitle: {
    text: 'May 11, 2012'
},
xAxis: {
    categories: [ '9:30 am', '10:00 am', '10:30 am',
                  '11:00 am', '11:30 am', '12:00 pm',
                  '12:30 pm', '1:00 pm', '1:30 pm',
                  '2:00 pm', '2:30 pm', '3:00 pm',
                  '3:30 pm', '4:00 pm' ],
    labels: {
        step: 3
    }
},
yAxis: {
    title: {
        text: null
    }
},
legend: {
    enabled: false
},
credits: {
    enabled: false
},
series: [{
    name: 'Nasdaq',
    color: '#4572A7',
    data: [ 2606.01, 2622.08, 2636.03, 2637.78, 2639.15,
            2637.09, 2633.38, 2632.23, 2632.33, 2632.59,
            2630.34, 2626.89, 2624.59, 2615.98 ]
}]
```

The preceding code snippet produces a graph that looks like the following screenshot:

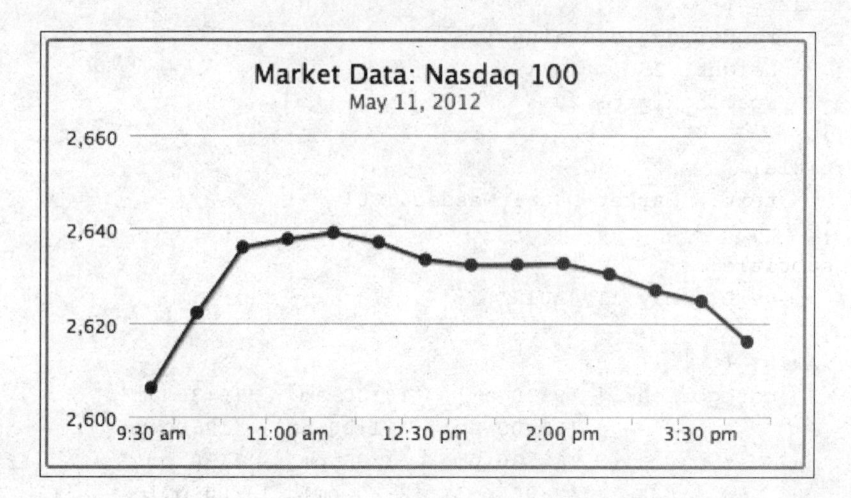

The first name in the categories field corresponds to the first value, **9:30 am**, 2606.01, in the series data array, and so on.

Alternatively, we can specify the time values inside the series data and use the `type` property of the *x* axis to format the time. The `type` property supports `'linear'` (default), `'logarithmic'`, or `'datetime'`. The `'datetime'` setting automatically interprets the time in the series data into human-readable form. Moreover, we can use the `dateTimeLabelFormats` property to predefine the custom format for the time unit. The option can also accept multiple time unit formats. This is for when we don't know in advance how long the time span is in the series data, so each unit in the resulting graph can be per hour, per day, and so on. The following example shows how the graph is specified with predefined hourly and minute formats. The syntax of the format string is based on the PHP `strftime` function:

```
xAxis: {
    type: 'datetime',
     // Format 24 hour time to AM/PM
     dateTimeLabelFormats: {

         hour: '%I:%M %P',
         minute: '%I %M'
     }
},
series: [{
    name: 'Nasdaq',
    color: '#4572A7',
```

```
            data: [ [ Date.UTC(2012, 4, 11, 9, 30), 2606.01 ],
                      [ Date.UTC(2012, 4, 11, 10), 2622.08 ],
                      [ Date.UTC(2012, 4, 11, 10, 30), 2636.03 ],
                      .....
                    ]
        }]
```

Note that the *x* axis is in the 12-hour time format, as shown in the following screenshot:

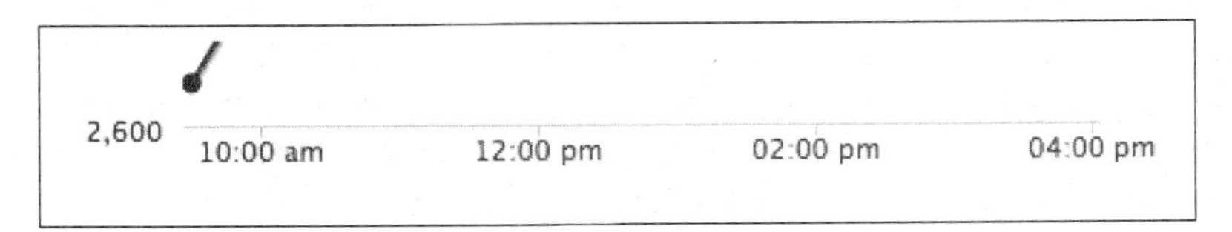

Instead, we can define the format handler for the `xAxis.labels.formatter` property to achieve a similar effect. Highcharts provides a utility routine, `Highcharts.dateFormat`, that converts the timestamp in milliseconds to a readable format. In the following code snippet, we define the `formatter` function using `dateFormat` and `this.value`. The keyword `this` is the axis's interval object, whereas `this.value` is the UTC time value for the instance of the interval:

```
        xAxis: {
            type: 'datetime',
            labels: {
                formatter: function() {
                    return Highcharts.dateFormat('%I:%M %P', this.value);
                }
            }
        },
```

Since the time values of our data points are in fixed intervals, they can also be arranged in a cut-down version. All we need is to define the starting point of time, `pointStart`, and the regular interval between them, `pointInterval`, in milliseconds:

```
    series: [{
        name: 'Nasdaq',
        color: '#4572A7',
        pointStart: Date.UTC(2012, 4, 11, 9, 30),
        pointInterval: 30 * 60 * 1000,
        data: [ 2606.01, 2622.08, 2636.03, 2637.78,
                2639.15, 2637.09, 2633.38, 2632.23,
                2632.33, 2632.59, 2630.34, 2626.89,
                2624.59, 2615.98 ]
    }]
```

Adjusting intervals and background

We have learned how to use axis categories and series data arrays in the last section. In this section, we will see how to format interval lines and the background style to produce a graph with more clarity.

We will continue from the previous example. First, let's create some interval lines along the *y* axis. In the chart, the interval is automatically set to 20. However, it would be clearer to double the number of interval lines. To do that, simply assign the `tickInterval` value to `10`. Then, we use `minorTickInterval` to put another line in between the intervals to indicate a semi-interval. In order to distinguish between interval and semi-interval lines, we set the semi-interval lines, `minorGridLineDashStyle`, to a dashed and dotted style.

There are nearly a dozen line style settings available in Highcharts, from `'Solid'` to `'LongDashDotDot'`. Readers can refer to the online manual for possible values.

The following is the first step to create the new settings:

```
yAxis: {
    title: {
        text: null
    },
    tickInterval: 10,
    minorTickInterval: 5,
    minorGridLineColor: '#ADADAD',
    minorGridLineDashStyle: 'dashdot'
}
```

The interval lines should look like the following screenshot:

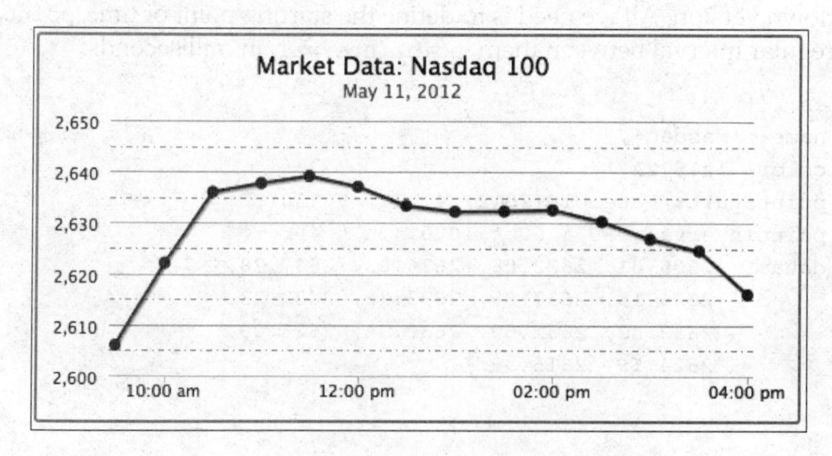

To make the graph even more presentable, we add a striping effect with shading using `alternateGridColor`. Then, we change the interval line color, `gridLineColor`, to a similar range with the stripes. The following code snippet is added into the `yAxis` configuration:

```
gridLineColor: '#8AB8E6',
alternateGridColor: {
    linearGradient: {
        x1: 0, y1: 1,
        x2: 1, y2: 1
    },
    stops: [ [0, '#FAFCFF' ],
             [0.5, '#F5FAFF'] ,
             [0.8, '#E0F0FF'] ,
             [1, '#D6EBFF'] ]
}
```

We will discuss the color gradient later in this chapter. The following is the graph with the new shading background:

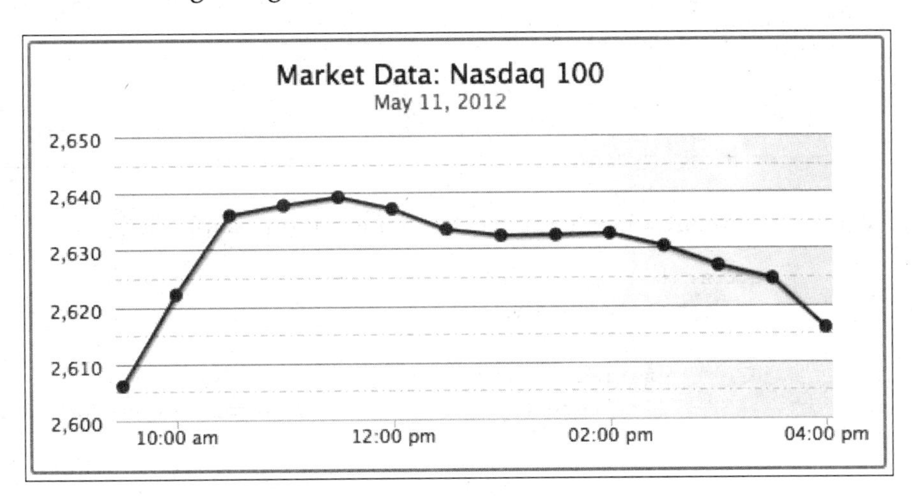

The next step is to apply a more professional look to the *y* axis line. We are going to draw a line on the *y* axis with the `lineWidth` property, and add some measurement marks along the interval lines with the following code snippet:

```
lineWidth: 2,
lineColor: '#92A8CD',
tickWidth: 3,
tickLength: 6,
```

```
            tickColor: '#92A8CD',
            minorTickLength: 3,
            minorTickWidth: 1,
            minorTickColor: '#D8D8D8'
```

The tickWidth and tickLength properties add the effect of little marks at the start of each interval line. We apply the same color on both the interval mark and the axis line. Then we add the ticks minorTickLength and minorTickWidth into the semi-interval lines in a smaller size. This gives a nice measurement mark effect along the axis, as shown in the following screenshot:

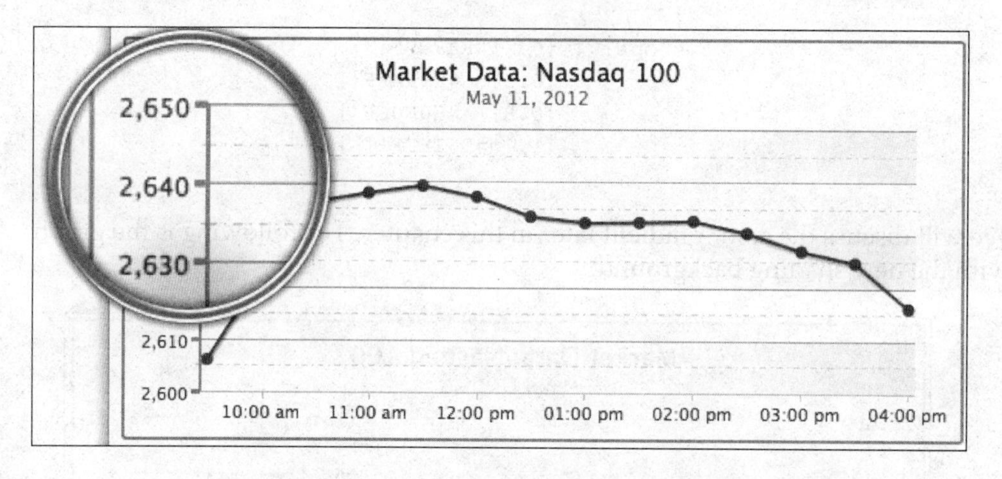

Now, we apply a similar polish to the xAxis configuration, as follows:

```
        xAxis: {
            type: 'datetime',
            labels: {
                formatter: function() {
                    return Highcharts.dateFormat('%I:%M %P', this.
value);
                },
            },
            gridLineDashStyle: 'dot',
            gridLineWidth: 1,
            tickInterval: 60 * 60 * 1000,
            lineWidth: 2,
            lineColor: '#92A8CD',
            tickWidth: 3,
            tickLength: 6,
            tickColor: '#92A8CD',
        },
```

We set the *x* axis interval lines to the hourly format and switch the line style to a dotted line. Then, we apply the same color, thickness, and interval ticks as on the *y* axis. The following is the resulting screenshot:

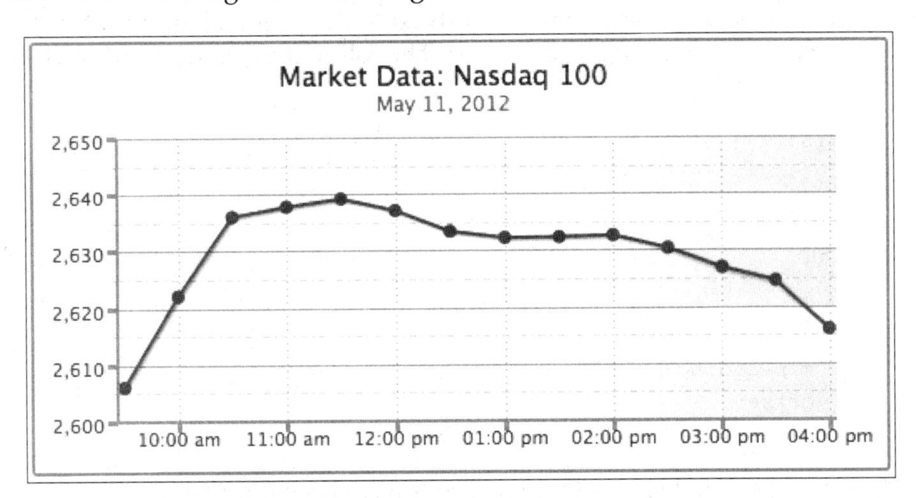

However, there are some defects along the *x* axis line. To begin with, the meeting point between the *x* axis and *y* axis lines does not align properly. Secondly, the interval labels at the *x* axis are touching the interval ticks. Finally, part of the first data point is covered by the y-axis line. The following is an enlarged screenshot showing the issues:

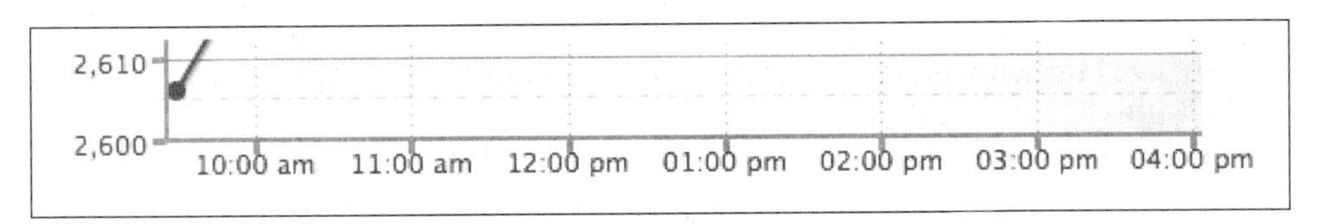

There are two ways to resolve the axis line alignment problem, as follows:

- Shift the plot area 1 pixel away from the x axis. This can be achieved by setting the `offset` property of `xAxis` to `1`.
- Increase the x-axis line width to 3 pixels, which is the same width as the y-axis tick interval.

As for the x-axis label, we can simply solve the problem by introducing the `y` offset value into the `labels` setting.

Finally, to avoid the first data point touching the y-axis line, we can impose `minPadding` on the *x* axis. What this does is to add padding space at the minimum value of the axis, the first point. The `minPadding` value is based on the ratio of the graph width. In this case, setting the property to `0.02` is equivalent to shifting along the *x* axis 5 pixels to the right (250 px * 0.02). The following are the additional settings to improve the chart:

```
xAxis: {
    ....
    labels: {
            formatter: ...,
            y: 17
    },
    .....
    minPadding: 0.02,
    offset: 1
}
```

The following screenshot shows that the issues have been addressed:

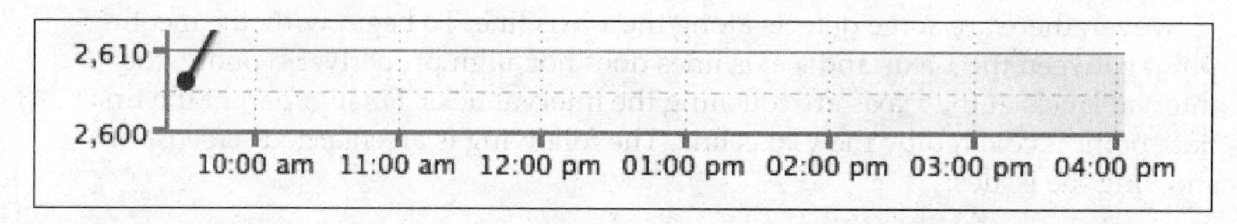

As we can see, Highcharts has a comprehensive set of configurable variables with great flexibility.

Using plot lines and plot bands

In this section, we are going to see how we can use Highcharts to place lines or bands along the axis. We will continue with the example from the previous section. Let's draw a couple of lines to indicate the day's highest and lowest index points on the y axis. The `plotLines` field accepts an array of object configurations for each plot line. There are no width and color default values for `plotLines`, so we need to specify them explicitly in order to see the line. The following is the code snippet for the plot lines:

```
yAxis: {
        ... ,
        plotLines: [{
             value: 2606.01,
             width: 2,
```

```
                color: '#821740',
                label: {
                    text: 'Lowest: 2606.01',
                    style: {
                        color: '#898989'
                    }
                }
            }, {
                value: 2639.15,
                width: 2,
                color: '#4A9338',
                label: {
                    text: 'Highest: 2639.15',
                    style: {
                        color: '#898989'
                    }
                }
            }]
        }
```

The following screenshot shows what it should look like:

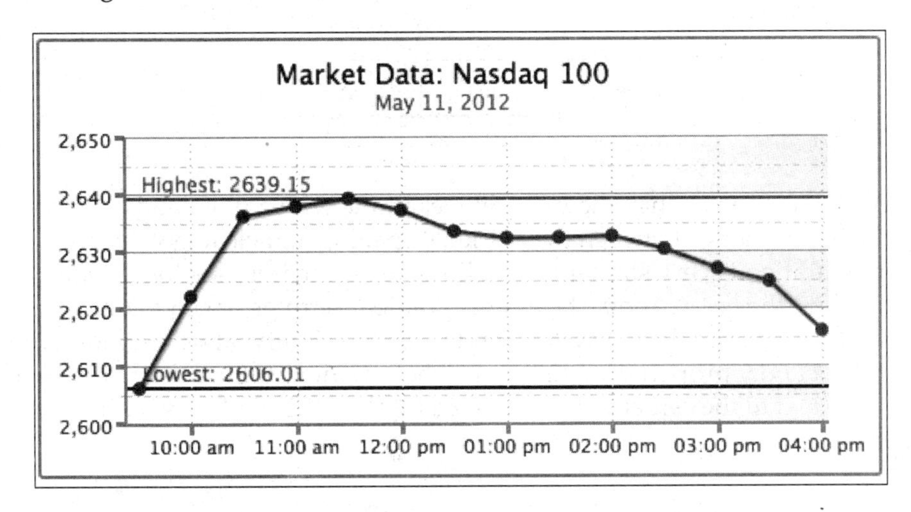

We can improve the look of the chart slightly. First, the text label for the top plot line should not be next to the highest point. Second, the label for the bottom line should be remotely covered by the series and interval lines, as follows:

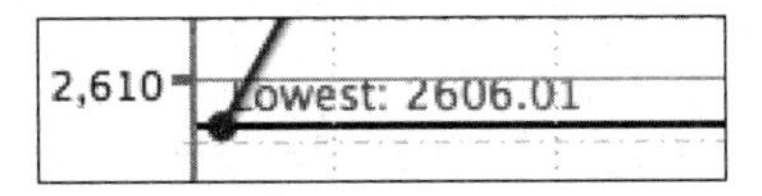

To resolve these issues, we can assign the plot line's `zIndex` to `1`, which brings the text label above the interval lines. We also set the `x` position of the label to shift the text next to the point. The following are the new changes:

```
plotLines: [{
        ... ,
        label: {
            ... ,
            x: 25
        },
        zIndex: 1
        }, {
        ... ,
        label: {
            ... ,
            x: 130
        },
        zIndex: 1
 }]
```

The following graph shows the label has been moved away from the plot line and over the interval line:

Now, we are going to change the preceding example with a plot band area that shows the index change between the market's opening and closing values. The plot band configuration is very similar to plot lines, except that it uses the `to` and `from` properties, and the `color` property accepts gradient settings or color code. We create a plot band with a triangle text symbol and values to signify a positive close. Instead of using the `x` and `y` properties to fine-tune label position, we use the `align` option to adjust the text to the center of the plot area (replace the plotLines setting from the above example):

```
plotBands: [{
     from: 2606.01,
     to: 2615.98,
     label: {
          text: '▲ 9.97 (0.38%)',
          align: 'center',
          style: {
               color: '#007A3D'
          }
```

```
            },
            zIndex: 1,
            color: {
                linearGradient: {
                    x1: 0, y1: 1,
                    x2: 1, y2: 1
                },
                stops: [ [0, '#EBFAEB' ],
                         [0.5, '#C2F0C2'] ,
                         [0.8, '#ADEBAD'] ,
                         [1, '#99E699']
                ]
            }
        }]
```

The triangle is an alt-code character; hold down the left *Alt* key and enter `30` in the number keypad. See `http://www.alt-codes.net` for more details.

This produces a chart with a green plot band highlighting a positive close in the market, as shown in the following screenshot:

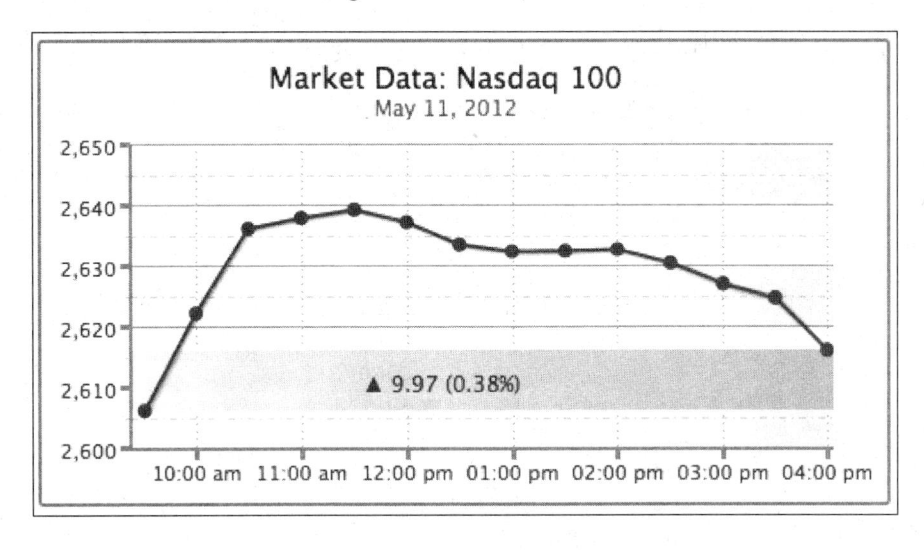

Extending to multiple axes

Previously, we ran through most of the axis configurations. Here, we explore how we can use multiple axes, which are just an array of objects containing axis configurations.

Continuing from the previous stock market example, suppose we now want to include another market index, Dow Jones, along with Nasdaq. However, both indices are different in nature, so their value ranges are vastly different. First, let's examine the outcome by displaying both indices with the common *y* axis. We change the title, remove the fixed interval setting on the y axis, and include data for another series:

```
chart: ... ,
title: {
    text: 'Market Data: Nasdaq & Dow Jones'
},
subtitle: ... ,
xAxis: ... ,
credits: ... ,
yAxis: {
    title: {
        text: null
    },
    minorGridLineColor: '#D8D8D8',
    minorGridLineDashStyle: 'dashdot',
    gridLineColor: '#8AB8E6',
    alternateGridColor: {
        linearGradient: {
            x1: 0, y1: 1,
            x2: 1, y2: 1
        },
        stops: [ [0, '#FAFCFF' ],
                 [0.5, '#F5FAFF'] ,
                 [0.8, '#E0F0FF'] ,
                 [1, '#D6EBFF'] ]
    },
    lineWidth: 2,
    lineColor: '#92A8CD',
    tickWidth: 3,
    tickLength: 6,
    tickColor: '#92A8CD',
    minorTickLength: 3,
    minorTickWidth: 1,
    minorTickColor: '#D8D8D8'
},
```

```
series: [{
  name: 'Nasdaq',
  color: '#4572A7',
  data: [ [ Date.UTC(2012, 4, 11, 9, 30), 2606.01 ],
          [ Date.UTC(2012, 4, 11, 10), 2622.08 ],
          [ Date.UTC(2012, 4, 11, 10, 30), 2636.03 ],
          ...
        ]
}, {
  name: 'Dow Jones',
  color: '#AA4643',
  data: [ [ Date.UTC(2012, 4, 11, 9, 30), 12598.32 ],
          [ Date.UTC(2012, 4, 11, 10), 12538.61 ],
          [ Date.UTC(2012, 4, 11, 10, 30), 12549.89 ],
          ...
        ]
}]
```

The following is the chart showing both market indices:

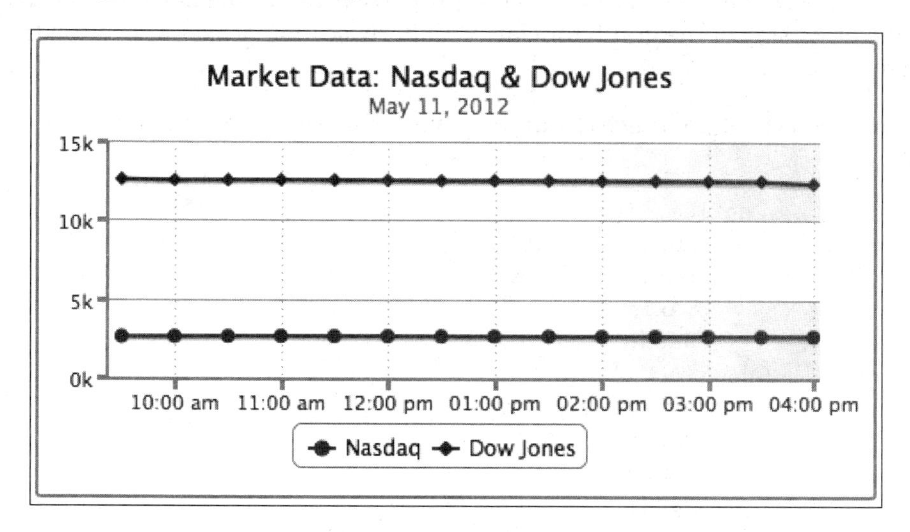

As expected, the index changes that occur during the day have been normalized by the vast differences in value. Both lines look roughly straight, which falsely implies that the indices have hardly changed.

Let us now explore putting both indices onto separate *y* axes. We should remove any background decoration on the *y* axis, because we now have a different range of data shared on the same background.

The following is the new setup for `yAxis`:

```
yAxis: [{
        title: {
            text: 'Nasdaq'
        },
    }, {
        title: {
            text: 'Dow Jones'
        },
        opposite: true
}],
```

Now `yAxis` is an array of axis configurations. The first entry in the array is for Nasdaq and the second is for Dow Jones. This time, we display the axis title to distinguish between them. The `opposite` property is to put the Dow Jones y axis onto the other side of the graph for clarity. Otherwise, both *y* axes appear on the left-hand side.

The next step is to align indices from the y-axis array to the series data array, as follows:

```
series: [{
    name: 'Nasdaq',
    color: '#4572A7',
    yAxis: 0,
    data: [ ... ]
}, {
    name: 'Dow Jones',
    color: '#AA4643',
    yAxis: 1,
    data: [ ... ]
}]
```

We can clearly see the movement of the indices in the new graph, as follows:

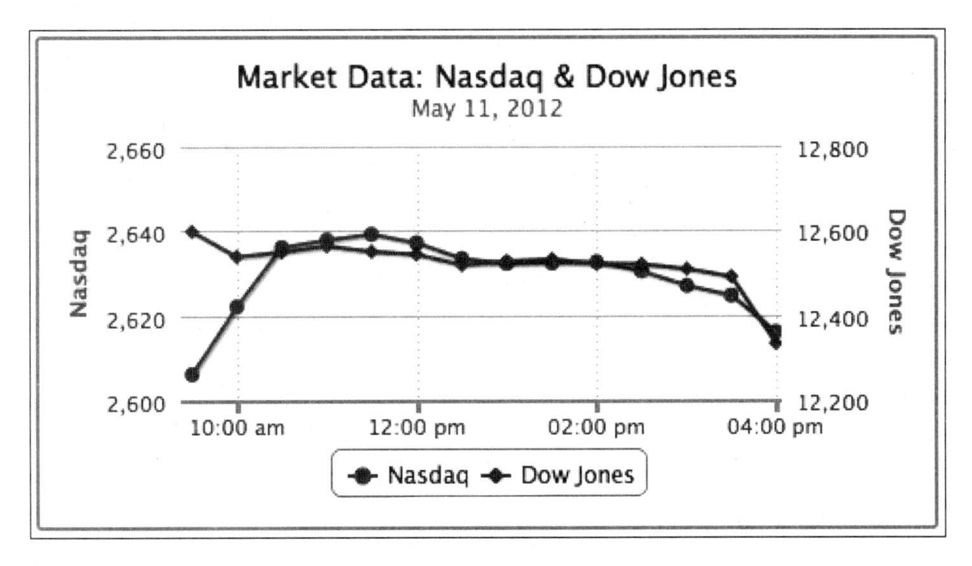

Moreover, we can improve the final view by color-matching the series to the axis lines. The `Highcharts.getOptions().colors` property contains a list of default colors for the series, so we use the first two entries for our indices. Another improvement is to set `maxPadding` for the *x* axis, because the new y-axis line covers parts of the data points at the high end of the *x* axis:

```
xAxis: {
    ... ,
    minPadding: 0.02,
    maxPadding: 0.02
    },
yAxis: [{
    title: {
        text: 'Nasdaq'
    },
    lineWidth: 2,
    lineColor: '#4572A7',
    tickWidth: 3,
    tickLength: 6,
    tickColor: '#4572A7'
}, {
    title: {
        text: 'Dow Jones'
    },
```

```
            opposite: true,
            lineWidth: 2,
            lineColor: '#AA4643',
            tickWidth: 3,
            tickLength: 6,
            tickColor: '#AA4643'
        }],
```

The following screenshot shows the improved look of the chart:

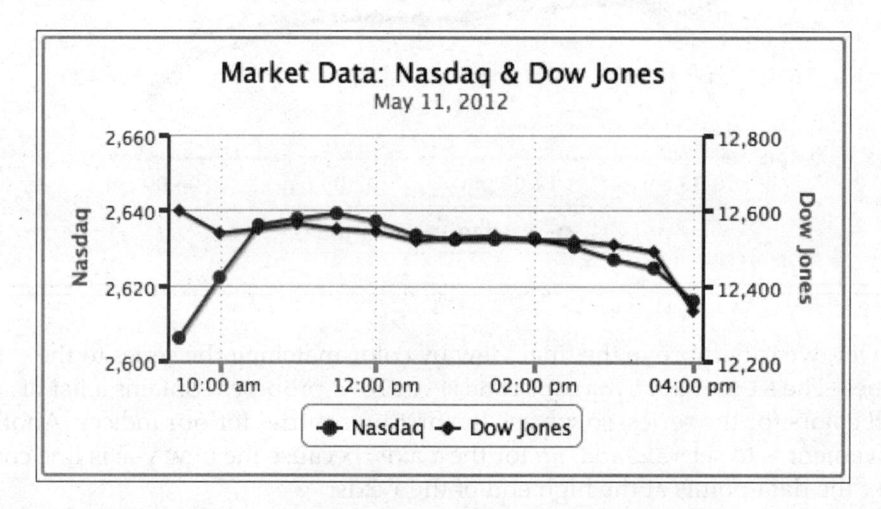

We can extend the preceding example and have more than a couple of axes, simply by adding entries into the `yAxis` and `series` arrays, and mapping both together. The following screenshot shows a 4-axis line graph:

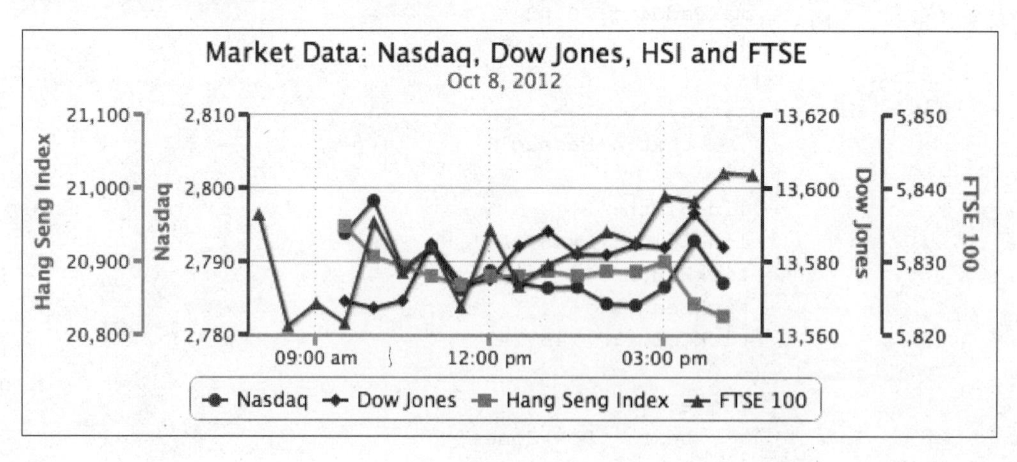

Revisiting the series config

By now, we should have an idea of what the `series` property does. In this section, we are going to examine it in more detail.

The `series` property is an array of series configuration objects that contain data- and series-specific options. It allows us to specify single-series data and multiple-series data. The purpose of series objects is to inform Highcharts of the format of the data and how the data is presented in the chart.

All data values in the chart are specified through the `data` field. The `data` field is highly flexible and can take an array in a number of forms, as follows:

- Numerical values
- An array with *x* and *y* values
- A point object with properties describing the data point

The first two options have already been examined in the *Accessing the axis data type* section. In this section, we will explore the third option. Let's use the single-series Nasdaq example and we will specify the series data through a mixture of numerical values and objects:

```
series: [{
    name: 'Nasdaq',
    pointStart: Date.UTC(2012, 4, 11, 9, 30),
    pointInterval: 30 * 60 * 1000,
    data: [{
        // First data point
        y: 2606.01,
        marker: {
            symbol: 'url(./sun.png)'
            }
        }, 2622.08, 2636.03, 2637.78,
      {
        // Highest data point
        y: 2639.15,
        dataLabels: {
            enabled: true
        },
        marker: {
            fillColor: '#33CC33',
            radius: 5
        }
      }, 2637.09, 2633.38, 2632.23, 2632.33,
```

```
            2632.59, 2630.34, 2626.89, 2624.59,
        {
          // Last data point
          y: 2615.98,
          marker: {
              symbol: 'url(./moon.png)'
              }
        }]
    }]
```

The first and last data points are objects that have *y* axis values and image files to indicate the opening and closing of the market. The highest data point is configured with a different color and data label. The size of the data point is also set slightly larger than the default. The rest of the data arrays are just numerical values, as shown in the following screenshot:

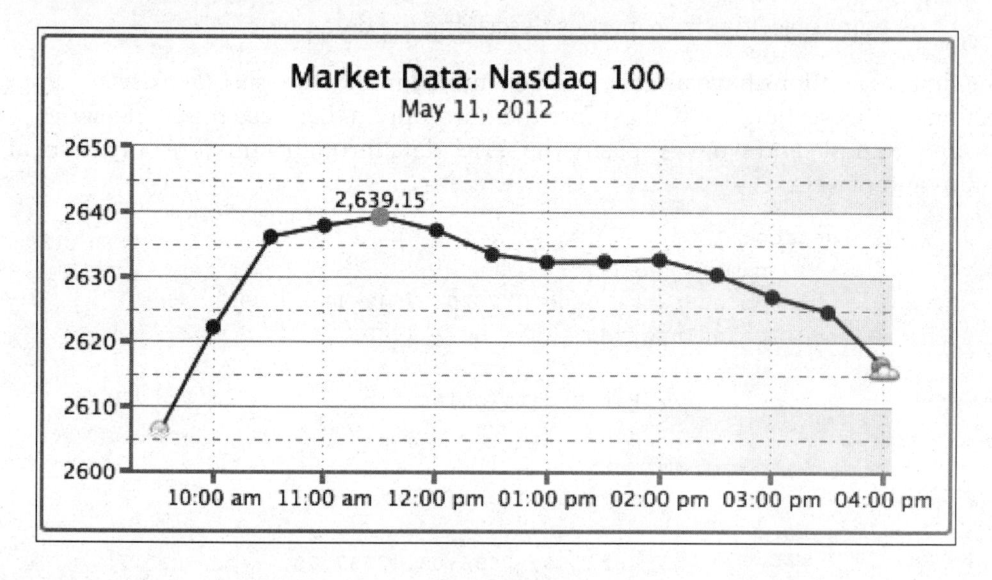

Exploring PlotOptions

The `plotOptions` object is a wrapper object for config objects for each series type supported in Highcharts. These configurations have properties such as `plotOptions.line.lineWidth`, common to other series types, as well as other configurations such as `plotOptions.pie.center` that is only specific to the pie series type. Among the specific series, there is `plotOptions.series`, which is used for common plotting options shared by the whole series.

The preceding `plotOptions` object can form a chain of precedence between `plotOptions.series`, `plotOptions.{series-type}`, and the series configuration. For example, `series[x].shadow` (where `series[x].type` is `'pie'`) has a higher precedence than `plotOptions.pie.shadow`, which in turn has a higher precedence than `plotOptions.series.shadow`.

The purpose of this is that the chart is composed of multiple different series types. For example, in a chart with multiple series of columns and a single line series, the common properties between column and line can be defined in `plotOptions.series.*`, whereas `plotOptions.column` and `plotOptions.line` hold their own specific property values. Moreover, properties in `plotOptions.{series-type}.*` can be further overridden by the same series type specified in the series array.

The following is a reference for the configurations in precedence. The higher-level ones have lower precedence, which means that configurations defined in the lower level of the chain can override properties defined in the higher level of the chain. For the series array, the preference is valid if `series[x].type` or the default series type value is the same as the series type in `plotOptions`:

```
chart.type
    series[x].type

plotOptions.series.{seriesProperty}
    plotOptions.{series-type}.{seriesProperty}
        series[x].{seriesProperty}

plotOptions.points.events.*
       series[x].data[y].events.*

plotOptions.series.marker.*
       series[x].data[y].marker.*
```

The `plotOptions` object contains properties controlling how a series type is presented in the chart—for example inverted charts, series colors, stacked column charts, user interactions with the series, and so on. All these options will be covered in detail when we study each type of chart. Meanwhile, we will explore the concept of `plotOptions` with a monthly Nasdaq graph. The graph has five different series data types: open, close, high, low, and volume. Normally, this data is used for plotting daily stock charts (OHLCV). We compact them into a single chart for the purpose of demonstrating `plotOptions`.

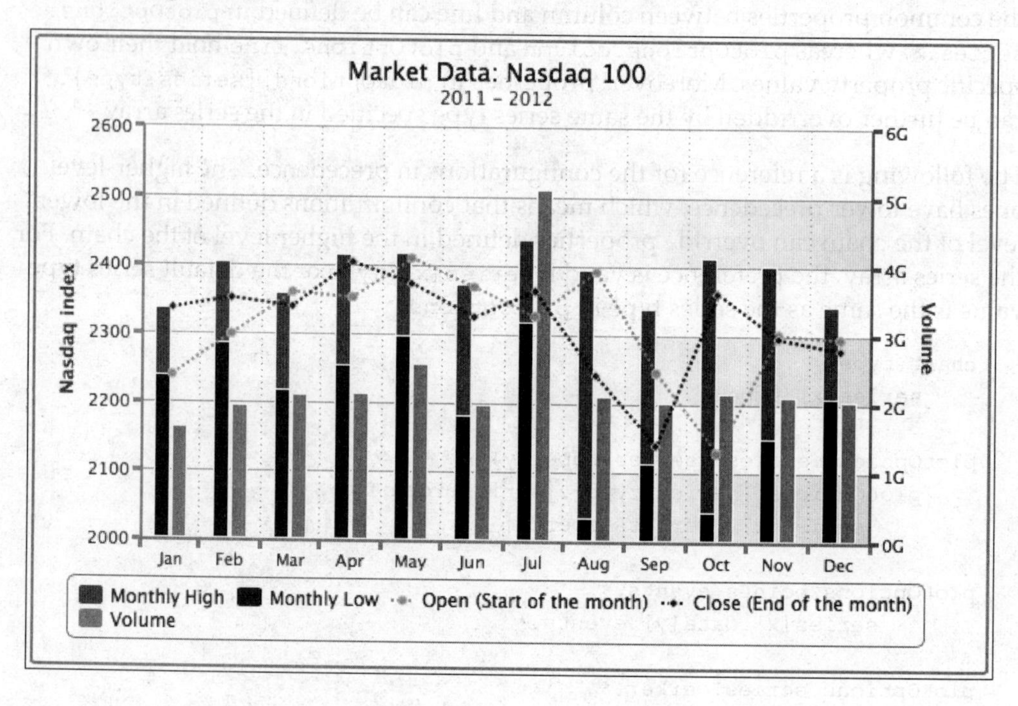

The following is the chart configuration code for generating the preceding graph:

```
chart: {
    renderTo: 'container',
    height: 250,
    spacingRight: 30
},
title: {
    text: 'Market Data: Nasdaq 100'
},
subtitle: {
    text: '2011 - 2012'
},
```

```
xAxis: {
    categories: [ 'Jan', 'Feb', 'Mar', 'Apr',
                  'May', 'Jun', 'Jul', 'Aug',
                  'Sep', 'Oct', 'Nov', 'Dec' ],
    labels: {
        y: 17
    },
    gridLineDashStyle: 'dot',
    gridLineWidth: 1,
    lineWidth: 2,
    lineColor: '#92A8CD',
    tickWidth: 3,
    tickLength: 6,
    tickColor: '#92A8CD',
    minPadding: 0.04,
    offset: 1
},
yAxis: [{
    title: {
        text: 'Nasdaq index'
    },
    min: 2000,
    minorGridLineColor: '#D8D8D8',
    minorGridLineDashStyle: 'dashdot',
    gridLineColor: '#8AB8E6',
    alternateGridColor: {
        linearGradient: {
            x1: 0, y1: 1,
            x2: 1, y2: 1
        },
        stops: [ [0, '#FAFCFF' ],
                 [0.5, '#F5FAFF'] ,
                 [0.8, '#E0F0FF'] ,
                 [1, '#D6EBFF'] ]
    },
    lineWidth: 2,
    lineColor: '#92A8CD',
    tickWidth: 3,
    tickLength: 6,
    tickColor: '#92A8CD'
}, {
    title: {
        text: 'Volume'
```

```
    },
    lineWidth: 2,
    lineColor: '#3D96AE',
    tickWidth: 3,
    tickLength: 6,
    tickColor: '#3D96AE',
    opposite: true
}],
credits: {
    enabled: false
},
plotOptions: {
    column: {
        stacking: 'normal'
    },
    line: {
        zIndex: 2,
        marker: {
            radius: 3,
            lineColor: '#D9D9D9',
            lineWidth: 1
        },
        dashStyle: 'ShortDot'
    }
},
series: [{
  name: 'Monthly High',
  // Use stacking column chart - values on
  // top of monthly low to simulate monthly
  // high
  data: [ 98.31, 118.08, 142.55, 160.68, ... ],
  type: 'column',
  color: '#4572A7'
}, {
  name: 'Monthly Low',
  data: [ 2237.73, 2285.44, 2217.43, ... ],
  type: 'column',
  color: '#AA4643'
}, {
  name: 'Open (Start of the month)',
  data: [ 2238.66, 2298.37, 2359.78, ... ],
  color: '#89A54E'
}, {
```

```
          name: 'Close (End of the month)',
          data: [ 2336.04, 2350.99, 2338.99, ... ],
          color: '#80699B'
        }, {
          name: 'Volume',
          data: [ 1630203800, 1944674700, 2121923300, ... ],
          yAxis: 1,
          type: 'column',
          stacking: null,
          color: '#3D96AE'
        }]
    }
```

Although the graph looks slightly complicated, we will go through the code step-by-step. First, there are two entries in the `yAxis` array: the first is for the Nasdaq index; the second y axis, displayed on the right-hand side (`opposite: true`), is for the volume trade. In the series array, the first and second series are specified as column series types (`type: 'column'`), which override the default series type `'line'`. Then the `stacking` option is defined as `'normal'` in `plotOptions.column`, which stacks the monthly high on top of the monthly low column (deep blue and black columns). Strictly speaking, the stacked column chart is used for displaying the ratio of data belonging to the same category. For the sake of demonstrating `plotOptions`, we used the stacked column chart to show the upper and lower ends of monthly trade. To do that, we take the difference between monthly high and monthly low and substitute the differences back into the monthly high series. So in the code, we can see that the data values in the monthly high series are much smaller than the monthly low.

The third and fourth series are the market open and market close index. Both take the default line series type and inherit options defined from `plotOptions.line`. The `zIndex` option is assigned to `2` to overlay both line series on top of the fifth volume series; otherwise, both lines are covered by the volume columns. The `marker` object configurations are to reduce the default data point size, as the whole graph is already compacted with columns and lines.

The last column series is the volume trade, and the `stacking` option in the series is manually set to `null`, which overrides the inherited option from `plotOptions.column`. This resets the series back to the non-stacking option, displaying as a separate column. Finally, the `yAxis` index option is set to align with the y axis of the volume series (`yAxis: 1`).

Styling tooltips

Tool tips in Highcharts are enabled by the `tooltip.enabled` Boolean option, which is `true` by default. In Highcharts 4, the default shape of the tooltip box has been changed to callout. The following shows the new style of tooltip:

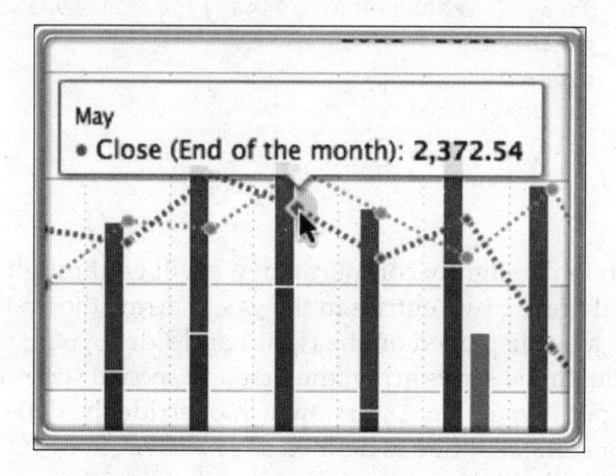

For the older style of tooltip shape, we can set the `tooltip.shape` option to `square`, which we will use in the following exercises.

Tooltip's content formats are flexible, which can be defined via a callback handler or in the HTML style. We will continue from the example in the previous section. As the chart is packed with multiple lines and columns, we can first enable the crosshair tool tip to help us align the data points onto the axes. The `crosshairs` configuration can take either a Boolean value to activate the feature or an object style for the crosshair line style. The following is the code snippet to set up crosshairs with an array of x- and y-axis configurations for the gray color and dash line styles:

```
tooltip : {
    shape: 'square',
    crosshairs: [{
        color: '#5D5D5D',
        dashStyle: 'dash',
        width: 2
    }, {
        color: '#5D5D5D',
        dashStyle: 'dash',
        width: 2
    }]
},
```

Again, the `dashStyle` option uses the same common line style values in Highcharts. See the crosshairs reference manual for all the possible values.

The following screenshot shows the view when hovering over a data point in the market close series. We see a tool tip box appear next to the pointer and gray crosshairs for both axes:

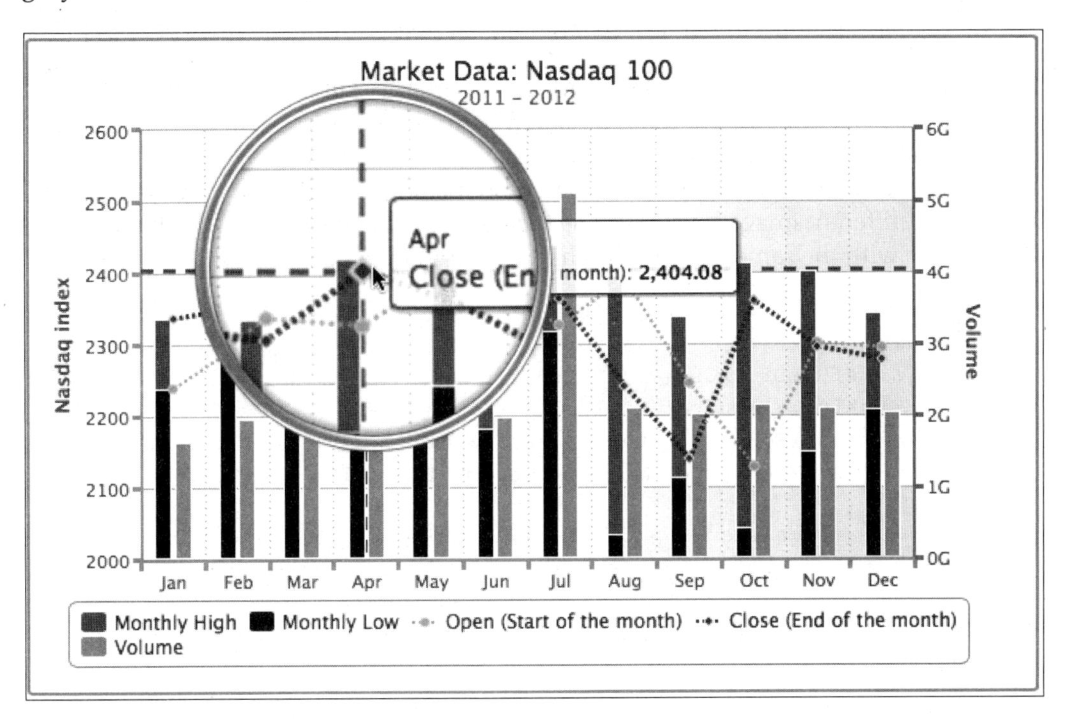

Formatting tooltips in HTML

Highcharts provides template options such as `headerFormat`, `pointFormat`, and `footerFormat` to construct the tool tip by specific template variables (or macros). These specific variables are series and point, and we can use their properties such as `point.x`, `point.y`, `series.name`, and `series.color` within the template. For instance, the default tool tip setting uses `pointFormat`, which has the default value of the following code snippet:

```
<span style="color:{series.color}">{series.name}</span>:
<b>{point.y}</b><br/>
```

Highcharts internally translates the preceding expression into SVG text markups, so only a subset of HTML syntax can be supported, which is `<b>`, `<br>`, `<strong>`, `<em>`, `<i>`, `<span>`, `<href>`, and font style attributes in CSS. However, if we want to have more flexibility in polishing the content, and the ability to include image files, we need to use the `useHTML` option for full HTML tool tips. This option allows us to do the following:

- Use other HTML tags such as `<img>` inside the tool tip
- Create a tool tip in real HTML content, so that it is outside the SVG markups

Here, we can format an HTML table inside a tool tip. We will use `headerFormat` to create a header column for the category and a bottom border to separate the header from the data. Then, we will use `pointFormat` to set up an icon image along with the series name and data. The image file is based on the `series.index` macro, so different series have different image icons. We use the `series.color` macro to highlight the series name with the same color in the chart and apply the `series.data` macro for the series value:

```
    tooltip : {
        useHTML: true,
        shape: 'square',
        headerFormat: '<table><thead><tr>' +
            '<th style="border-bottom: 2px solid #6678b1; color:
#039" ' +
            'colspan=2 >{point.key}</th></tr></thead><tbody>',
        pointFormat: '<tr><td style="color: {series.color}">' +
            '<img src="./series_{series.index}.png" ' +
            'style="vertical-align:text-bottom; margin-right: 5px" >'
+
            '{series.name}: </td><td style="text-align: right; color:
#669;">' +
            '<b>{point.y}</b></td></tr>',
        footerFormat: '</tbody></table>'
    },
```

When we hover over a data point, the template variable `point` is substituted internally for the hovered point object, and the series is replaced by the `series` object containing the data point.

The following is the screenshot of the new tool tip. The icon next to the series name indicates market close:

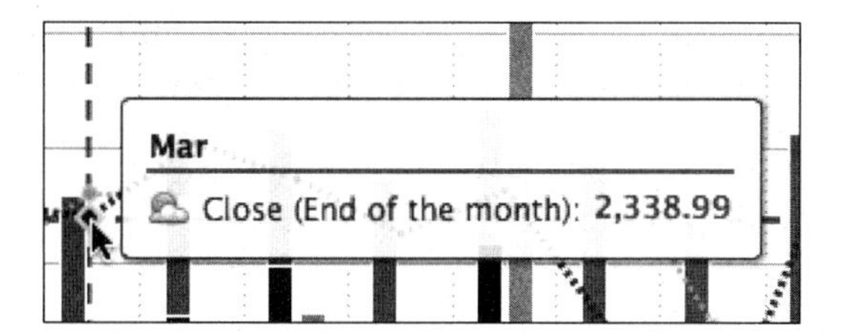

Using the callback handler

Alternatively, we can implement the tool tip through the callback handler in JavaScript. The tool tip handler is declared through the `formatter` option. The major difference between template options and the handler is that we can disable the tool tip display for certain points by setting conditions and returning the Boolean to `false`, whereas for template options we cannot. In the callback example, we use the `this.series` and `this.point` variables for the series name and values for the data point that is hovered over.

The following is an example of the handler:

```
formatter: function() {
    return '<span style="color:#039;font-weight:bold">' +
           this.point.category +
           '</span><br/><span style="color:' +
           this.series.color + '">' + this.series.name +
           '</span>: <span style="color:#669;font-weight:bold">'
+
           this.point.y + '</span>';
}
```

The preceding handler code returns an SVG text tool tip with the series name, category, and value, as shown in the following screenshot:

Applying a multiple-series tooltip

Another flexible tooltip feature is to allow all the series data to be displayed inside the same tooltip. This simplifies user interaction by looking up multiple series data in one action. To enable this feature, we need to set the `shared` option to `true`.

We will continue with the previous example for a multiple series tooltip. The following is the new tooltip code:

```
shared: true,
useHTML: true,
shape: 'square',
headerFormat: '<table><thead><tr><th colspan=2 >' +
              '{point.key}</th></tr></thead><tbody>',
pointFormat:  '<tr><td style="color: {series.color}">' +
              '{series.name}: </td>' +
              '<td style="text-align: right; color: #669;"> '
+
              '<b>{point.y}</b></td></tr>',
footerFormat: '</tbody></table>'
```

The preceding code snippet will produce the following screenshot:

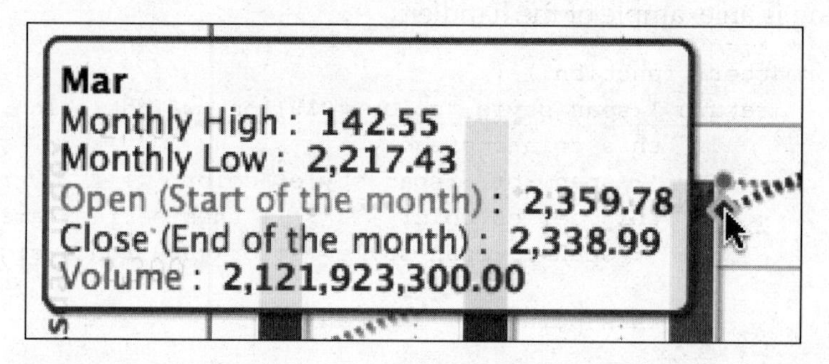

As previously discussed, we will use the monthly high and monthly low series to plot stacked columns that are actually used for plotting data within the same category. Therefore, the tooltip for the monthly high series is showing the subtracted values that we previously put in. To correct this within the tooltip, we can use the handler to apply different properties for the monthly high series, as follows:

```
shared: true,
shape: 'square',
formatter: function() {
    return '<span style="color:#039;font-weight:bold">' +
        this.x + '</span><br/>' +
        this.points.map(function(point, idx) {
            return '<span style="color:' + point.series.color +
```

```
                    '">' + point.series.name +
                    '</span>: <span style="color:#669;font-
weight:bold">' +
                     Highcharts.numberFormat((idx == 0) ? point.total
: point.y) + '</span>';
               }).join('<br/>');
        }
```

`point.total` is the total of the difference and the monthly low series value. The following screenshot shows the new corrected monthly high value:

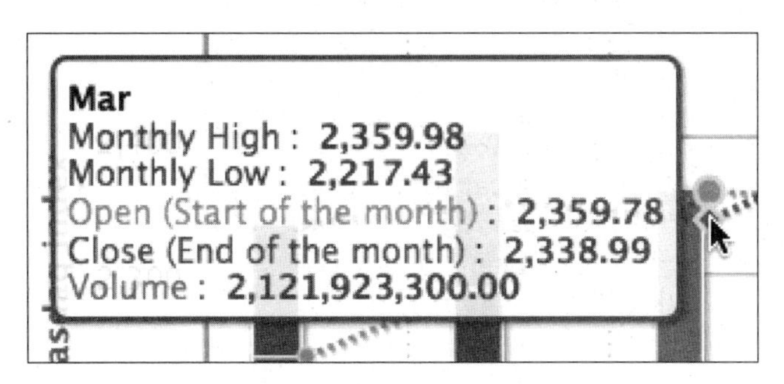

Animating charts

There are two types of animations in Highcharts: initial and update animations. An initial animation is the animation that happens when the series data is ready and the chart is displayed. An update animation occurs after the initial animation, when the series data or any parts of the chart anatomy have been changed.

The initial animation configurations can be specified through `plotOptions.series.animation` or `plotOptions.{series-type}.animation`, whereas the update animation is configured via the `chart.animation` property.

All Highcharts animations use jQuery implementation. The `animation` property can be a Boolean value or a set of options. For Boolean values, it is `true`. Highcharts can use jQuery for swing animation. These are the options:

- `duration`: This is the time, in milliseconds, to complete the animation.
- `easing`: This is the type of animation jQuery provides. The variety of animations can be extended by importing the jQuery UI plugin. A good reference can be found at `http://plugindetector.com/demo/easing-jquery-ui/`.

Here, we continue the example from the previous section. We will apply the animation settings to `plotOptions.column` and `plotOptions.line`, as follows:

```
plotOptions: {
    column: {
        ... ,
        animation: {
            duration: 2000,
            easing: 'swing'
        }
    },
    line: {
        .... ,
        animation: {
            duration: 3000,
            easing: 'linear'
        }
    }
},
```

The animations are tuned into at a much slower pace, so we can see the difference between linear and swing animations. The line series appears at a linear speed along the x axis, whereas the column series expands upwards at a linear speed and then decelerates sharply when approaching the end of the display. The following is a screenshot showing an ongoing linear animation:

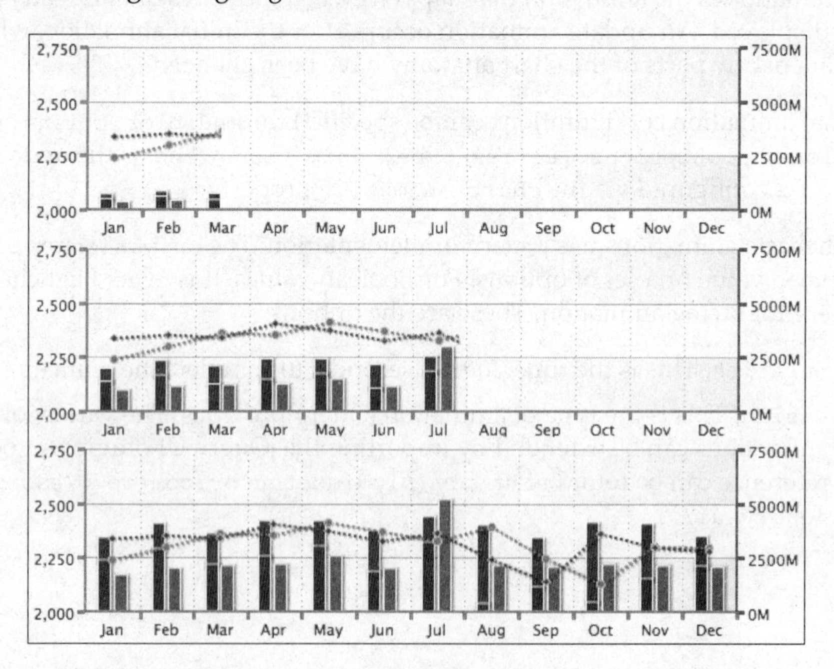

Expanding colors with gradients

Highcharts not only supports single color values, but also allows complex color gradient definitions. In Highcharts, the color gradient is based on the SVG linear color gradient standard, which is composed of two sets of information as follows:

- `linearGradient`: This gives a gradient direction for a color spectrum made up of two sets of x and y coordinates; ratio values are between 0 and 1, or in percentages
- `stops`: This gives a sequence of colors to be filled in the spectrum, and their ratio positions within the gradient direction

We can use the previous stock market example with only the volume series, and redefine `yAxis alternateGridColor` as follows:

```
yAxis: [{
     title: { text: 'Nasdaq index' },
     ....

     alternateGridColor: {
         linearGradient: [ 10, 250, 400, 250 ],
             stops: [
                    [ 0, 'red' ],
                 [ 0.2, 'orange' ],
                 [ 0.5, 'yellow' ] ,
                 [ 0.8, 'green' ] ,
                 [ 1, 'lime' ] ]
     }
```

`linearGradient` is an array of coordinate values that are arranged in the x1, y1, x2, y2 order. The values can be absolute coordinates, percentage strings, or ratio values between 0 and 1. The difference is that colors defined in coordinate values can be affected by the chart size, whereas percentage and ratio values avoid that.

The array syntax for absolute position gradients is deprecated because it doesn't work in the same way in both SVG and VML, it also doesn't scale well with varying chart sizes.

The stops property has an array of tuples: the first value is the offset ranging from 0 to 1 and the second value is the color definition. The offset and color values define where the color is positioned within the spectrum. For example, `[0, 'red' ]` and `[0.2, 'orange' ]` mean starting with red at the beginning and gradually changing the color to orange in a horizontal direction towards the position at *x = 80 (0.2 * 400)*, before changing from orange at *x = 80* to yellow at *x = 200*, and so on. The following is a screenshot of the multicolor gradient:

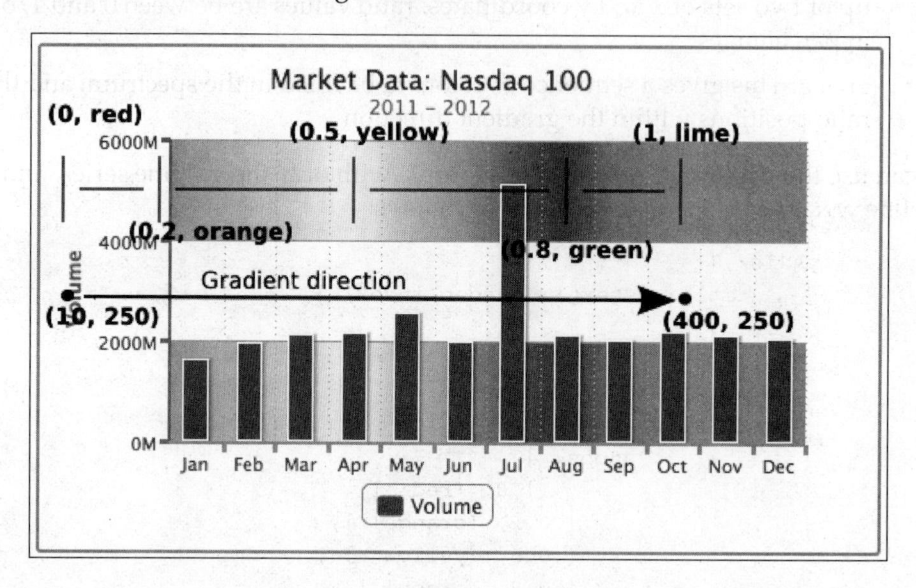

As we can see, the red and orange colors do not appear on the chart because the gradient is based on coordinates. Hence, depending on the size of the chart, the position of the y axis exceeds the red and orange coordinates in this example. Alternatively, we can specify `linearGradient` in terms of percentage, as follows:

```
linearGradient: [ '20%', 250, '90%', 250 ]
```

This means `linearGradient` stretches from `20%` of the width of the chart to `90%`, so that the color bands are not limited to the size of the chart. The following screenshot shows the effect of the new `linearGradient` setting:

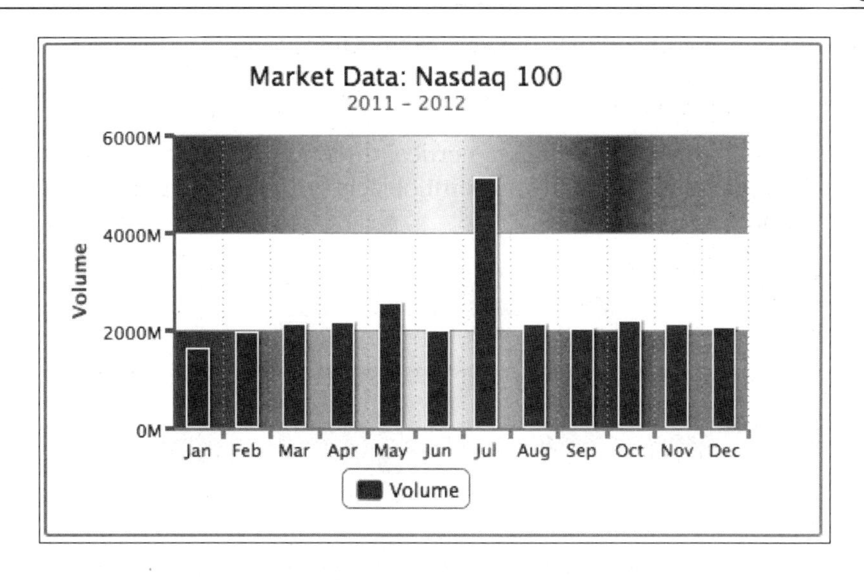

The chart background now has the complete color spectrum. As for specifying ratio values between 0 and 1, `linearGradient` must be defined in an object style, otherwise the values will be treated as coordinates. Note that the ratio values are referred to as the fraction over the plot area only, and not the whole chart.

```
linearGradient: { x1: 0, y1: 0, x2: 1, y2: 0 }
```

The preceding line of code is an alternative way to set the horizontal gradient.

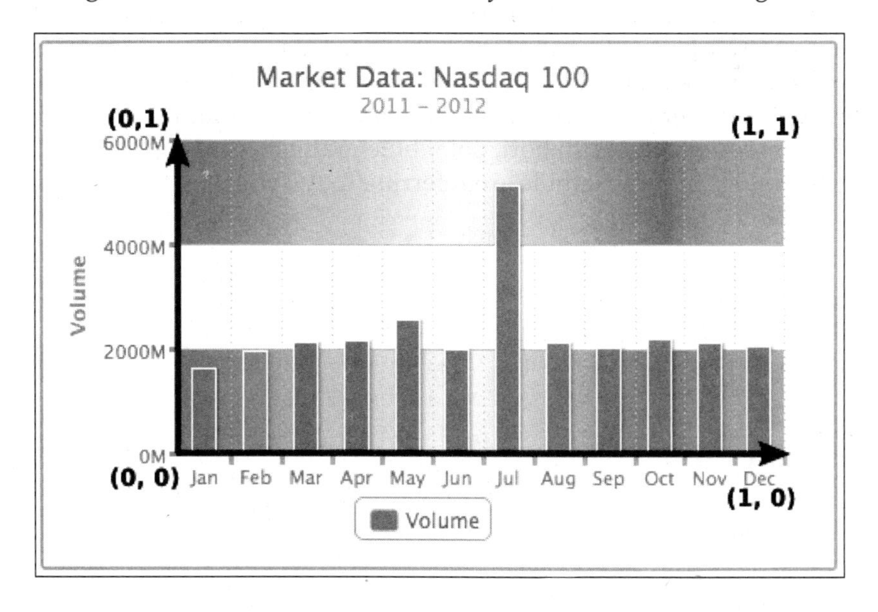

The following line of code adjusts the vertical gradient:

```
linearGradient: { x1: 0, y1: 0, x2: 0, y2: 1 }
```

It produces a gradient background in a vertical direction. We also set the `'Jan'` and `'Jul'` data points individually as point objects with linear shading in a vertical direction:

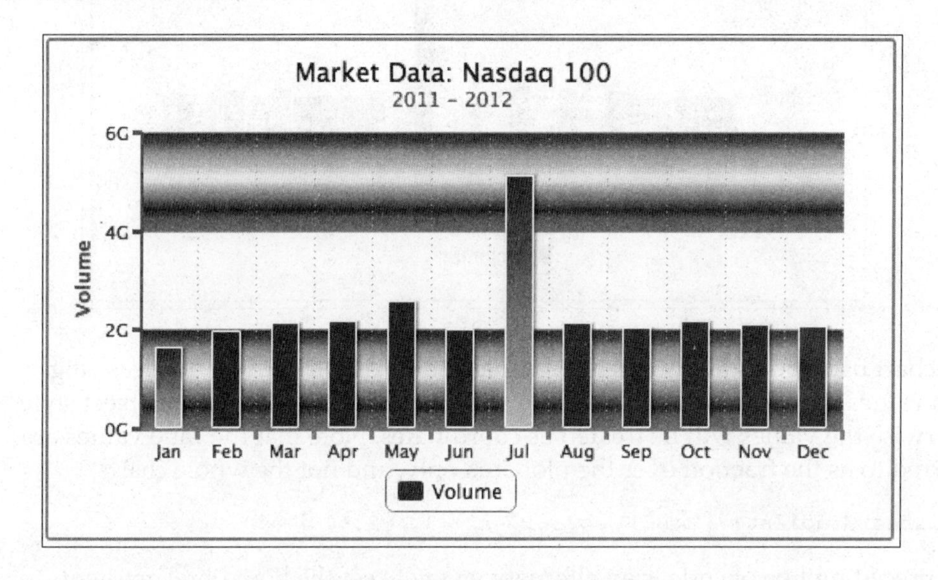

Moreover, we can manipulate Highcharts standard colors to trigger a color gradient in the series plot. This approach is taken from a post in a Highcharts forum experimenting with the look of 3D charts. Before plotting a chart, we need to overwrite the default series color with a gradient color. The following code snippet replaces the first series color with horizontal blue gradient shading. Note that the ratio gradient values in this example are referring to the width of the series column:

```
$(document).ready(function() {

    Highcharts.getOptions().colors[0] = {
            linearGradient: { x1: 0, y1: 0, x2: 1, y2: 0 },
            stops: [ [ 0, '#4572A7' ],
                     [ 0.7, '#CCFFFF' ],
                     [ 1, '#4572A7' ] ]
    };

    var chart = new Highcharts.Chart({  ...
```

The following is a screenshot of a column chart with color shading:

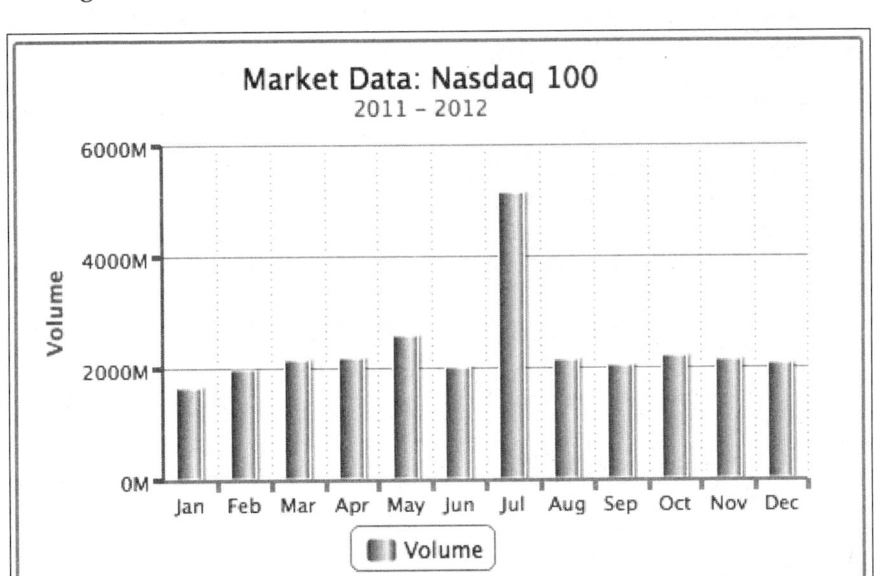

Zooming data with the drilldown feature

Highcharts provides an easy way to zoom data interactively between two series of data. This feature is implemented as a separate module. To use the feature, users need to include the following JavaScript:

```
<script type="text/javascript"
 src="http://code.highcharts.com/modules/drilldown.js"></script>
```

There are two ways to build an interactive drilldown chart: synchronous and asynchronous. The former method requires users to supply both top and detail levels of data in advance, and arrange both levels of data inside the configuration. Both levels of data are specified as standard Highcharts `series` configs, except that the zoom-in series is located inside the `drilldown` property. To join both levels, the top-level series must provide the option `series.data[x].drilldown`, with a matching name to the option `drilldown.series.[y].id` in the detail level.

Let's revisit the web browser statistic example and plot the data in a column chart. We will go through column series in more detail later in the book. The following is the `series` config for the top-level data:

```
series: [{
    type: 'column',
    name: 'Web browsers',
    colorByPoint: true,
    data: [{
        name: 'Chrome', y: 55.8, drilldown: 'chrome'
    }, {
        name: 'Firefox', y: 26.8, drilldown: 'firefox',
    }, {
        name: 'Internet Explorer', y: 9, drilldown: 'ie'
    }, {
        name: 'Safari', y: 3.8
    }]
}],
```

Each column is defined with a `drilldown` option and assigned a specific name. Note that it is not mandatory to make all the data points zoomable. The preceding example demonstrates the `Safari` column without the `drilldown` property.

To correspond to each of the `drilldown` assignments, an array of series data (detail level) is configured to match against the values. The following shows the setup of the drilldown data:

```
drilldown: {
   series: [{
       id: 'chrome',
       name: 'Chrome',
       data: [{
           name: 'C 32', y: 1
       }, {
           name: 'C 31', y: 49
       }, {

       }]
   }, {
       id: 'firefox',
       name: 'Firfox',
       data: [{
```

```
                        name: 'FF 26', y: 6.9
                    }, {
                        name: 'FF 25', y: 13.3
                    }, {
                                                              w.
                    }]
                                                  ....
                }]
            }
```

This produces the following screenshot:

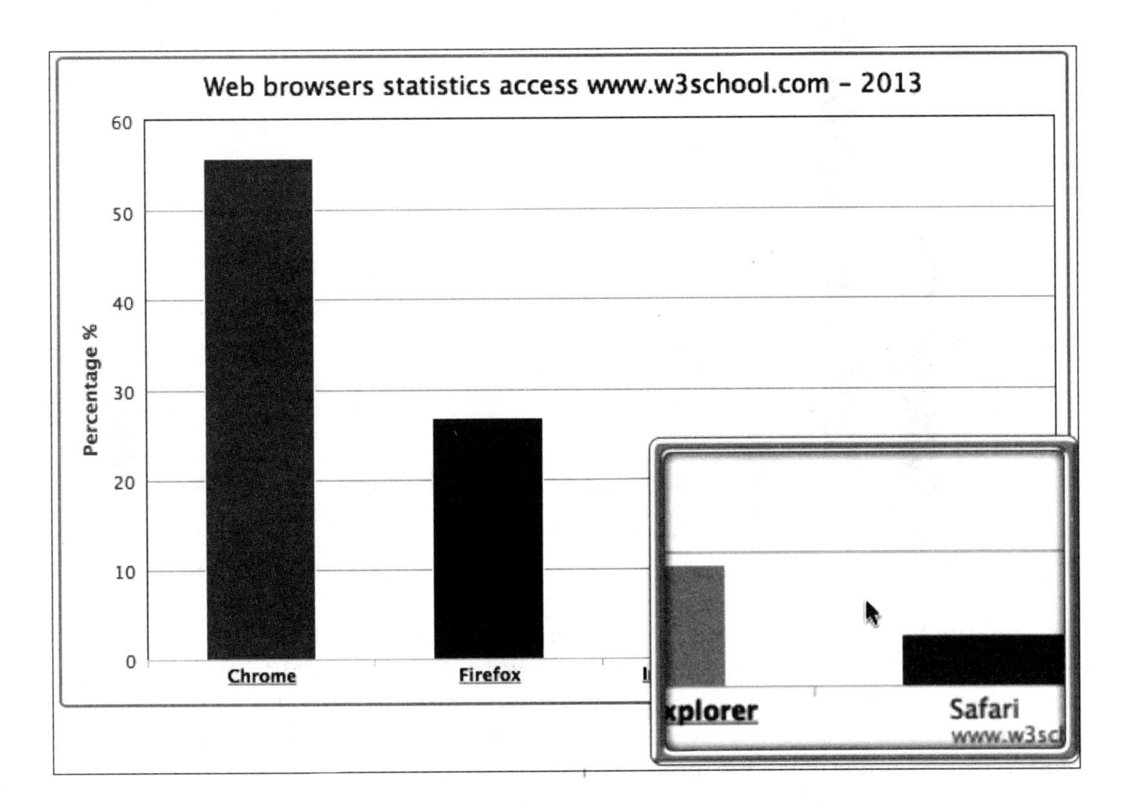

As we can see, the category labels along the x-axis are decorated in a dark color and underlined (the default style). They are clickable, and appear different when compared to the non-drilldown column, Safari. When we click on the Firefox column, or the label, the column animates and zooms into multiple columns along with a back button at the top right corner, as shown in the following screenshot:

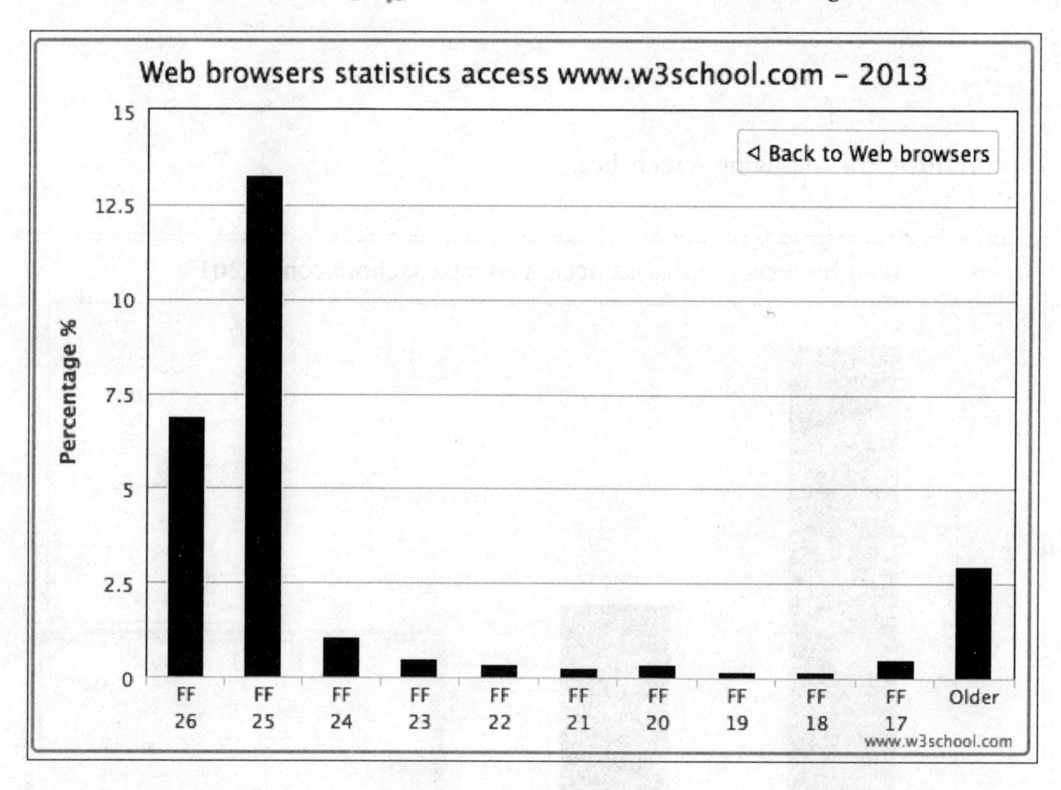

Let's modify the chart with more styles. There are properties within the `drilldown` option for smartening the data labels such as `activeAxisLabelStyle` and `activeDataLabelStyle` that take a configuration object of CSS styles. In the following code, we polish the category labels with features such as hyperlinks, blue color, and underlines. Additionally, we change the cursor style into the browser-specific zoom-in icon:

webkit-zoom-in is the browser-specific cursor name for Chrome and Safari; moz-zoom-in/out is for Firefox.

```
plotOptions: {
    column: {
        dataLabels: {
```

```
                enabled: true
            }
        }
    },
    drilldown: {
        activeAxisLabelStyle: {
            textDecoration: 'none',
            cursor: '-webkit-zoom-in',
            color: 'blue'
        },
        activeDataLabelStyle: {
            cursor: '-webkit-zoom-in',
            color: 'blue',
            fontStyle: 'italic'
        },
        series: [{
            ....
        }]
    }
```

The following screenshot shows the new text label style as well as the new cursor icon over the category label:

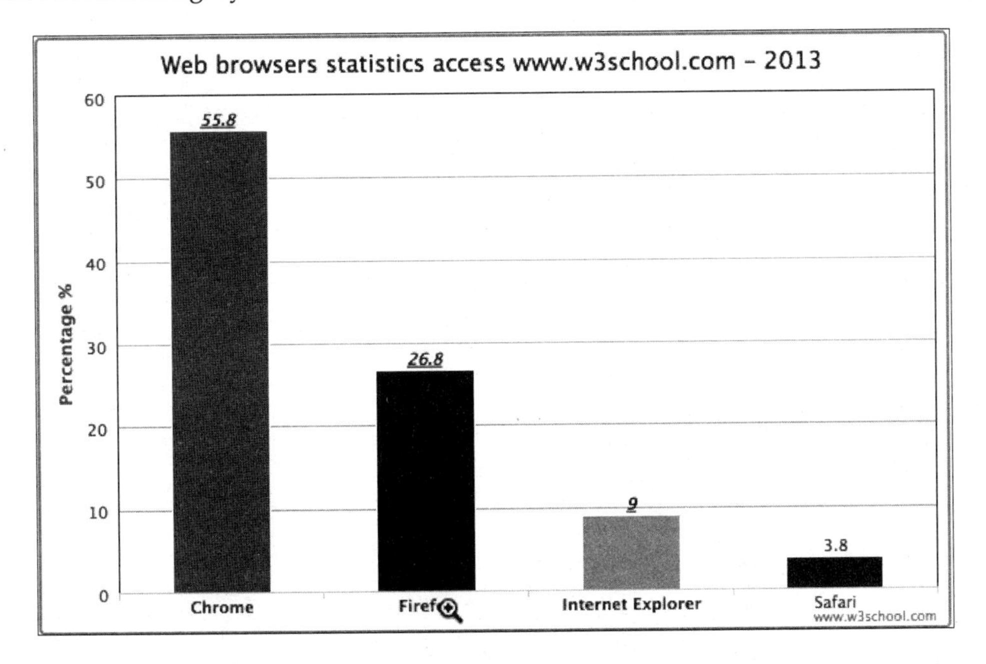

Another great flexibility in Highcharts is that the top and detail level charts are not limited to the same type of series. The drilldown series can be any series best suited for presenting the detail data. This can be accomplished by assigning a different series `type` in the detail level. The following is the further code update for the example, in which the Internet Explorer column is zoomed in and presented in a donut (or ring) chart. We also set the data labels for the pie chart to show both the version name and percentage value:

```
plotOptions: {
    column: {
        ....
    },
    pie: {
        dataLabels: {
            enabled: true,
            format: '{point.name} - {y}%'
        }
    }
},
drilldown: {
    activeAxisLabelStyle: {
        ....
    },
    activeDataLabelStyle: {
        ....
    },
    series: [{
        id: 'ie',
        name: 'Internet Explorer',
        type: 'pie',
        innerSize: '30%',
        data: [{
            ....
        }],
    }, {
        ....
    }]
}
```

The following is the screenshot of the donut chart:

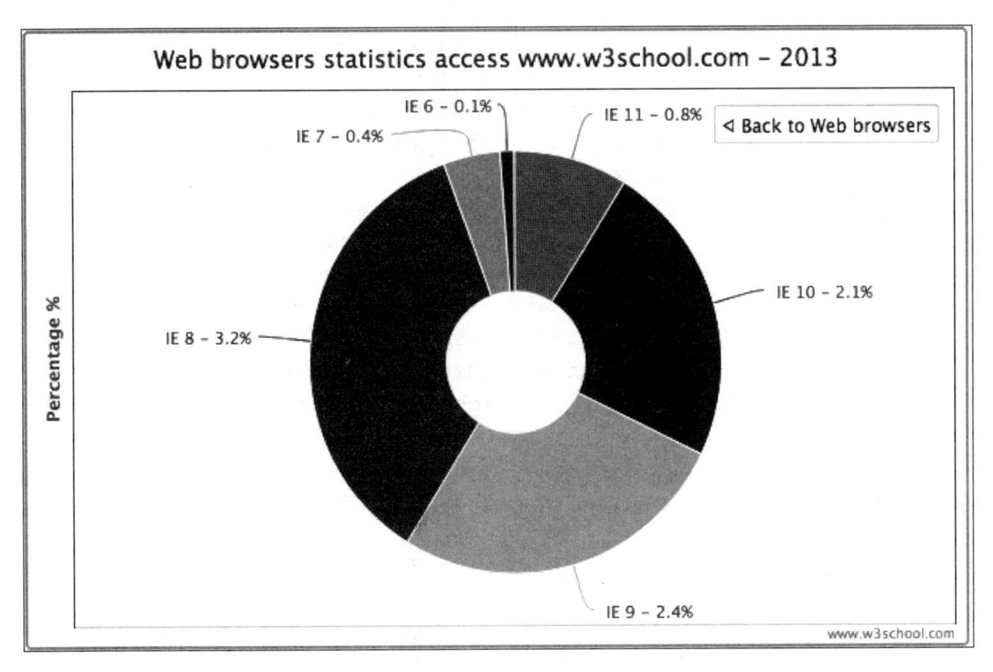

The preceding donut chart is slightly misleading because the total value of all the data labels is 9 percent, but the presentation of the donut chart gives the impression of 100 percent in total. In order to clarify that 9 percent is the total value, we can take advantage of the Highcharts Renderer engine (we will explore this further in *Chapter 5, Pie Charts*) to display an SVG text box in the middle of the ring. However, the `drilldown` option only allows us to declare series-specific options. Moreover, we would like the 9 percent text box to appear only when clicking on the 'Internet Explorer' column.

One trick to achieve that is to use Highcharts events, which will be examined further in *Chapter 11, Highcharts Events*. Here, we use the events specific to the `drilldown` action. The basic idea is that, when the `drilldown` event is triggered, we check the data point being clicked is **Internet Explorer**. If so, then we create a textbox using the chart's renderer engine. In the `drillup` event (triggered when the **Back to ...** button is clicked), we remove the text box if it exists:

```
chart = new Highcharts.Chart({
    chart: {
        renderTo: 'container',
        events: {
```

```
            drilldown: function(event) {
                // According to Highcharts documentation,
                // the clicked data point is embedded inside
                // the event object
                var point = event.point;
                var chart = point.series.chart;
                var renderer = chart.renderer;
                if (point.name == 'Internet Explorer') {
                    // Create SVG text box from
                    // Highcharts Renderer
                    // chart center position based on
                    // it's dimension
                    pTxt = renderer.text('9%',
                            (chart.plotWidth / 2) +
                            chart.plotLeft - 43,
                            (chart.plotHeight / 2) +
                            chart.plotTop + 25)
                        .css({
                            fontSize: '45px',
                            // Google Font
                            fontFamily: 'Old Standard TT',
                            fontStyle: 'italic',
                            color: 'blue'
                        }).add();
                }
            }, // drilldown
            drillup: function() {
                pTxt && (pTxt = pTxt.destroy());
            }
        }
```

Here is the refined chart with the textbox centered in the chart:

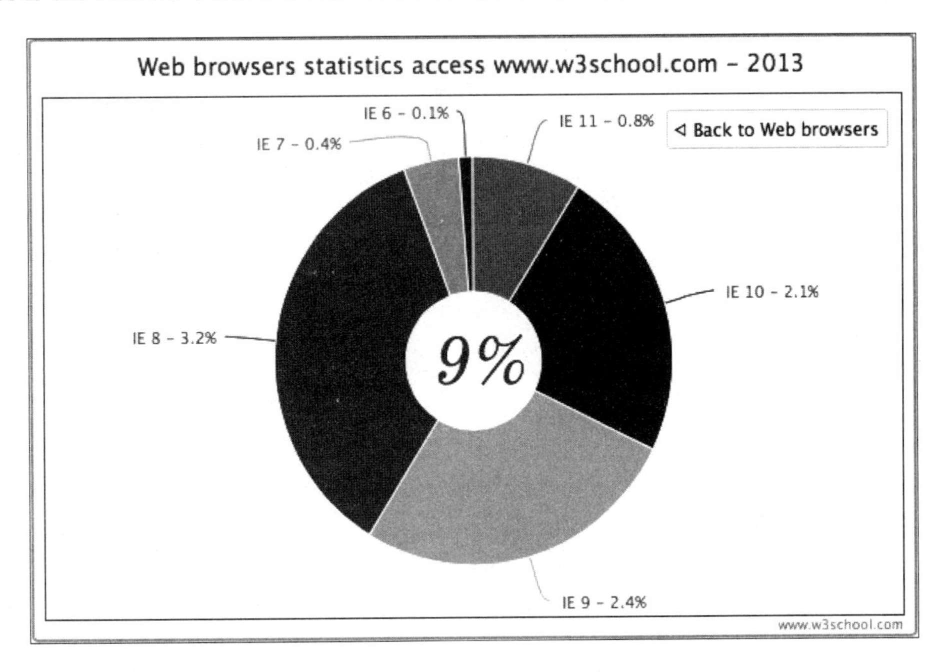

Summary

In this chapter, major configuration components were discussed and experimented with, and examples shown. By now, we should be comfortable with what we have covered already and ready to plot some of the basic graphs with more elaborate styles.

In the next chapter, we will explore the line, area, and scatter graphs supported by Highcharts. We will apply configurations that we have learned in this chapter and explore the series-specific style options to plot charts in an artistic style.

3

Line, Area, and Scatter Charts

In this chapter, we will learn about line, area, and scatter charts and explore their plotting options in more details. We will also learn how to create a stacked area chart and projection charts. Then, we will attempt to plot the charts in a slightly more artistic style. The reason for that is to provide us with an opportunity to utilize various plotting options. In this chapter, we will cover the following:

- Introducing line charts
- Sketching an area chart
- Highlighting and raising the base level
- Mixing line and area series
- Combining scatter and area series

Introducing line charts

First let's start with a single-series line chart. We will use one of the many data sets provided by *The World Bank* organization at `www.worldbank.org`. The following is the code snippet to create a simple line chart that shows the percentage of population aged 65 and above in Japan for the past three decades:

```
var chart = new Highcharts.Chart({
    chart: {
        renderTo: 'container'
    },
    title: {
        text: 'Population ages 65 and over (% of total)'
    },
```

```
                credits: {
                            position: {
                            align: 'left',
                            x: 20
                         },
                         text: 'Data from The World Bank'
                },
                yAxis: {
                          title: {
                          text: 'Percentage %'
                        }
                },
                xAxis: {
                    categories: ['1980', '1981', '1982', ... ],
                    labels: {
                               step: 5
                           }
                        },
                series: [{
                            name: 'Japan - 65 and over',
                            data: [  9, 9, 9, 10, 10, 10, 10 ... ]
                         }]
        });
```

The following is the display of the simple chart:

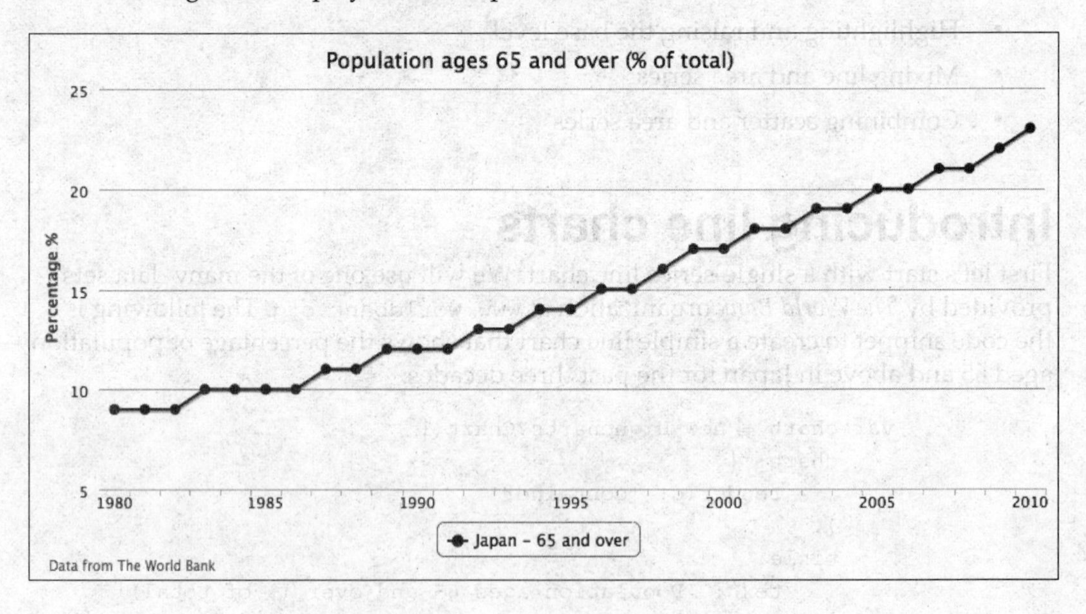

Instead of specifying the year number manually as strings in categories, we can use the `pointStart` option in the `series` config to initiate the *x* axis value for the first point. So we have an empty `xAxis` config and `series` config, as follows:

```
xAxis: {
},
series: [{
    pointStart: 1980,
    name: 'Japan - 65 and over',
    data: [  9, 9, 9, 10, 10, 10, 10 ... ]
}]
```

Extending to multiple-series line charts

We can include several more line series and emphasize the Japan series by increasing the line width to be `6`-pixels wide, as follows:

```
series: [{
    lineWidth: 6,
    name: 'Japan',
    data: [  9, 9, 9, 10, 10, 10, 10 ... ]
}, {
    Name: 'Singapore',
    data: [ 5, 5, 5, 5, ... ]
}, {
        // Norway, Brazil, and South Africa series data...
                    ...
}]
```

By making that line thicker, the line series for the Japanese population becomes the focus in the chart, as shown in the following screenshot:

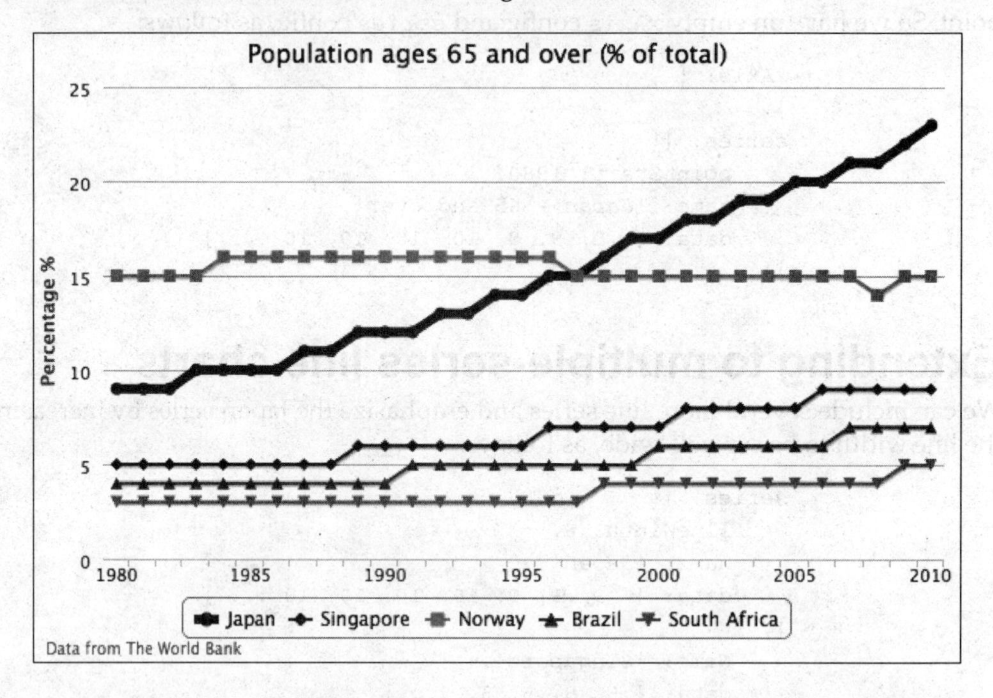

Let's move on to a more complicated line graph. For the sake of demonstrating inverted line graphs, we use the `chart.inverted` option to flip the y and x axes to opposite orientations. Then, we change the line colors of the axes to match the same series colors as in the previous chapter. We also disable data point markers for all the series, and finally align the second series to the second entry on the y-axis array, as follows:

```
chart: {
    renderTo: 'container',
    inverted: true
},
yAxis: [{
    title: {
        text: 'Percentage %'
    },
    lineWidth: 2,
    lineColor: '#4572A7'
}, {
    title: {
        text: 'Age'
    },
    opposite: true,
```

```
            lineWidth: 2,
            lineColor: '#AA4643'
        }],
        plotOptions: {
            series: {
                marker: {
                    enabled: false
                }
            }
        },
        xAxis: {
           categories: [ '1980', '1981', '1982', ...,
                         '2009', '2010' ],
           labels: {
               step: 5
           }
        },
        series: [{
            name: 'Japan - 65 and over',
            type: 'spline',
            data: [ 9, 9, 9, ... ]
        }, {
            name: 'Japan - Life Expectancy',
            yAxis: 1,
            data: [ 76, 76, 77, ... ]
        }]
```

The following is the inverted graph with double y axes:

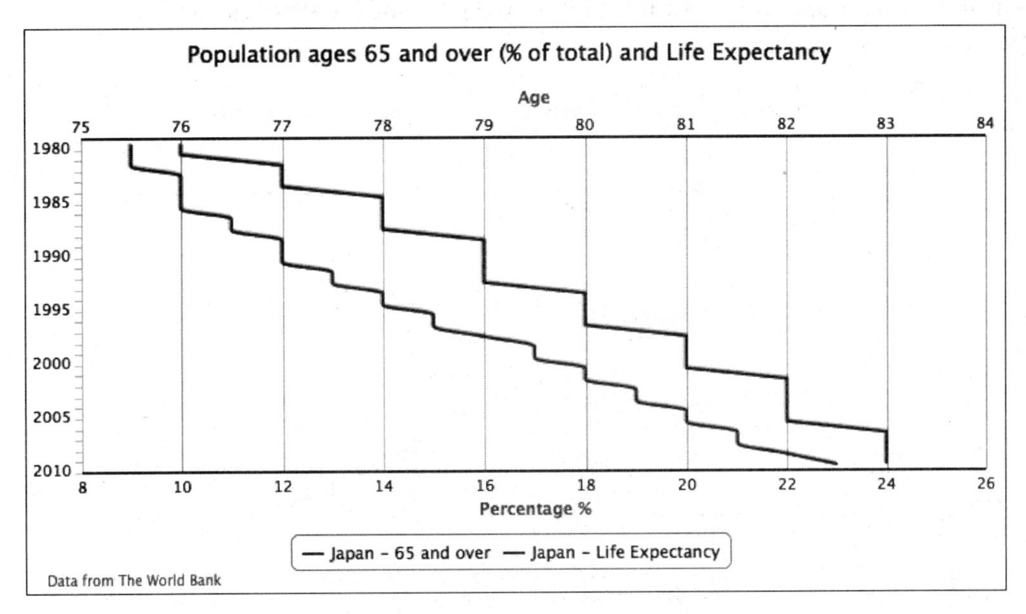

The data representation of the chart may look slightly odd as the usual time labels are swapped to the *y* axis and the data trend is difficult to comprehend. The `inverted` option is normally used for showing data in a non-continuous form and in a bar format. If we interpret the data from the graph, 12 percent of the population is 65 or over in 1990, and the life expectancy is 79.

Setting `plotOptions.series.marker.enabled` to `false` switches off all the data point markers. If we want to display a point marker for a particular series, we can either switch off the marker globally and then turn the marker on for individual series, or the other way round:

```
plotOptions: {
    series: {
        marker: {
            enabled: false
        }
    }
},
series: [{
    marker: {
        enabled: true
    },
    name: 'Japan - 65 and over',
    type: 'spline',
    data: [ 9, 9, 9, ... ]
}, {
```

The following graph demonstrates that only the 65-and-over series has point markers:

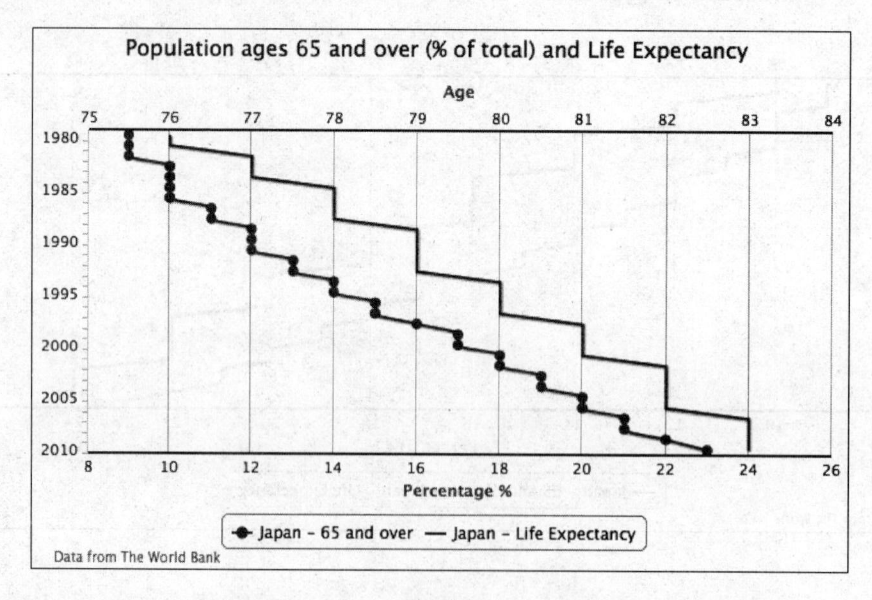

Highlighting negative values and raising the base level

Sometimes, we may want to highlight both positive and negative regions in different colors. In such cases, we can specify the series color for negative values with the series option, `negativeColor`. Let's create a simple example with inflation data, containing both positive and negative data. Here is the series configuration:

```
plotOptions: {
    line: {
        negativeColor: 'red'
    }
},
series: [{
    type: 'line',
    color: '#0D3377',
    marker: {
        enabled: false
    },
    pointStart: 2004,
    data:[ 2.9, 2.8, 2.4, 3.3, 4.7,
           2.3, 1.1, 1.0, -0.3, -2.1
    ]
}]
```

We assign the color red for negative inflation values and disable the markers in the line series. The line series color is defined by another color, blue, which is used for positive values. The following is the graph showing the series in separate colors:

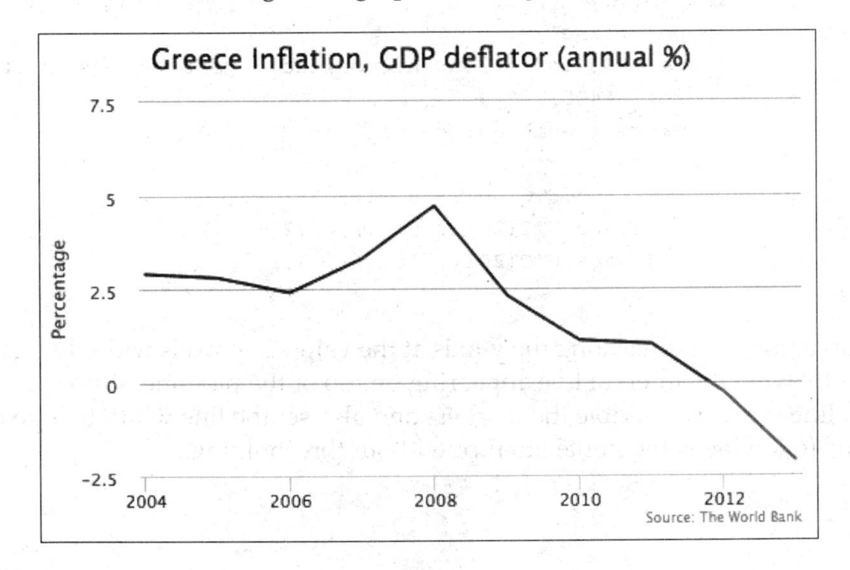

Let's create another slightly more advanced example, in which we define the negative territory in terms of the subject. We plot another inflation chart, but based on the **European Central Bank** (**ECB**) definition of healthy inflation that is 2 percent. Anything below that level is regarded as unhealthy for the economy, so we set the color below that threshold to red. Beside the color threshold, we also set up a plotline on the y-axis to indicate the cut-off level. The following is our first try to set up the chart:

```
yAxis: {
    title: { text: null },
    min: 0,
    max: 4,
    plotLines: [{
        value: 2,
        width: 3,
        color: '#6FA031',
        zIndex: 1,
        label: {
            text: 'ECB....',
            ....
        }
    }]
},
xAxis: { type: 'datetime' },
plotOptions: {
    line: { lineWidth: 3 }
},
series: [{
    type: 'line',
    name: 'EU Inflation (harmonized), year-over-year (%)',
    color: '#0D3377',
    marker: { enabled: false },
    data:[
        [ Date.UTC(2011, 8, 1), 3.3 ],
        [ Date.UTC(2011, 9, 1), 3.3 ],
        [ Date.UTC(2011, 10, 1), 3.3 ],
        ....
```

We set up a green plotline along the y-axis at the value 2, 3 pixels wide. The `zIndex` option is to avoid the interval line appearing on top of the plot line. With the inflation line series, we disable the markers and also set the line width to 3 pixels wide. The following is the initial attempt without thresholding:

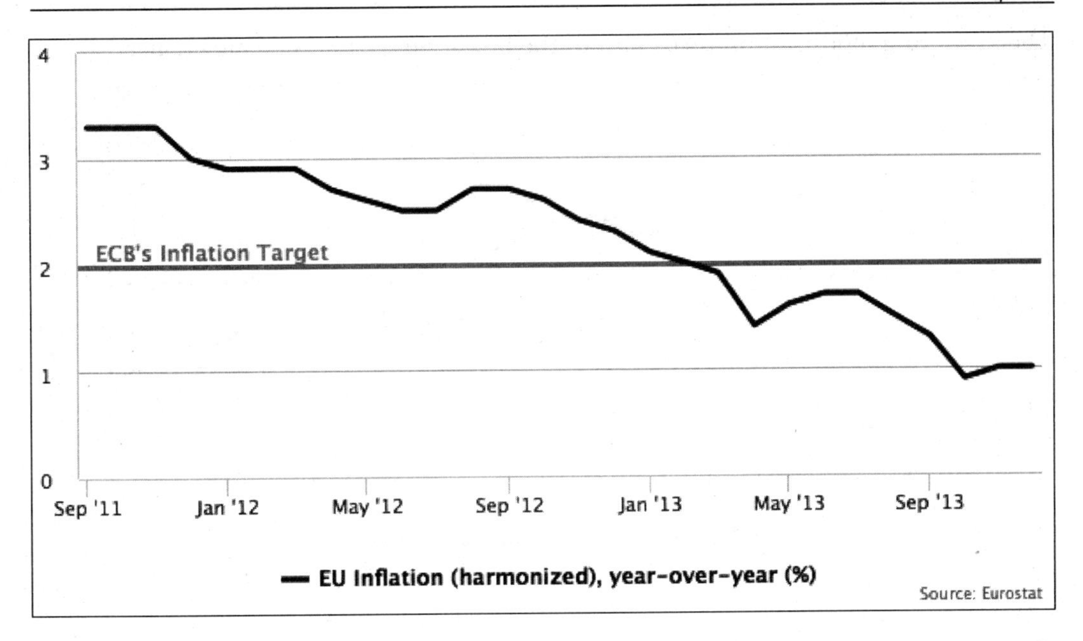

Let's apply the threshold level to the lines series. The default negative color on the y-axis level is at 0 value. As for this particular example, the base level for a negative color would be 2. To raise the base level to 2, we set the `threshold` property along with the `negativeColor` option:

```
plotOptions: {
    line: {
        lineWidth: 3,
        negativeColor: 'red',
        threshold: 2
    }
},
```

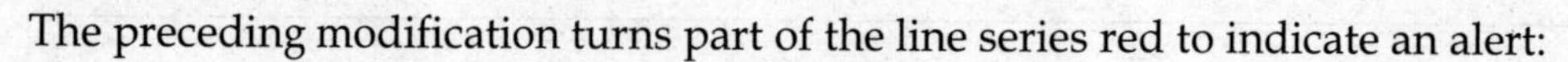

The preceding modification turns part of the line series red to indicate an alert:

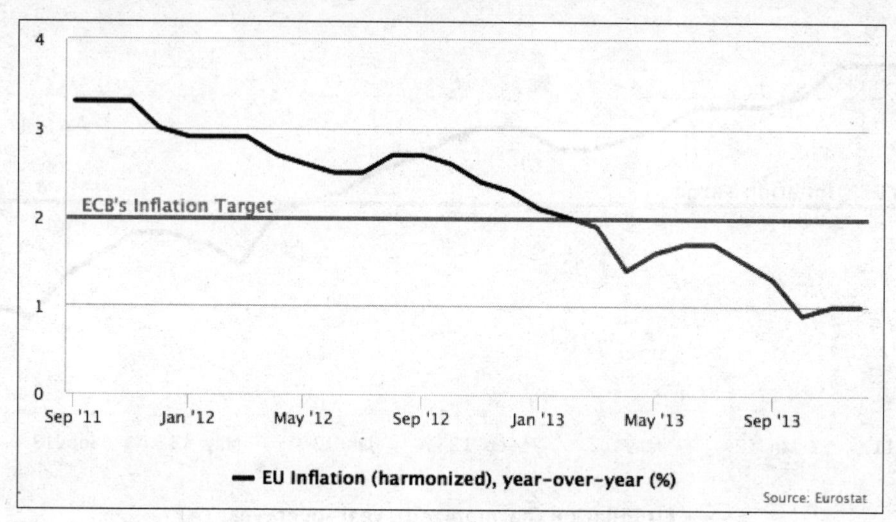

Sketching an area chart

In this section, we are going to use our very first example and turn it into a more stylish graph (based on the design of a wind energy poster by Kristin Clute): an area spline chart. An **area spline chart** is generated using the combined properties of area and spline charts. The main data line is plotted as a spline curve and the region underneath the line is filled in a similar color with a gradient and opaque style:

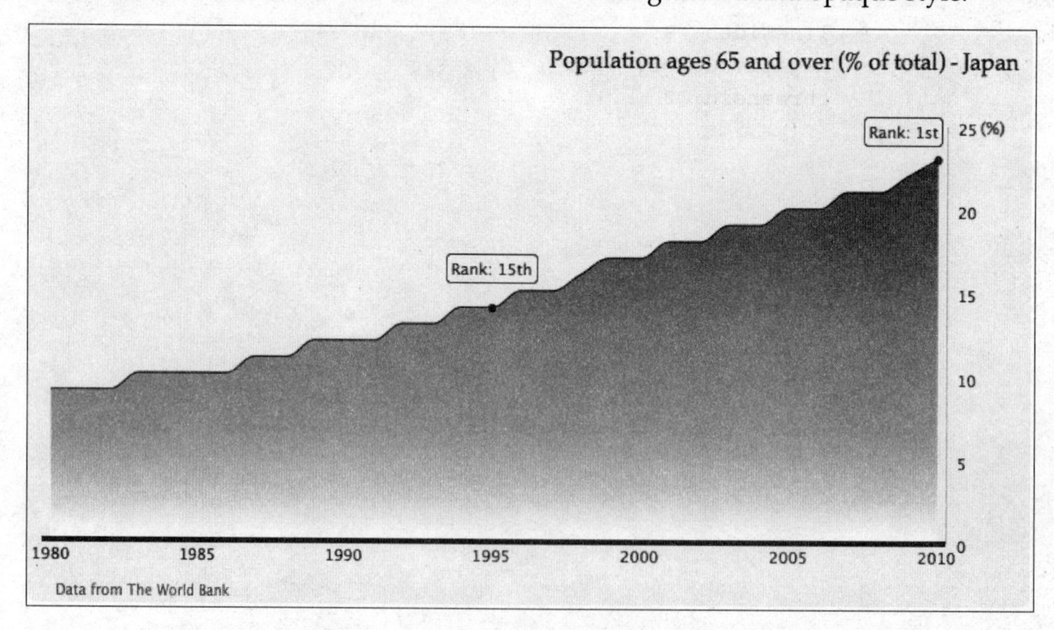

First, we want to make the graph easier for viewers to look up the values for the current trend, so we move the y-axis values to the right side of the chart, where they will be closest to the most recent year:

```
yAxis: { ....
    opposite: true
}
```

The next thing is to remove the interval lines and have a thin axis line along the y axis:

```
yAxis: { ....
    gridLineWidth: 0,
    lineWidth: 1,
}
```

Then, we simplify the y-axis title with a percentage sign and align to the top of the axis:

```
yAxis: { ....
    title: {
        text: '(%)',
        rotation: 0,
        x: 10,
        y: 5,
        align: 'high'
    },
}
```

As for the x axis, we thicken the axis line with red and remove the interval ticks:

```
xAxis: { ....
    lineColor: '#CC2929',
    lineWidth: 4,
    tickWidth: 0,
    offset: 2
}
```

For the chart title, we move the title to the right of the chart, increase the margin between the chart and the title, and then adopt a different font for the title:

```
title: {
    text: 'Population ages 65 and over (% of total) -
Japan ',

    margin: 40,
    align: 'right',
    style: {
```

```
            fontFamily: 'palatino'
        }
    }
```

After that, we are going to modify the whole series presentation, so we first change the `chart.type` property from `'line'` to `'areaspline'`. Notice that setting the properties inside this `series` object will overwrite the same properties defined in `plotOptions.areaspline` and so on in `plotOptions.series`.

Since so far there is only one series in the graph, there is no need to display the legend box. We can disable it with the `showInLegend` property. We then smarten the area part with a gradient color and the spline with a darker color:

```
series: [{
    showInLegend: false,
    lineColor: '#145252',
    fillColor: {
        linearGradient: {
            x1: 0, y1: 0,
            x2: 0, y2: 1
        },
        stops:[ [ 0.0, '#248F8F' ] ,
                [ 0.7, '#70DBDB' ],
                [ 1.0, '#EBFAFA' ] ]
    },
    data: [ ... ]
}]
```

After that, we introduce a couple of data labels along the line to indicate that the ranking of old age population has increased over time. We use the values in the series data array corresponding to the year 1995 and 2010, and then convert the numerical value entries into data point objects. Since we only want to show point markers for these two years, we turn off markers globally in `plotOptions.series.marker.enabled` and set the marker on individually inside the point objects, accompanied by style settings:

```
plotOptions: {
    series: {
        marker: {
            enabled: false
        }
    }
},
series: [{ ...,
    data:[ 9, 9, 9, ...,
```

```
                { marker: {
                        radius: 2,
                        lineColor: '#CC2929',
                        lineWidth: 2,
                        fillColor: '#CC2929',
                        enabled: true
                      },
                  y: 14
                }, 15, 15, 16, ... ]
}]
```

We then set a bounding box around the data labels with round corners (`borderRadius`) in the same border color (`borderColor`) as the x axis. The data label positions are then finely adjusted with the `x` and `y` options. Finally, we change the default implementation of the data label formatter. Instead of returning the point value, we print the country ranking:

```
series: [{ ...,
     data:[ 9, 9, 9, ...,
             { marker: {
                     ...
                 },
                 dataLabels: {
                     enabled: true,
                     borderRadius: 3,
                     borderColor: '#CC2929',
                     borderWidth: 1,
                     y: -23,
                     formatter: function() {
                          return "Rank: 15th";
                     }
                 },
                 y: 14
             }, 15, 15, 16, ... ]
}]
```

The final touch is to apply a gray background to the chart and add extra space for `spacingBottom`. The extra space for `spacingBottom` is to avoid the credit label and x-axis label getting too close together, because we have disabled the legend box:

```
chart: {
     renderTo: 'container',
     spacingBottom: 30,
     backgroundColor: '#EAEAEA'
},
```

When all these configurations are put together, it produces the chart shown in the screenshot at the start of this section.

Mixing line and area series

In this section, we are going to explore different plots including line and area series together, as follows:

- Creating a projection chart, where a single trend line is joined with two series in different line styles
- Plotting an area spline chart with another step line series
- Exploring a stacked area spline chart, where two area spline series are stacked on top of each other

Simulating a projection chart

The projection chart has a spline area with the section of real data, and continues in a dashed line with the projection data. To do that we separate the data into two series, one for real data and the other for projection data. The following is the series configuration code for the data from the year 2015 to 2024. This data is based on the National Institute of Population and Social Security Research report (`http://www.ipss.go.jp/pp-newest/e/ppfj02/ppfj02.pdf`):

```
series: [{
    name: 'project data',
    type: 'spline',
    showInLegend: false,
    lineColor: '#145252',
    dashStyle: 'Dash',
    data: [ [ 2015, 26 ], [ 2016, 26.5 ],
              ... [ 2024, 28.5 ] ]
}]
```

The future series is configured as a spline in a dashed line style and the legend box is disabled, because we want to show both series as being part of the same series. Then we set the future (second) series color the same as the first series. The final part is to construct the series data. As we specify the x-axis time data with the `pointStart` property, we need to align the projection data after 2014. There are two approaches that we can use to specify the time data in a continuous form, as follows:

- Insert null values into the second series data array for padding, to align with the real data series

- Specify the second series data in `tuples`, an array with both time and projection data

Here we are going to use the second approach because the series presentation is simpler. The following is the screenshot for the future data series only:

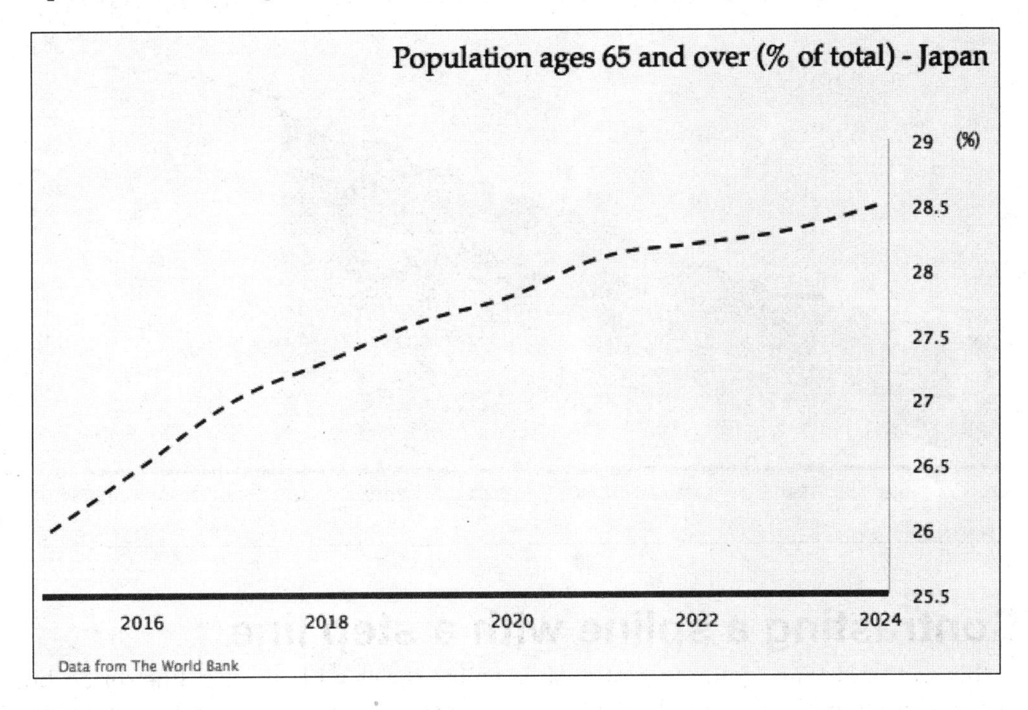

The real data series is exactly the same as the graph in the screenshot at the start of the *Sketching an area chart* section, except without the point markers and data labels decorations. The next step is to join both series together, as follows:

```
series: [{
    name: 'real data',
    type: 'areaspline',
    ....
}, {
    name: 'project data',
    type: 'spline',
    ....
}]
```

Since there is no overlap between both series data, they produce a smooth projection graph:

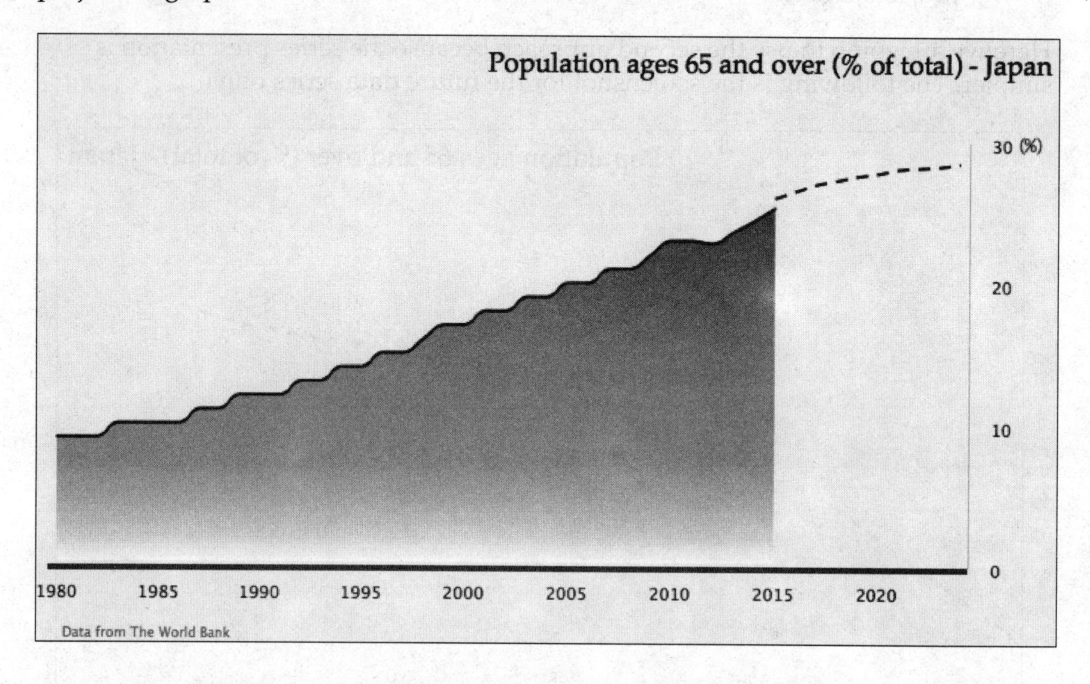

Contrasting a spline with a step line

In this section, we are going to plot an area spline series with another line series, but in step presentation. The step line transverses vertically and horizontally according to changes in series data only. It is generally used for presenting discrete data: data without continuous/gradual movement.

For the purpose of showing a step line, we will continue from the first area spline example. First of all, we need to enable the legend by removing the disabled `showInLegend` setting and also remove `dataLabels` in the series data.

Next is to include a new series—**Ages 0 to 14**—in the chart with the default line type. Then, we change the line style slightly into different steps. The following is the configuration for both series:

```
series: [{
    name: 'Ages 65 and over',
    type: 'areaspline',
    lineColor: '#145252',
```

```
            pointStart: 1980,
            fillColor: {
                ....
            },
            data: [ 9, 9, 9, 10, ...., 23 ]
        }, {
          name: 'Ages 0 to 14',
          // default type is line series
          step: true,
          pointStart: 1980,
          data: [ 24, 23, 23, 23, 22, 22, 21,
                  20, 20, 19, 18, 18, 17, 17, 16, 16, 16,
                  15, 15, 15, 15, 14, 14, 14, 14, 14, 14,
                  14, 14, 13, 13 ]
        }]
```

The following screenshot shows the second series in line step style:

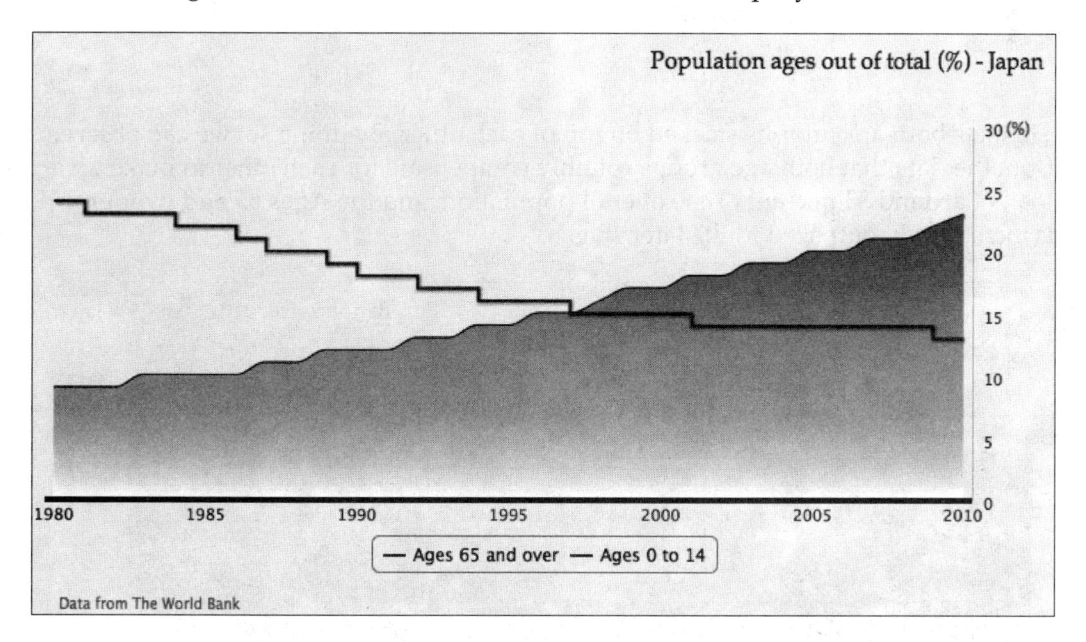

Extending to the stacked area chart

In this section, we are going to turn both series into area splines and stack them on top of each other to create a stacked area chart. As the data series are stacked together, we can observe the series quantities as individual, proportional, and total amounts.

Let's change the second series into another `'areaspline'` type:

```
name: 'Ages 0 to 14',
type: 'areaspline',
pointStart: 1980,
data: [ 24, 23, 23, ... ]
```

Set the `stacking` option to `'normal'` as a default setting for `areaspline`, as follows:

```
plotOptions: {
    areaspline: {
      stacking: 'normal'
    }
}
```

This sets both area graphs stacked on top of each other. By doing so, we can observe from the data that both age groups roughly compensate for each other to make up a total of around 33 percent of the overall population, and the **Ages 65 and over** group is increasingly outpaced in the later stage:

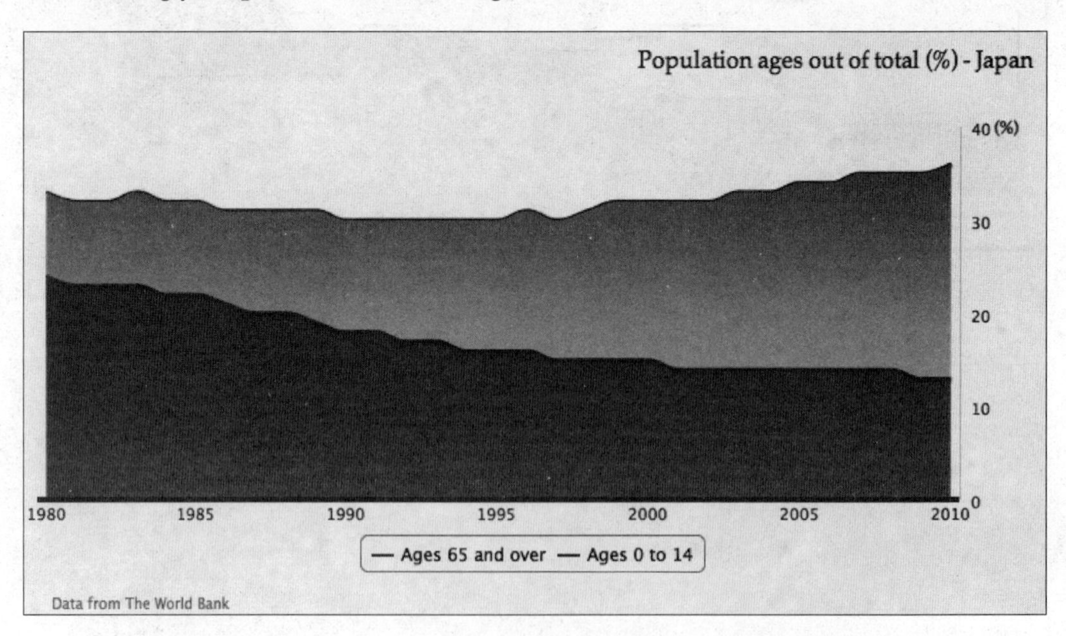

Suppose we have three area spline series and we only want to stack two of them (although it is clearer to do that in a column chart rather than in an area spline chart). As described in the *Exploring PlotOptions* section in *Chapter 2, Highcharts Configurations,* we can set the `stacking` option in `plotOptions.series` to `'normal'`, and manually turn off `stacking` in the third series configuration. The following is the series configuration with another series:

```
plotOptions: {
    series: {
      marker: {
         enabled: false
      },
      stacking: 'normal'
    }
},
series: [{
    name: 'Ages 65 and over',
    ....
}, {
    name: 'Ages 0 to 14',
    ....
}, {
    name: 'Ages 15 to 64',
    type: 'areaspline',
    pointStart: 1980,
    stacking: null,
    data: [ 67, 67, 68, 68, .... ]
}]
```

This creates an area spline graph with the third series **Ages 15 to 64** covering the other two stacked series, as shown in the following screenshot:

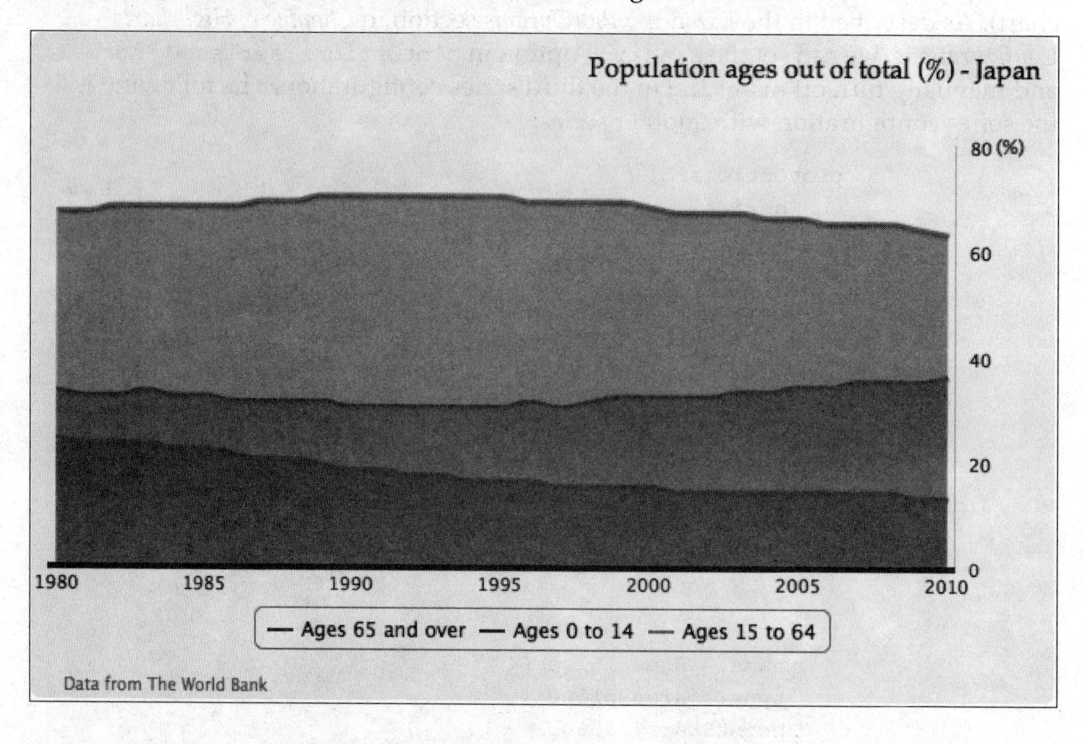

Plotting charts with missing data

If a series has missing data, then the default action of Highcharts is to display the series as a broken line. There is an option—`connectNulls`—that allows a series line to continue even if there is missing data. The default value for this option is `false`. Let's examine the default behavior by setting two spline series with null data points. We also enable the point markers, so that we can clearly view the missing data points:

```
series: [{
    name: 'Ages 65 and over',
    connectNulls: true,
    ....,
    // Missing data from 2004 - 2009
    data: [ 9, 9, 9, ..., null, null, null, 22, 23 ]

}, {
    name: 'Ages 0 to 14',
    ....,
```

```
            // Missing data from 1989 - 1994
            data: [ 24, 23, 23, ..., 19, null, null, ..., 13 ]
        }]
```

The following is a chart with the spline series presenting missing points in different styles:

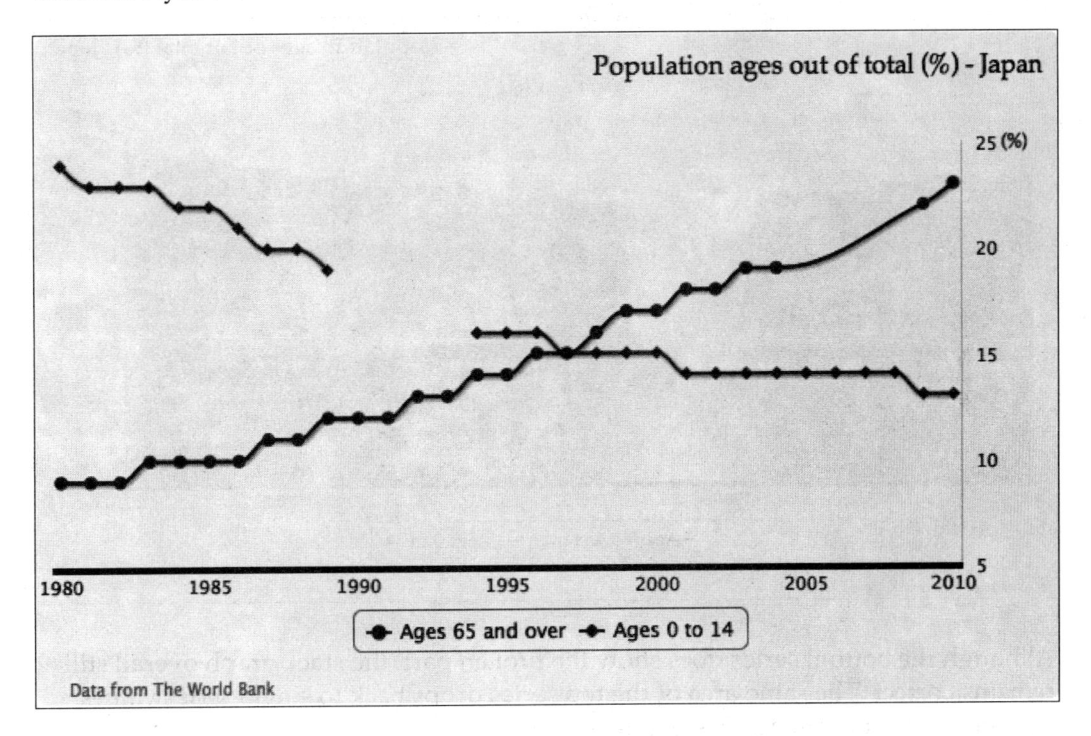

As we can see, the **Ages 0 to 14** series has a clear broken line, whereas **Ages 65 and over** is configured by setting `connectNulls` to `true`, which joins the missing points with a spline curve. If the point marker is not enabled, we wouldn't be able to notice the difference.

However, we should use this option with caution and it should certainly never be enabled with the `stacking` option. Suppose we have a stacked area chart with both series and there is missing data only in the **Ages 0 to 14** series, which is the bottom series. The default action for the missing data will make the graph look like the following screenshot:

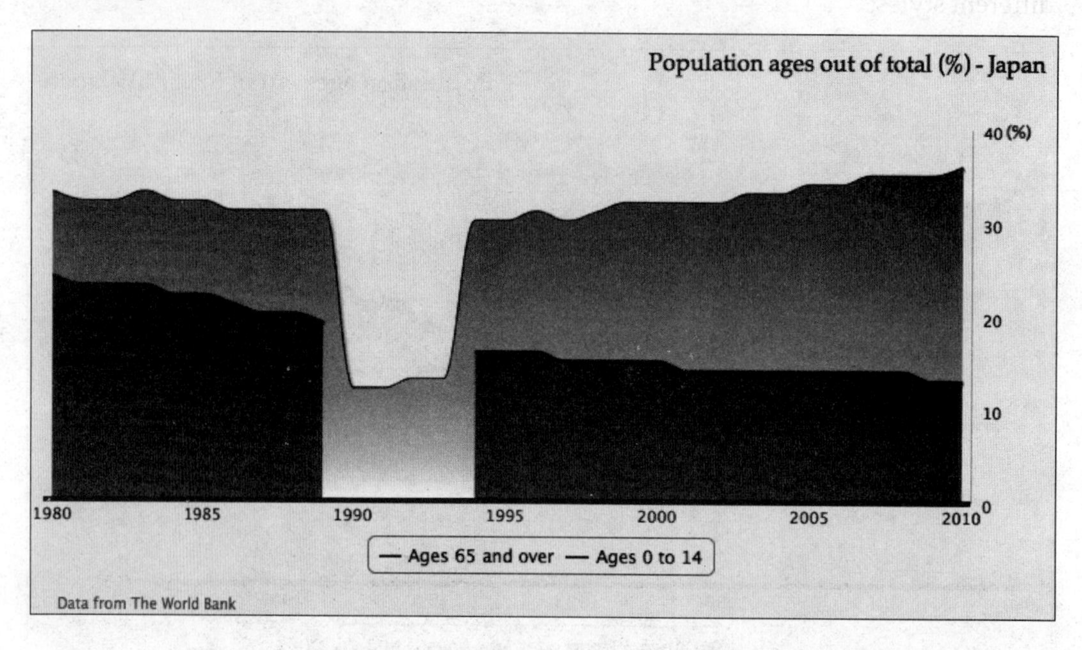

Although the bottom series does show the broken part, the stack graph overall still remains correct. The same area of the top series drops back to single-series values and the overall percentage is still intact.

The problem arises when we set the `connectNulls` option to `true` and do not realize that there is missing data in the series. This results in an inconsistent graph, as follows:

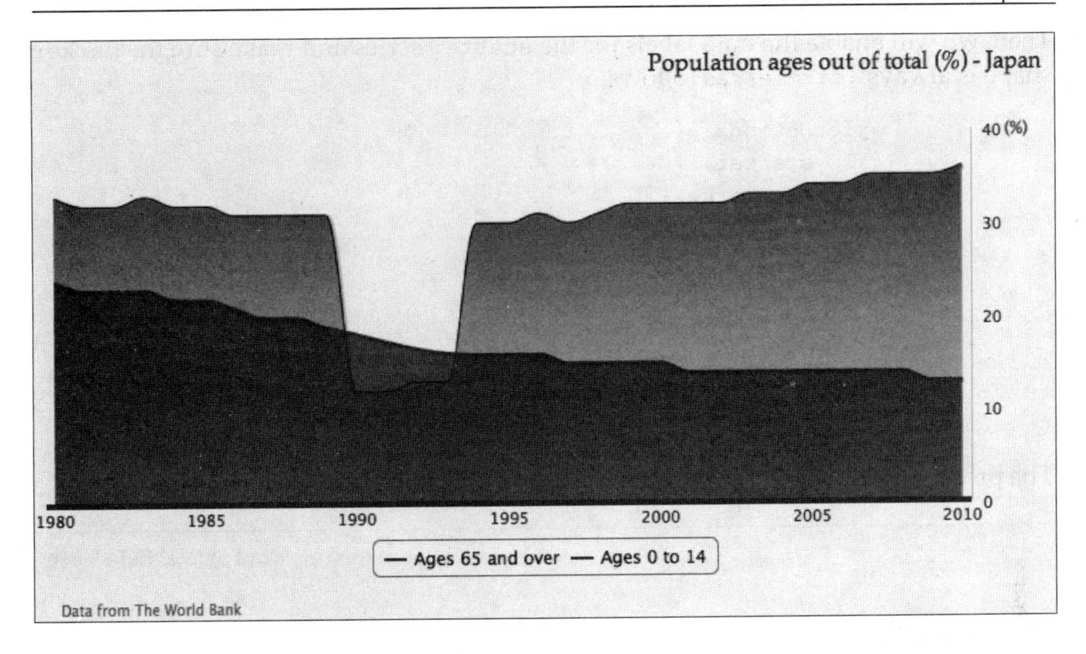

The bottom series covers a hole left from the top series that contradicts the stack graph's overall percentage.

Combining the scatter and area series

Highcharts also supports a scatter chart that enables us to plot the data trend from a large set of data samples. Here we are going to use the scatter series differently, which makes our chart a bit like a poster chart.

First, we are going to use a subset of the `'Ages 0 to 14'` data and set the series to the `scatter` type:

```
name: 'Ages 0 to 14',
type: 'scatter',
data: [ [ 1982, 23 ], [ 1989, 19 ],
        [ 2007, 14 ], [ 2004, 14 ],
        [ 1997, 15 ], [ 2002, 14 ],
        [ 2009, 13 ], [ 2010, 13 ] ]
```

Then, we will enable the data labels for the `scatter` series and make sure the `marker` shape is always `'circle'`, as follows:

```
plotOptions: {
    scatter: {
        marker: {
            symbol: 'circle'
        },
        dataLabels: {
            enabled: true
        }
    }
}
```

The preceding code snippet gives us the following graph:

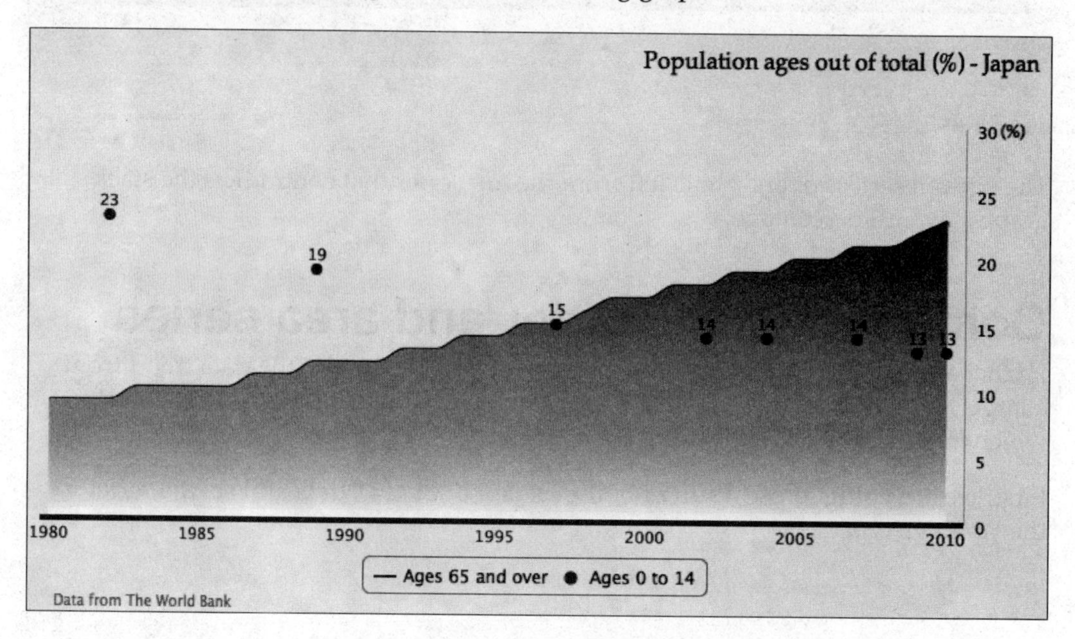

Highcharts provides a list of marker symbols as well as allowing users to supply their own marker icons (see *Chapter 2, Highcharts Configurations*). The list of supported symbols contains: circle, square, diamond, triangle, and triangle-down.

Polishing a chart with an artistic style

The next step is to format each scatter point into a bubble style with the `radius` property and manually set the data label font size proportional to the percentage value.

> The reason we used the scatter series instead of the bubble series is because most of the material in this chapter was written for the first edition; this chart was created with an earlier version of Highcharts that didn't support the bubble series.

Then use the `verticalAlign` property to adjust the labels to center inside the enlarged scatter points. The various sizes of scatter points require us to present each data point with different attributes. Therefore, we need to change the series data definition into an array of point object configurations, such as:

```
plotOptions: {
  scatter: {
      marker: {
          symbol: 'circle'
      },
      dataLabels: {
          enabled: true,
          verticalAlign: 'middle'
      }
  }
},
data: [ {
    dataLabels: {
        style: {
            fontSize: '25px'
        }
    },
    marker: { radius: 31 },
    y: 23,
    x: 1982
    }, {
    dataLabels: {
        style: {
            fontSize: '22px'
        }
    },
```

```
            marker: { radius: 23 },
            y: 19,
            x: 1989
        }, .....
```

The following screenshot shows a graph with a sequence of data points, starting with a large marker size and font, then gradually becoming smaller according to their percentage values:

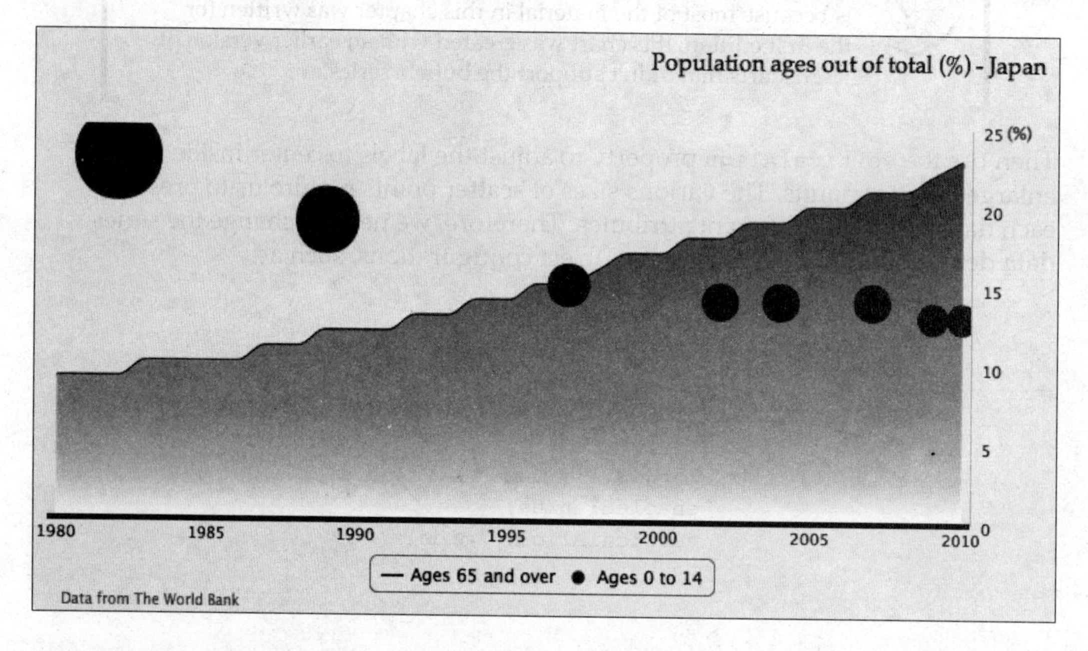

Now, we have two issues with the preceding graph. First, the scatter series color (the default second series color) clashes with the gray text labels inside the markers, making them hard to read.

To resolve this issue, we will change the scatter series to a lighter color with the gradient setting:

```
color: {
    linearGradient: { x1: 0, y1: 0, x2: 0, y2: 1 },
    stops: [ [ 0, '#FF944D' ],
             [ 1, '#FFC299' ] ]
},
```

Then, we give the scatter points a darker outline in `plotOptions`, as follows:

```
plotOptions: {
    scatter: {
        marker: {
            symbol: 'circle',
            lineColor: '#E65C00',
            lineWidth: 1
        },
```

Secondly, the data points are blocked by the end of the axes' ranges. The issue can be resolved by introducing extra padding spaces into both axes:

```
yAxis: {
    .....,
    maxPadding: 0.09
},
xAxis: {
    .....,
    maxPadding: 0.02
}
```

The following is the new look for the graph:

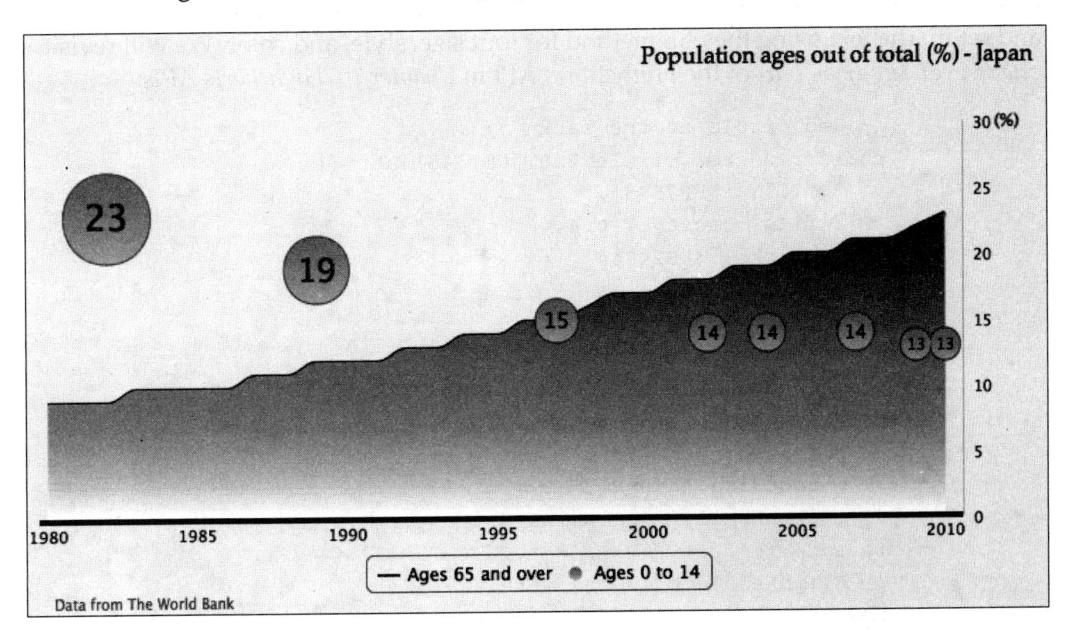

For the next part, we will put up a logo and some decorative text. There are two ways to import an image into a chart—the `plotBackgroundImage` option or the `renderer.image` API call. The `plotBackgroundImage` option brings the whole image into the chart background, which is not what we intend to do. The `renderer.image` method offers more control over the location and size of the image. The following is the call after the chart is created:

```
var chart = new Highcharts.Chart({
                ...
            });
chart.renderer.image('logo.png', 240, 10, 187, 92).add();
```

`logo.png` is the URL path for the logo image file. The next two parameters are the x and y positions (starting from 0, where 0 is the upper-left corner) of the chart where the image will be displayed. The last two parameters are the width and height of the image file. The `image` call basically returns an `element` object and the subsequent `.add` call puts the returned image object into the renderer.

As for the decorative text, it is a red circle with white bold text in a different size. They are all created from the renderer. In the following code snippet, the first renderer call is to create a red circle with x and y locations, and radius size. Then SVG attributes are immediately set with the `attr` method that configures the transparency and outline in a darker color. The next three renderer calls are to create text inside the red circle and set up the text using the `css` method for font size, style, and color. We will revisit `chart.renderer` as part of the Highcharts API in *Chapter 10, Highcharts APIs*:

```
// Red circle at the back
chart.renderer.circle(220, 65, 45).attr({
     fill: '#FF7575',
    'fill-opacity': 0.6,
     stroke: '#B24747',
    'stroke-width': 1
}).add();
// Large percentage text with special font
chart.renderer.text('37.5%', 182, 63).css({
     fontWeight: 'bold',
     color: '#FFFFFF',
     fontSize: '30px',
     fontFamily: 'palatino'
}).add();
// Align subject in the circle
chart.renderer.text('65 and over', 184, 82).css({
    'fontWeight': 'bold',
```

```
})).add();
chart.renderer.text('by 2050', 193, 96).css({
    'fontWeight': 'bold',
}).add();
```

Finally, we move the legend box to the top of the chart. In order to locate the legend inside the plot area, we need to set the `floating` property to `true`, which forces the legend into a fixed layout mode. Then, we remove the default border line and set the legend items list into a vertical direction:

```
legend: {
    floating: true,
    verticalAlign: 'top',
    align: 'center',
    x: 130,
    y: 40,
    borderWidth: 0,
    layout: 'vertical'
},
```

The following is our final graph with the decorations:

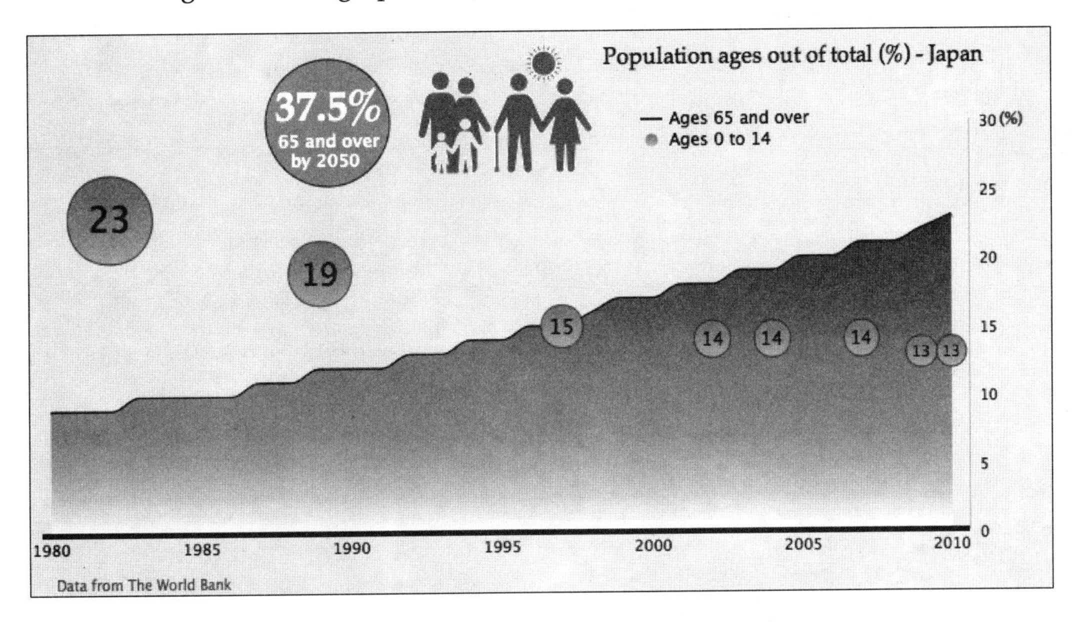

Summary

In this chapter, we have explored the usage of line, area, and scatter charts. We have seen how much flexibility Highcharts can offer to make a poster-like chart.

In the next chapter, we will learn how to plot column and bar charts with their plotting options.

4

Bar and Column Charts

In this chapter, we will start off by learning about column charts and their plotting options. Then, we will apply more advanced options for stacking and grouping columns together. After that, we will move on to bar charts by following the same example. Then, we will learn how to polish up a bar chart and apply tricks to turn a bar chart into mirror and gauge charts. Finally, a web page of multiple charts will be put together as a concluding exercise. In this chapter, we will cover the following:

- Introducing column charts
- Stacking and grouping a column chart
- Adjusting column colors and data labels
- Introducing bar charts
- Constructing a mirror chart
- Converting a single bar chart into a horizontal gauge chart
- Sticking the charts together

Most of the content in this chapter remains the same as the first edition. All the examples have been run through with the latest version of Highcharts. Apart from the different color schemes, the resulting charts look the same as the screenshots presented in this chapter.

Introducing column charts

The difference between column and bar charts is trivial. The data in column charts is aligned vertically, whereas it is aligned horizontally in bar charts. Column and bar charts are generally used for plotting data with categories along the x axis. In this section, we are going to demonstrate plotting column charts. The dataset we are going to use is from the U.S. Patent and Trademark Office.

The graph just after the following code snippet shows a column chart for the number of patents granted in the United Kingdom over the last 10 years. The following is the chart configuration code:

```
chart: {
    renderTo: 'container',
    type: 'column',
    borderWidth: 1
},
title: {
    text: 'Number of Patents Granted',
},
credits: {
    position: {
        align: 'left',
        x: 20
    },
    href: 'http://www.uspto.gov',
    text: 'Source: U.S. Patent & Trademark Office'
},
xAxis: {
    categories: [
        '2001', '2002', '2003', '2004', '2005',
        '2006', '2007', '2008', '2009', '2010',
        '2011' ]
},
yAxis: {
    title: {
        text: 'No. of Patents'
    }
},
series: [{
    name: 'UK',
    data: [ 4351, 4190, 4028, 3895, 3553,
            4323, 4029, 3834, 4009, 5038, 4924 ]
}]
```

The following is the result that we get from the preceding code snippet:

This data can be found in the online report *All Patents, All Types Report* by the Patent Technology Monitoring Team at `http://www.uspto.gov/web/offices/ac/ido/oeip/taf/apat.htm`.

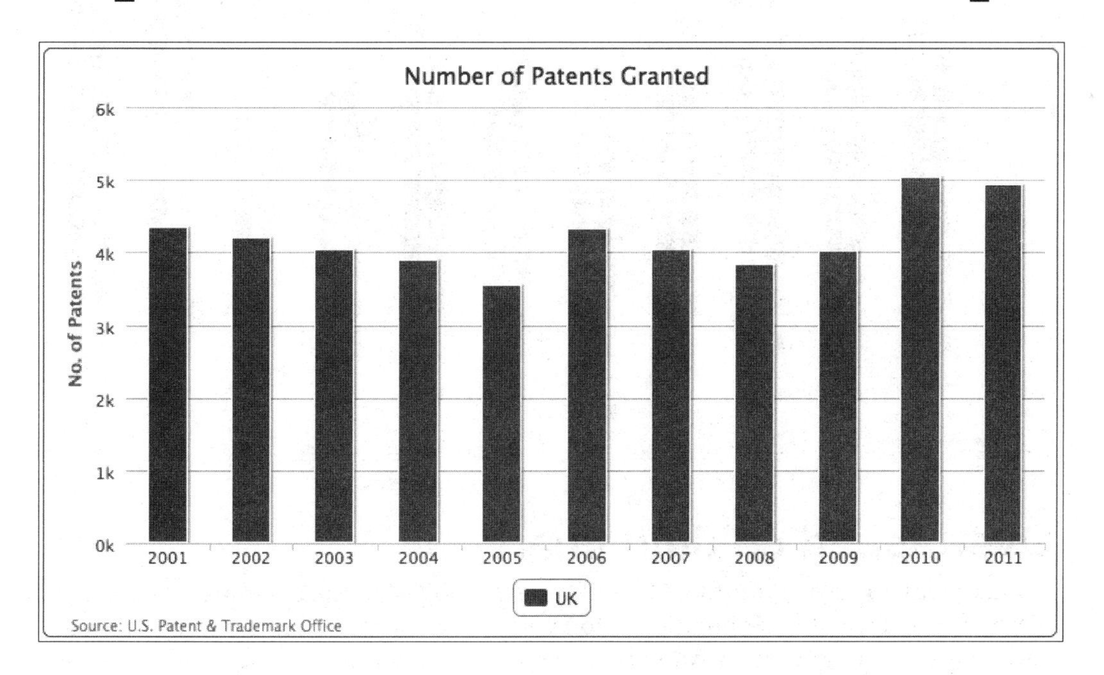

Let's add another series, France:

```
series: [{
    name: 'UK',
    data: [ 4351, 4190, 4028, .... ]
}, {
    name: 'France',
    data: [ 4456, 4421, 4126, 3686, 3106,
            3856, 3720, 3813, 3805, 5100, 5022 ]
}]
```

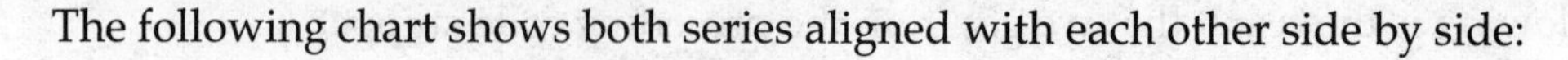

The following chart shows both series aligned with each other side by side:

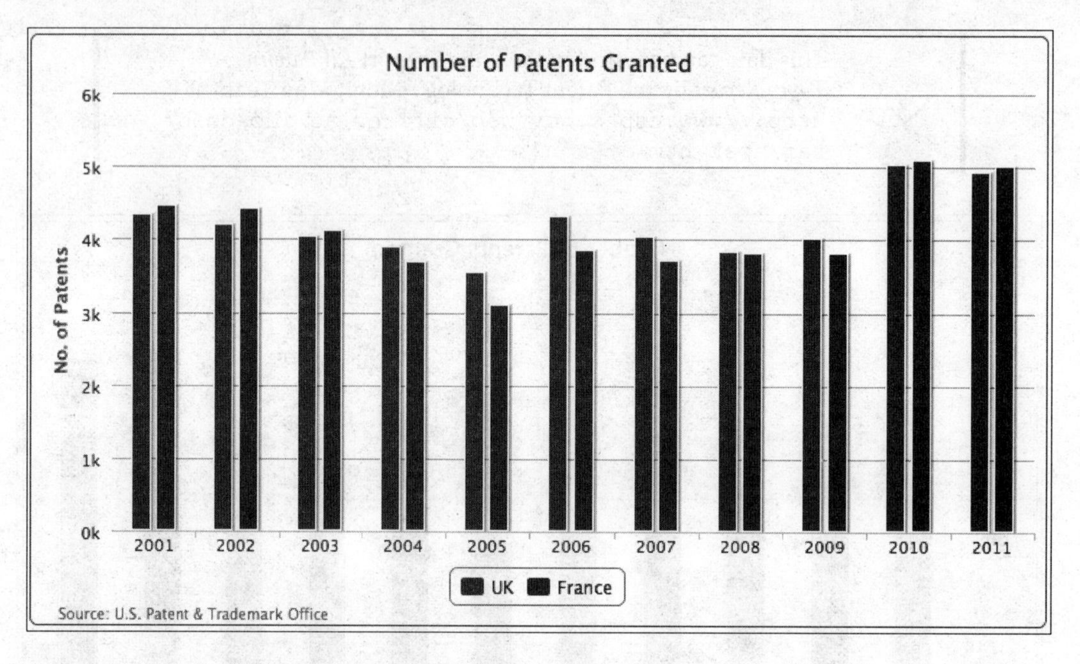

Overlapped column chart

Another way to present multi-series columns is to overlap the columns. The main reason for this type of presentation is to avoid columns becoming too thin and over-packed if there are too many categories in the chart. As a result, it becomes difficult to observe the values and compare them. Overlapping the columns provides more space between each category, so each column can still retain its width.

We can make both series partially overlap each other with the padding options, as follows:

```
plotOptions: {
    series: {
        pointPadding: -0.2,
        groupPadding: 0.3
    }
},
```

The default setting for padding between columns (also for bars) is 0.2, which is a fraction value of the width of each category. In this example, we are going to set `pointPadding` to a negative value, which means that, instead of having padding distance between neighboring columns, we bring the columns together to overlap each other. `groupPadding` is the distance of group values relative to each category width, so the distance between the pair of UK and France columns in 2005 and 2006. In this example, we have set it to 0.3 to make sure the columns don't automatically become wider, because overlapping produces more space between each group. The following is the screenshot of the overlapping columns:

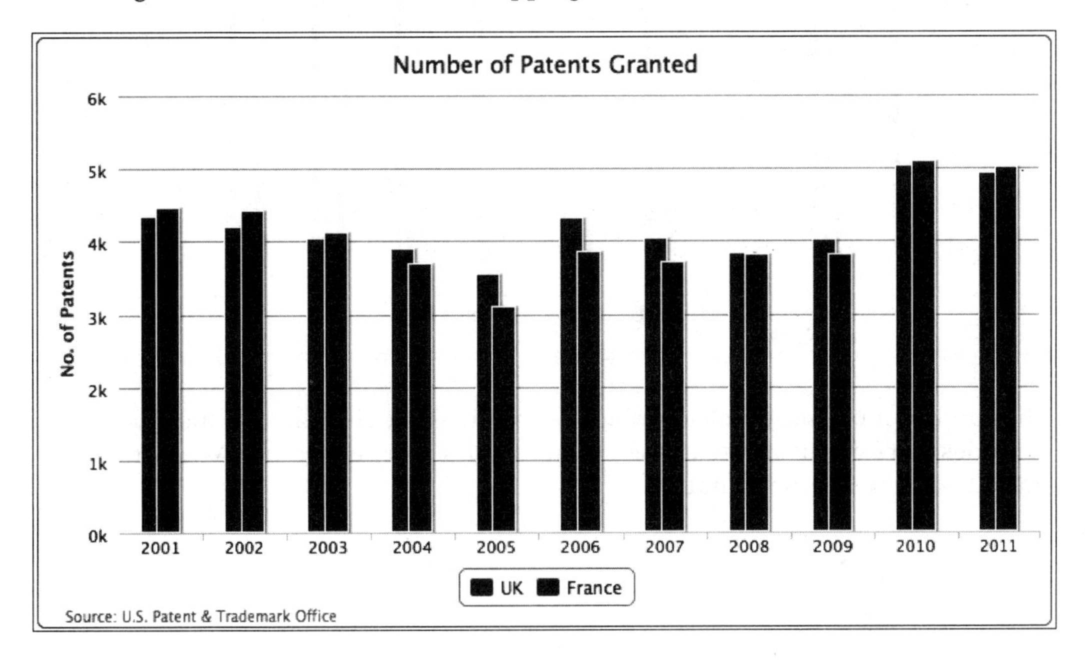

Stacking and grouping a column chart

Instead of aligning columns side by side, we can stack the columns on top of each other. Although this will make it slightly harder to visualize each column's values, we can instantly observe the total values of each category and the change of ratios between the series. Another powerful feature with stacked columns is to group them selectively when we have more than a couple of series. This can give a sense of proportion between multiple groups of stacked series.

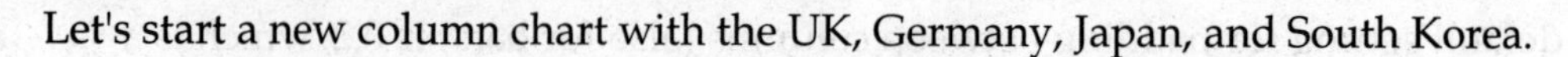

Let's start a new column chart with the UK, Germany, Japan, and South Korea.

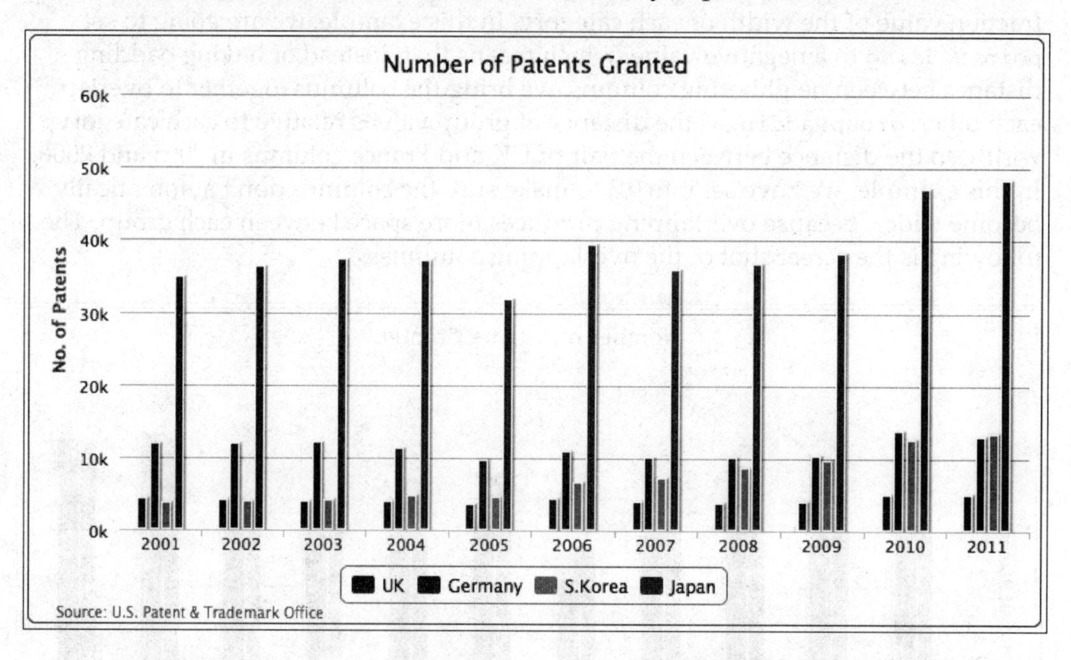

The number of patents granted for Japan is off-the-scale compared to the other countries. Let's group and stack the multiple series into Europe and Asia with the following series configuration:

```
plotOptions: {
      column: {
          stacking: 'normal'
      }
  },
series: [{
      name: 'UK',
      data: [ 4351, 4190, 4028, .... ],
      stack: 'Europe'
   }, {
      name: 'Germany',
      data: [ 11894, 11957, 12140, ... ],
      stack: 'Europe'
   }, {
      name: 'S.Korea',
      data: [ 3763, 4009, 4132, ... ],
      stack: 'Asia'
   }, {
```

```
        name: 'Japan',
        data: [ 34890, 36339, 37248, ... ],
        stack: 'Asia'
}]
```

We declare column `stacking` in `plotOptions` as `'normal'` and then for each column series assign a stack group name, `'Europe'` and `'Asia'`, which produces the following graph:

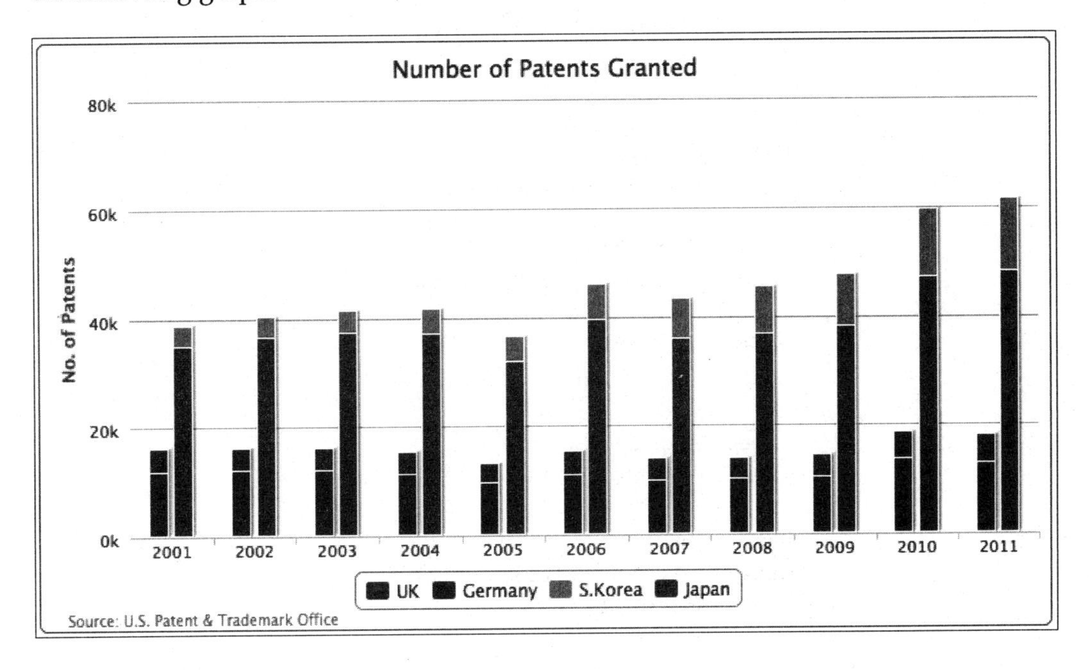

As we can see, the chart reduces four vertical bars into two and each column comprises two series. The first vertical bar is the `'Europe'` group and the second one is `'Asia'`.

Mixing the stacked and single columns

In the previous section, we acknowledged the benefit of grouping and stacking multiple series. There are also occasions when multiple series can belong to a group but there are also individual series in their own groups. Highcharts offers the flexibility to mix stacked and grouped series with single series.

Let's look at an example of mixing a stacked column and a single column together. First, we remove the stack group assignment in each series, as the default action for all the column series is to remain stacked together. Then, we introduce a new column series, US, and manually declare the `stacking` option as `null` in the series configuration to override the default `plotOptions` setting:

```
plotOptions: {
        column: {
            stacking: 'normal'
        }
},
series: [{
        name: 'UK',
        data: [ 4351, 4190, 4028, .... ]
    }, {
        name: 'Germany',
        data: [ 11894, 11957, 12140, ... ]
    }, {
        name: 'S.Korea',
        data: [ 3763, 4009, 4132, ... ]
    }, {
        name: 'Japan',
        data: [ 34890, 36339, 37248, ... ]
    }, {
        name: 'US',
        data: [ 98655, 97125, 98590, ... ],
        stacking: null
    }

}]
```

The new series array produces the following graph:

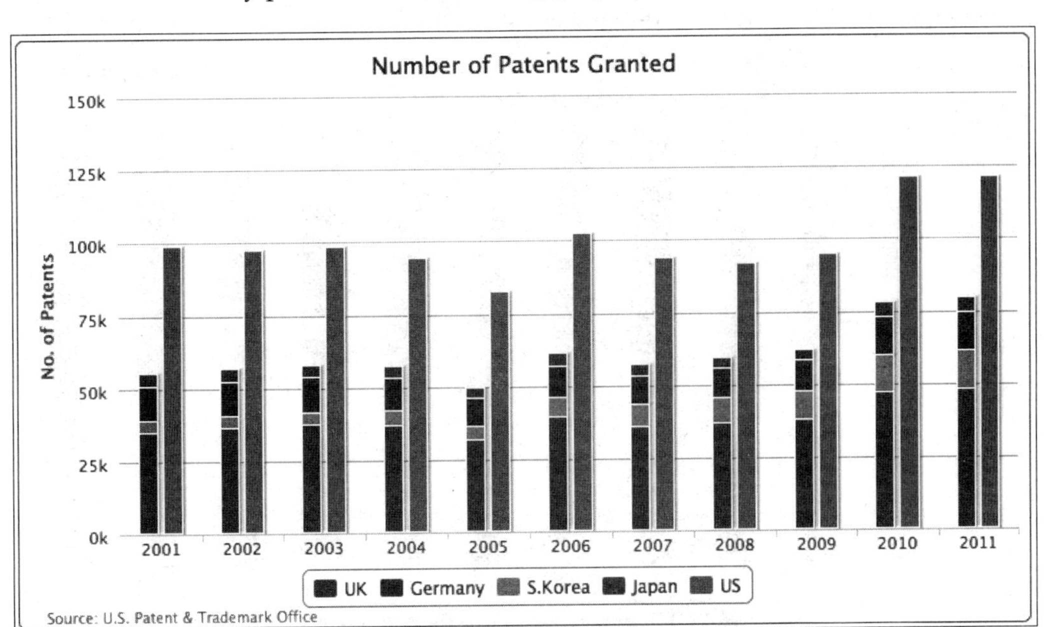

The first four series, **UK**, **Germany**, **S. Korea**, and **Japan**, are stacked together as a single column and **US** is displayed as a separate column. We can easily observe by stacking the series together that the number of patents from the other four countries put together is less than two-thirds of the number of patents from the **US** (the US is nearly 25 times that of the UK).

Comparing the columns in stacked percentages

Alternatively, we can see how each country compares in columns by normalizing the values into percentages and stacking them together. This can be achieved by removing the manual `stacking` setting in the US series and setting the global column `stacking` as `'percent'`:

```
plotOptions: {
    column: {
        stacking: 'percent'
    }
}
```

All the series are put into a single column and their values are normalized into percentages, as shown in the following screenshot:

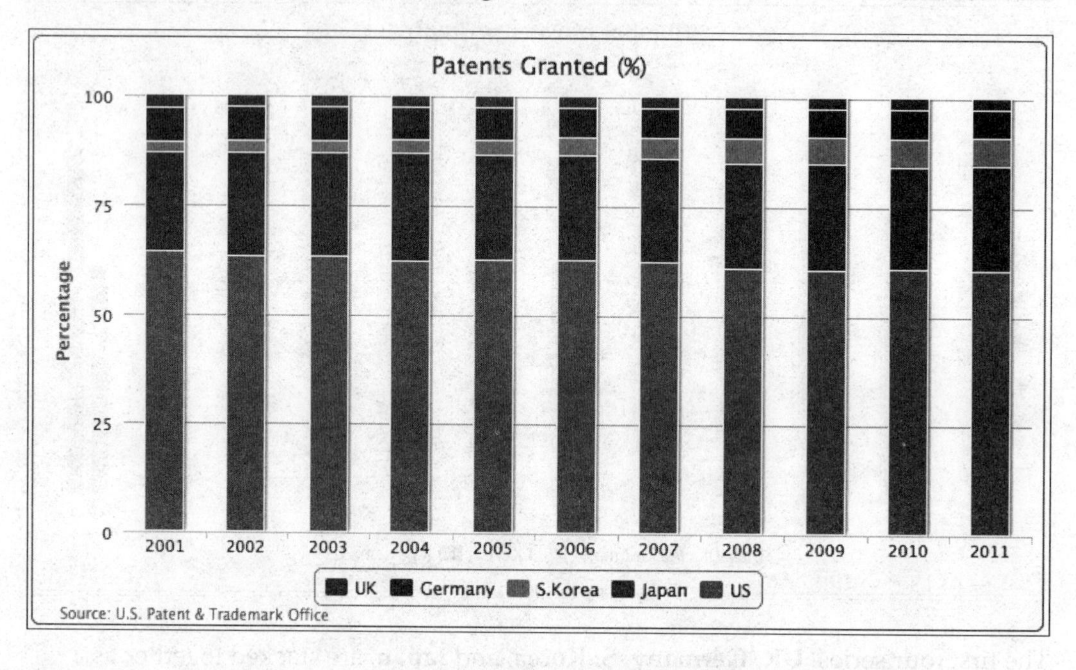

Adjusting column colors and data labels

Let's make another chart; this time we will plot the top ten countries or areas with patents granted. The following is the code to produce the chart:

```
chart: {
    renderTo: 'container',
    type: 'column',
    borderWidth: 1
},
title: {
    text: 'Number of Patents Granted in 2011'
},
credits: { ... },
xAxis: {
    categories: [
        'United States', 'Japan',
        'South Korea', 'Germany', 'Chinese Taiwan',
        'Canada', 'France', 'United Kingdom',
        'China', 'Italy' ]
```

```
        },
        yAxis: {
            title: {
                text: 'No. of Patents'
            }
        },
        series: [{
            showInLegend: false,
            data: [ 121261, 48256, 13239, 12968, 9907,
                    5754, 5022, 4924, 3786, 2333 ]
        }]
```

The preceding code snippet generates the following graph:

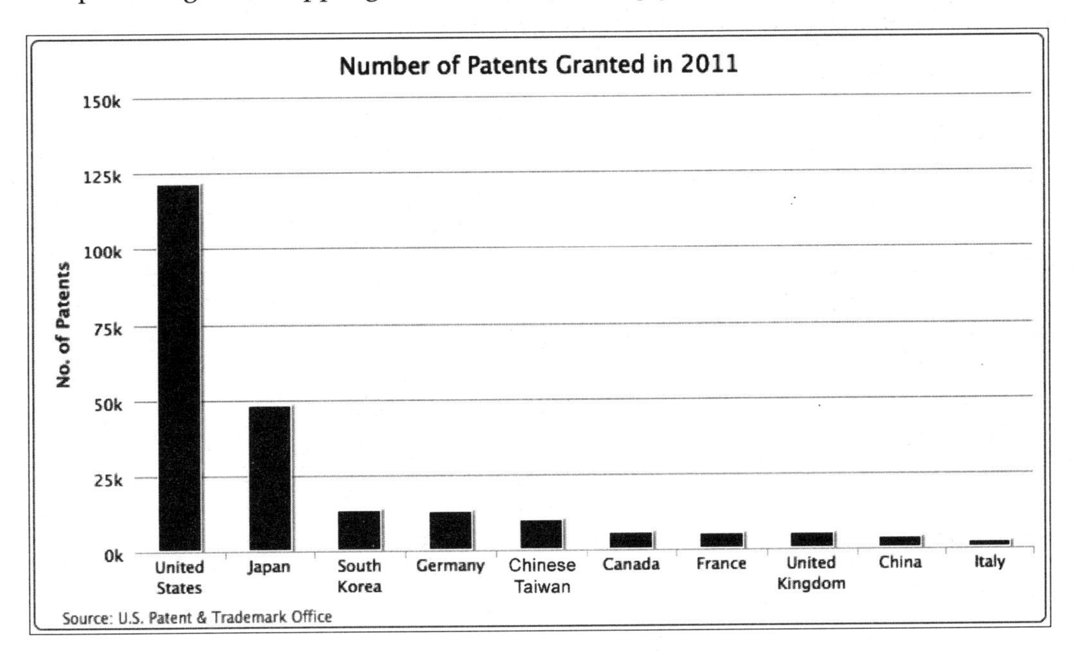

There are several areas that we would like to change in the preceding graph. First, there are word wraps in the country/area names. In order to avoid that, we can apply rotation to the x-axis labels, as follows:

```
        xAxis: {
            categories: [
                'United States', 'Japan',
                'South Korea',  ... ],
            labels: {
                rotation: -45,
```

```
            align: 'right'
        }
    },
```

Secondly, the large value from `'United States'` has gone off the scale compared to values from other countries/areas, so we cannot easily identify their values. To resolve this issue we can apply a logarithmic scale onto the y axis, as follows:

```
    yAxis: {
        title: ... ,
        type: 'logarithmic'
    },
```

Finally, we would like to print the value labels along the columns and decorate the chart with different colors for each column, as follows:

```
plotOptions: {
    column: {
        colorByPoint: true,
        dataLabels: {
            enabled: true,
            rotation: -90,
            y: 25,
            color: '#F4F4F4',
                formatter: function() {
                    return
                     Highcharts.numberFormat(this.y, 0);
                },
                x: 10,
                style: {
                    fontWeight: 'bold'
                }
            }
        }
    },
```

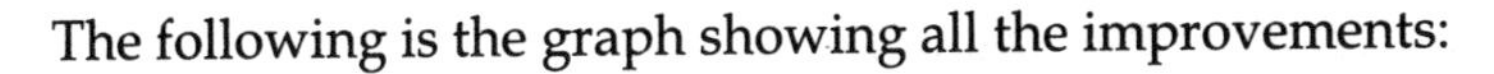

The following is the graph showing all the improvements:

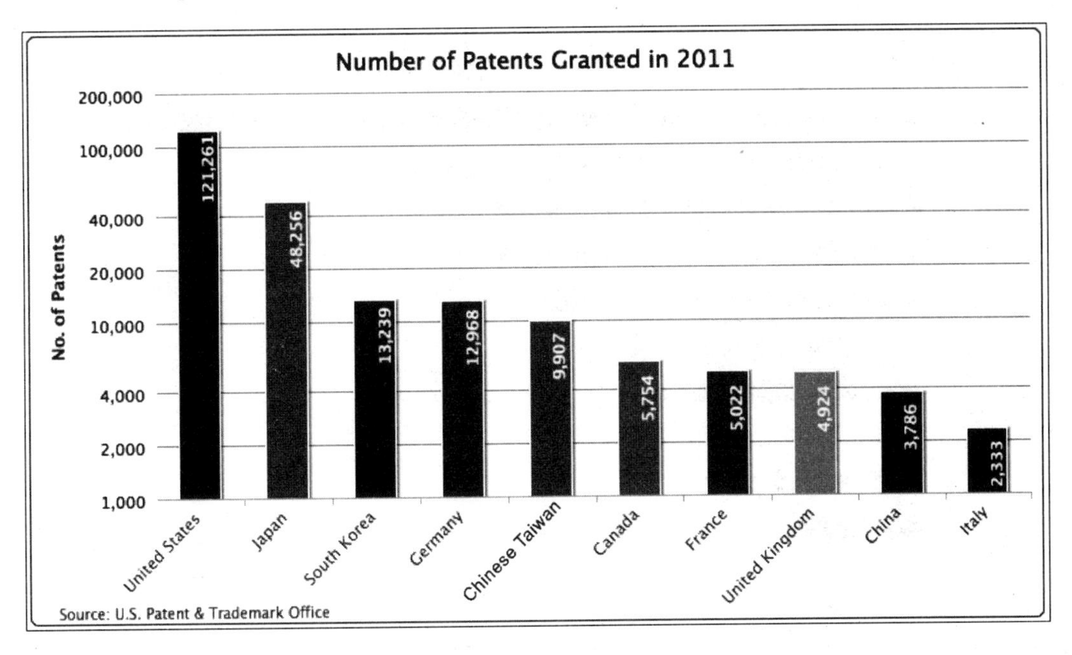

Introducing bar charts

In Highcharts, there are two ways to specify bar charts—setting the series `type` to `'bar'` or setting the `chart.inverted` option to `true` with column series (also true for switching from bar to column). Switching between column and bar is simply a case of swapping the display orientation between the y and x axes; all the label rotations are still intact. Moreover, the actual configurations still remain in the x and y axes. To demonstrate this, we will use the previous example along with the `inverted` option set to `true`, as follows:

```
chart: {
    .... ,
    type: 'column',
    inverted: true
},
```

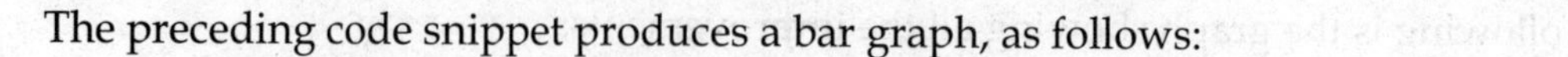

The preceding code snippet produces a bar graph, as follows:

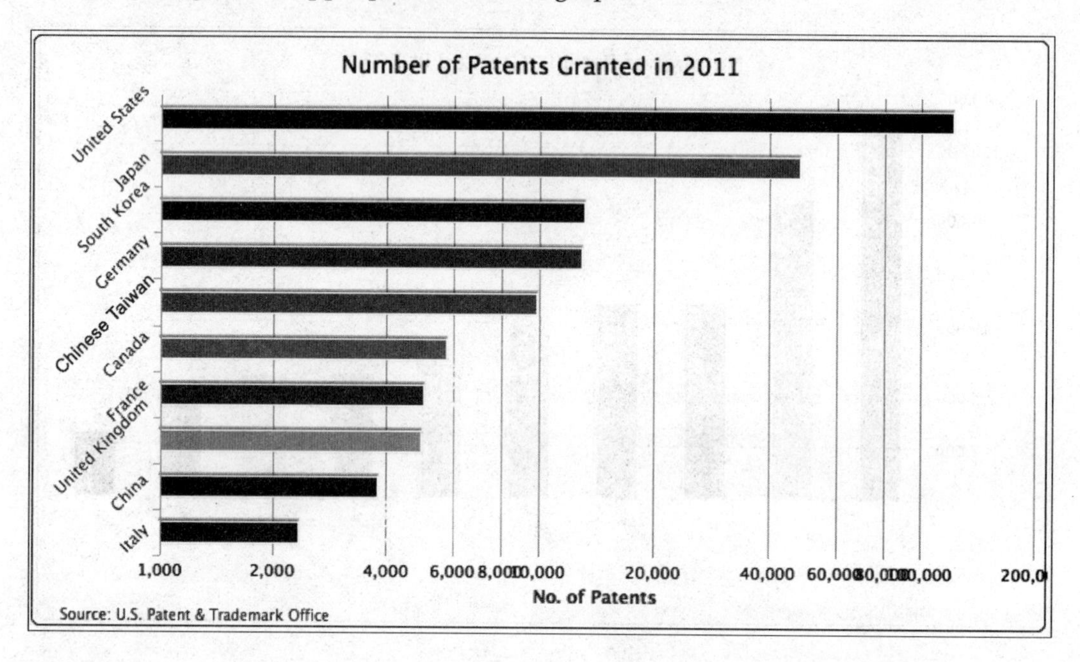

The rotation of the country/area name and the logarithmic axis labels still remains the same. In fact, now the value labels are muddled together and the category names are not aligned properly with the bars. The next step is to reset the label orientations to restore the graph to a readable form. We will simply swap the label setting from the y axis to the x axis:

```
xAxis: {
    categories: [ 'United States',
                  'Japan', 'South Korea', ... ]
},
yAxis: {
     .... ,
    labels: {
        rotation: -45,
        align: 'right'
    }
},
```

Then we will reset the default column `dataLabel` settings by removing the rotation option and re-adjusting the x and y positioning to align inside the bars:

```
plotOptions: {
    column: {
```

```
        ..... ,
        dataLabels: {
            enabled: true,
            color: '#F4F4F4',
            x: -50,
            y: 0,
            formatter: ....,
            style: ...
        }
    }
```

The following is the graph with fixed data labels:

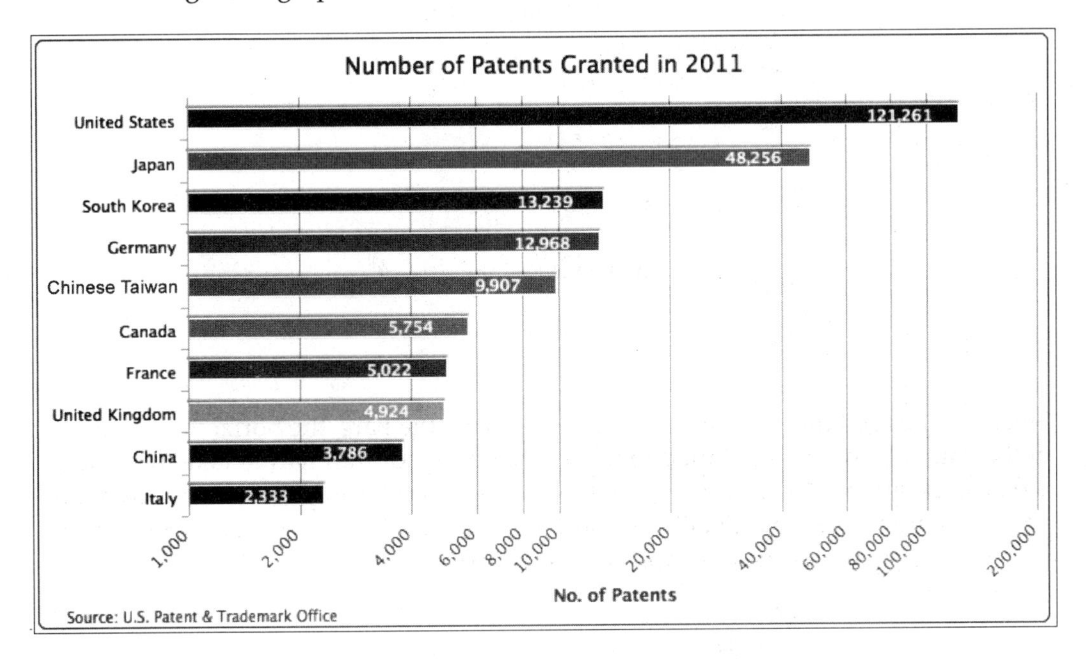

Giving the bar chart a simpler look

Here, we are going to strip the axes back to a minimal, bare presentation. We remove the whole y axis and adjust the category name above the bar. To strip off the y axis we will use the following code snippet:

```
yAxis: {
    title: {
        text: null
    },
    labels: {
```

```
            enabled: false
        },
        gridLineWidth: 0,
        type: 'logarithmic'
    },
```

Then, we move the country/area labels above the bars. This is accompanied by removing the axis line and the interval tick line, then changing the label alignments and their x and y positioning as follows:

```
    xAxis: {
        categories: [ 'United States', 'Japan',
                      'South Korea', ... ],
        lineWidth: 0,
        tickLength: 0,
        labels: {
            align: 'left',
            x: 0,
            y: -13,
            style: {
                fontWeight: 'bold'
            }
        }
    },
```

Since we changed the label alignments to go above the bars, the horizontal position of the bars (plot area) has shifted to the left-hand side of the chart to take over the old label positions. Therefore, we need to increase the spacing on the left to avoid the chart looking too packed. Finally, we add a background image, `chartBg.png`, to the plot area just to fill up the empty space, as follows:

```
    chart: {
        renderTo: 'container',
        type: 'column',
        spacingLeft: 20,
        plotBackgroundImage: 'chartBg.png',
        inverted: true
    },
    title: {
        text:  null
    },
```

The following screenshot shows the new simple look of our bar chart:

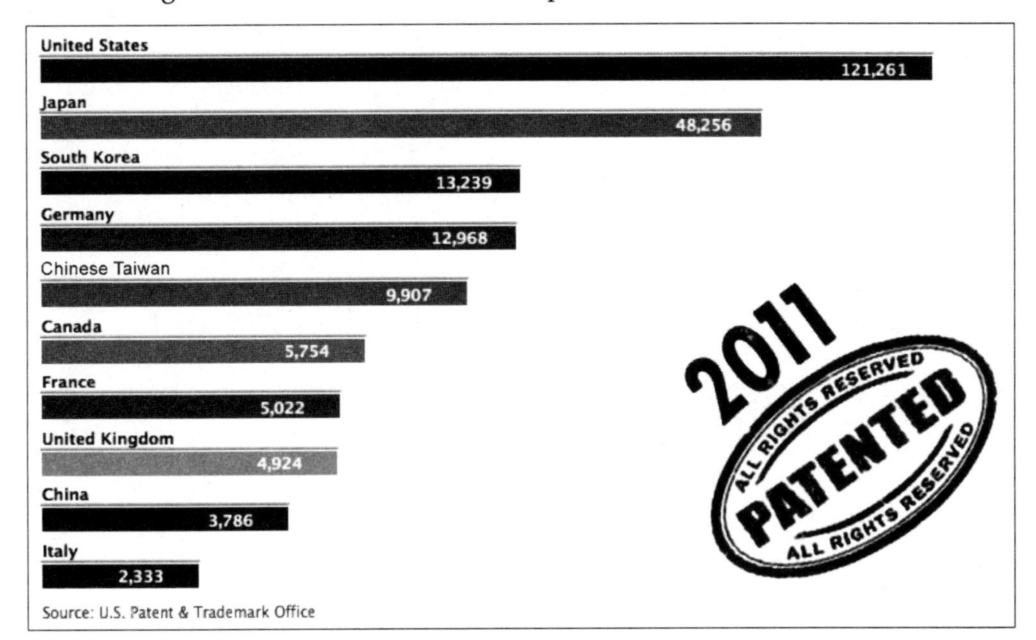

Constructing a mirror chart

Using a mirror chart is another way of comparing two column series. Instead of aligning the two series as columns adjacent to each other, mirror charts align them in bars opposite to each other. Sometimes, this is used as a preferred way for presenting the trend between the two series.

In Highcharts, we can make use of a stacked bar chart and change it slightly into a mirror chart for comparing two sets of data horizontally side by side. To do that, let's start with a new data series from **Patents Granted**, which shows the comparison between the United Kingdom and China with respect to the number of patents granted for the past decade.

The way we configure the chart is really a stacked-column bar chart, with one set of data being positive and another set being manually converted to negative values, so that the zero value axis is in the middle of the chart. Then we invert the column chart into a bar chart and label the negative range as positive. To demonstrate this concept, let's create a stacked column chart first, with both positive and self-made negative ranges, as follows:

```
chart: {
    renderTo: 'container',
    type: 'column',
```

```
        borderWidth: 1
    },
    title: {
        text: 'Number of Patents Granted'
    },
    credits: { ... },
    xAxis: {
        categories: [ '2001', '2002', '2003', ... ]
    },
    yAxis: {
        title: {
            text: 'No. of Patents'
        }
    },
    plotOptions: {
        series: {
          stacking: 'normal'
        }
    },
    series: [{
        name: 'UK',
        data: [ 4351, 4190, 4028, ... ]
        }, {
        name: 'China',
        data: [ -265, -391, -424, ... ]
    }]
```

The following screenshot shows the stacked-column chart with the zero value in the middle of the y axis:

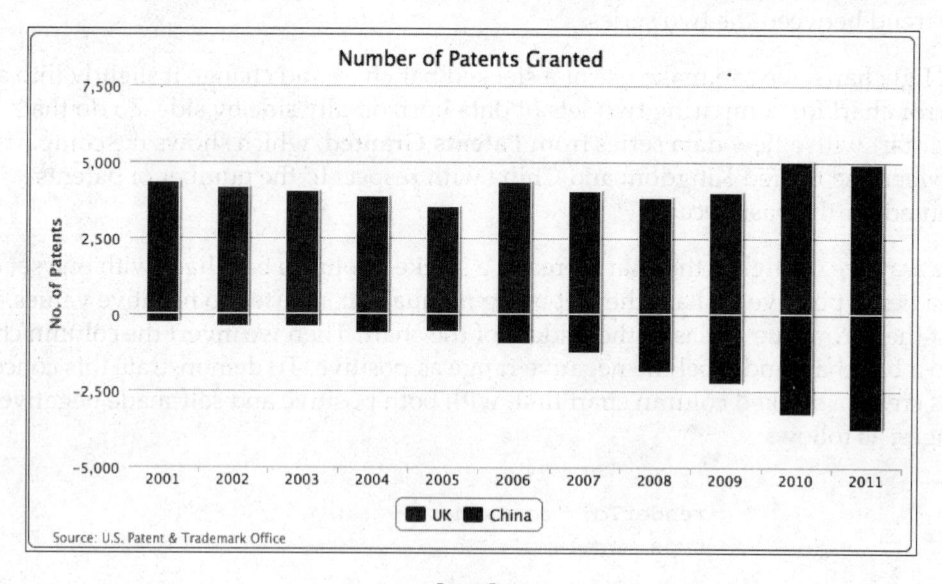

Then, we change the configuration into a bar chart with two x axes showing on each side with the same range. The last step is to define the y-axis label's `formatter` function to turn the negative labels into positive ones, as follows:

```
chart: {
    .... ,
    type: 'bar'
},
xAxis: [{
    categories: [ '2001', '2002', '2003', ... ]
}, {
    categories: [ '2001', '2002', '2003', ... ],
    opposite: true,
    linkedTo: 0
}],
yAxis: {
    .... ,
    labels: {
        formatter: function() {
            return
          Highcharts.numberFormat(Math.abs(this.value), 0);
        }
    }
},
```

The following is the final bar chart for comparing the number of patents granted between the UK and China for the past decade:

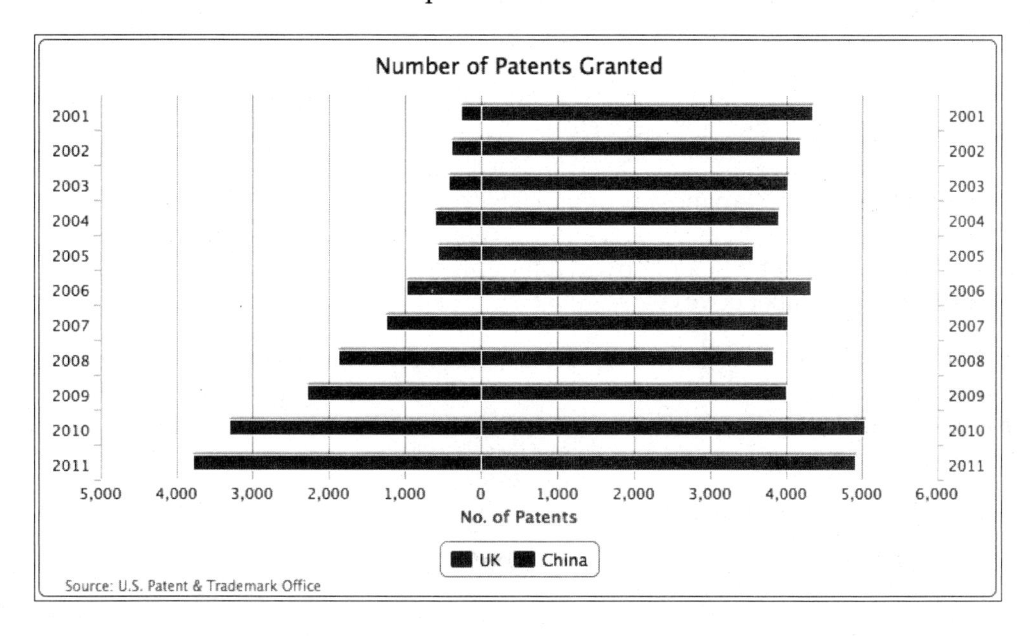

Extending to a stacked mirror chart

We can also apply the same principle from the column example to stacked and grouped series charts. Instead of having two groups of stacked columns displayed next to each other, we can have all the series stacked together with the zero value to divide both groups. The following screenshot demonstrates the comparison between the European and Asian stacked groups in a bar chart:

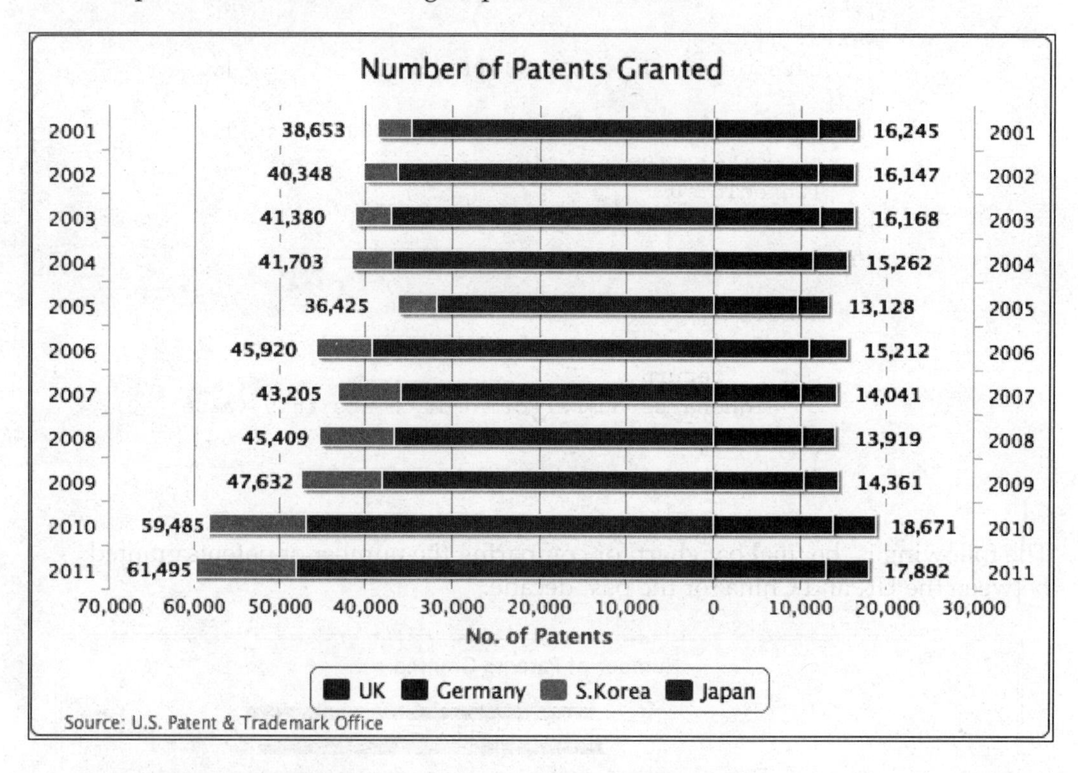

The South Korean and Japanese series are stacked together on the left-hand side (the negative side), whereas the UK and Germany are grouped on the right-hand side (the positive side). The only tricky bit in producing the preceding graph is how to output the data label boxes.

First of all, the South Korean and Japanese series data are manually set to negative values. Second, since South Korea and the UK are both the outer series of their own group, we enable the data label for these series. The following code snippet shows the series array configuration:

```
series: [{
    name: 'UK',
    data: [ 4351, 4190, 4028, ... ],
```

```
            dataLabels : {
                enabled: true,
                backgroundColor: '#FFFFFF',
                x: 40,
                formatter: function() {
                    return
        Highcharts.numberFormat(Math.abs(this.total), 0);
                },
                style: {
                    fontWeight: 'bold'
                }
            }
        }, {
            name: 'Germany',
            data: [ 11894, 11957, 12140, ... ]
        }, {
            name: 'S.Korea',
            data: [ -3763, -4009, -4132, ... ],
            dataLabels : {
                enabled: true,
                x: -48,
                backgroundColor: '#FFFFFF',
                formatter: function() {
                    return
        Highcharts.numberFormat(Math.abs(this.total), 0);
                },
                style: {
                    fontWeight: 'bold'
                }
            }
        }, {
            name: 'Japan',
            data: [ -34890, -36339, -37248, ... ]
        }]
```

Note that the definition for the `formatter` function is using `this.total` and not `this.y`, because we are using the position of the outer series to print the group's total value. The white background settings for the data labels are to avoid interfering with the y-axis interval lines.

Converting a single bar chart into a horizontal gauge chart

A gauge chart is generally used as an indicator for the current threshold level, meaning the extreme values in the y axis are fixed. Another characteristic is the single value (one dimension) in the x axis that is the current time.

Next, we are going to learn how to turn a chart with a single bar into a gauge-level chart. The basic idea is to diminish the plot area to the same size as the bar. This means we have to fix the size of both the plot area and the bar, disregarding the dimensions of the container. To do that, we set `chart.width` and `chart.height` to some values. Then, we decorate the plot area with a border and background color to make it resemble a container for the gauge:

```
chart: {
    renderTo: 'container',
    type: 'bar',
    plotBorderWidth: 2,
    plotBackgroundColor: '#D6D6EB',
    plotBorderColor: '#D8D8D8',
    plotShadow: true,
    spacingBottom: 43,
    width: 350,
    height: 120
},
```

We then switch off the y-axis title and set up a regular interval within the percentage, as follows:

```
xAxis: {
    categories: [ 'US' ],
    tickLength: 0
},
yAxis: {
    title: {
        text: null
    },
    labels: {
        y: 20
    },
    min: 0,
    max: 100,
    tickInterval: 20,
```

```
        minorTickInterval: 10,
        tickWidth: 1,
        tickLength: 8,
        minorTickLength: 5,
        minorTickWidth: 1,
        minorGridLineWidth: 0
    },
```

The final part is to configure the bar series, so that the bar width fits perfectly within the plot area. The rest of the series configuration is to brush up the bar with an SVG gradient effect, as follows:

```
    series: [{
        borderColor: '#7070B8',
        borderRadius: 3,
        borderWidth: 1,
        color: {
            linearGradient:
                { x1: 0, y1: 0, x2: 1, y2: 0 },
            stops: [ [ 0, '#D6D6EB' ],
                     [ 0.3, '#5C5CAD' ],
                     [ 0.45, '#5C5C9C' ],
                     [ 0.55, '#5C5C9C' ],
                     [ 0.7, '#5C5CAD' ],
                     [ 1, '#D6D6EB'] ]
        },
        pointWidth: 50,
        data: [ 48.9 ]
    }]
```

Multiple stop gradients are supported by SVG, but not by VML. For VML browsers, such as Internet Explorer 8, the number of stop gradients should be restricted to two.

The following is the final polished look of the gauge chart:

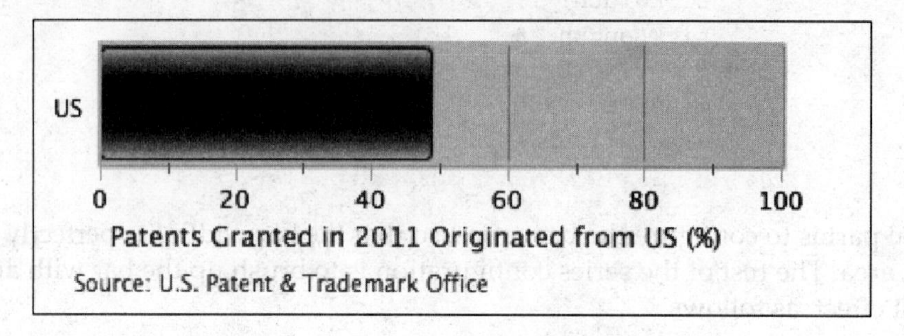

Sticking the charts together

In this section, we are building a page with a mixture of charts. The main chart is displayed in the left-hand side panel and three mini charts are displayed in the right-hand side panel in top-down order. The layout is achieved by HTML `div` boxes and CSS styles.

The left-hand side chart is from the multicolored column chart example that we discussed previously. All the axis lines and labels are disabled in the mini charts.

The first mini chart from the top is a two-series line chart with `dataLabels` enabled for the last point in each series only: the last point in the data array is a data object instead. The label color is set to the same color as its series. Then, `plotLine` is inserted into the y axis at the 50 percent value mark. The following is a sample of one of the series configurations:

```
pointStart: 2001,
marker: {
    enabled: false
},
data: [ 53.6, 52.7, 52.7, 51.9, 52.4,
        52.1, 51.2, 49.7, 49.5, 49.6,
        { y: 48.9,
          name: 'US',
          dataLabels: {
              color: '#4572A7',
              enabled: true,
              x: -10,
              y: 14,
              formatter: function() {
```

```
                    return
            this.point.name + ": " + this.y + '%';
                        }
                    }
                }]
```

The second mini chart is a simple bar with data labels outside the categories. The style for the data label is set to a larger, bold font.

The last mini chart is basically a scatter chart where each series has a single point, so that each series can appear in the right-hand side legend. Moreover, we set the x value for each series to zero, so that we can have different sizes of data points, and stacked on top of each other as well. The following is an example for one of the scatter series configurations:

```
zIndex: 1,
legendIndex: 0,
color: {
    linearGradient:
        { x1: 0, y1: 0, x2: 0, y2: 1 },
          stops: [ [ 0, '#FF6600' ],
                   [ 0.6, '#FFB280'  ] ]
},
name: 'America - 49%',
marker: {
    symbol: 'circle',
        lineColor: '#B24700',
        lineWidth: 1
},
data: [
    { x: 0, y: 49, name: 'America',
      marker: { radius: 74 }
} ]
```

The following is the screenshot of these multiple charts displayed next to each other:

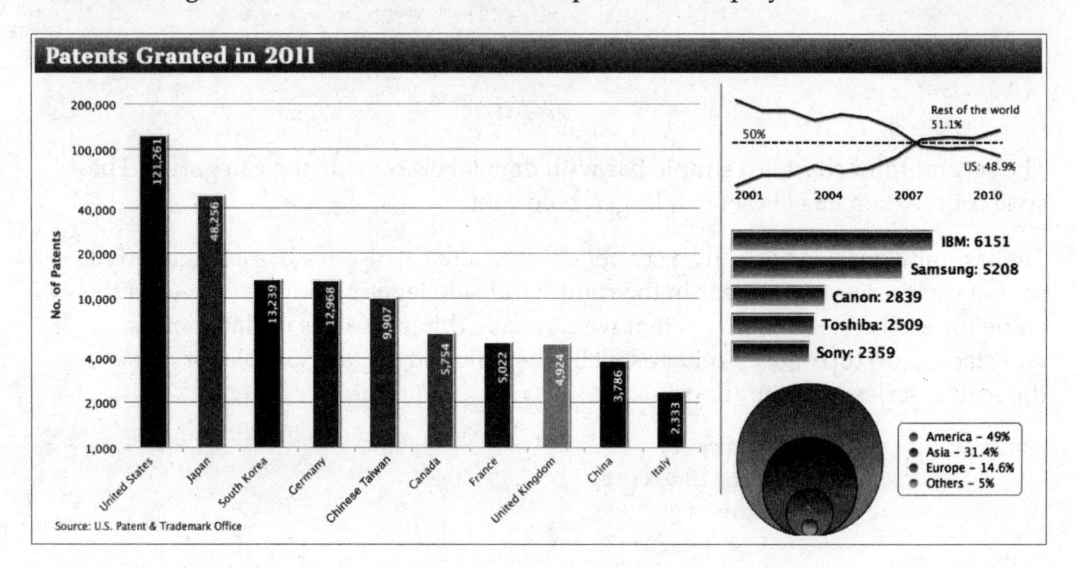

Summary

In this chapter, we have learned how to use both column and bar charts. We utilized their options to achieve various presentation configurations of columns and bars for ease of comparison between sets of data. We also learned advanced configurations for different chart appearances such as mirror and gauge charts.

In the next chapter, we will explore the pie chart series.

5

Pie Charts

In this chapter, you will learn how to plot pie charts and explore their various options. We will then examine how to put multiple pies inside a chart. After that, we will find out how to create a donut chart. We will then end the chapter by sketching a chart that contains all the series types that we have learned so far—column, line, and pie. In this chapter, we will be covering the following topics:

- Understanding the relationship between chart, pie, and series
- Plotting simple pie charts - single series
- Plotting multiple pies in a chart - multiple series
- Preparing a donut chart - multiple series
- Building a chart with multiple series types
- Understanding the `startAngle` and `endAngle` options
- Creating a simplified version of the stock picking wheel chart

Understanding the relationship between chart, pie, and series

Pie charts are simple to plot: they have no axes to configure and all they need is data with categories. Generally, the term **pie chart** refers to a chart with a single pie series. In Highcharts, a chart can handle multiple pie series. In this case, a chart can display more than one pie: each pie is associated with a series of data. Instead of showing multiple pies, Highcharts can display a donut chart that is basically a pie chart with multiple concentric rings lying on top of each other. Each concentric ring is a pie series, similar to a stacked pie chart. We will first learn how to plot a chart with a single pie, and then later on in the chapter we will explore plotting with multiple pie series in separate pies and a donut chart.

Plotting simple pie charts – single series

In this chapter, we are going to use video gaming data supplied by **vgchartz** (`www.vgchartz.com`). The following is the pie chart configuration, and the data is the number of games sold in 2011 according to publishers, based on the top 100 games sold. Wii Sports is taken out of the dataset because it is free with the Wii console:

```
chart: {
  renderTo: 'container',
  type: 'pie',
  borderWidth: 1,
  borderRadius: 5
},
title: {
  text: 'Number of Software Games Sold in 2011 Grouped by 
Publishers',
},
credits: {
  ...
},
series: [{
  data: [ [ 'Nintendo', 54030288 ],
       [ 'Electronic Arts', 31367739 ],
       ... ]
}]
```

Here is a simple pie chart screenshot with the first data point (Nintendo) starting at the 12 o'clock position. The first slice always starts at the 12 o'clock position unless we specify a different start point with the `startAngle` option, which we will explore later in the chapter.

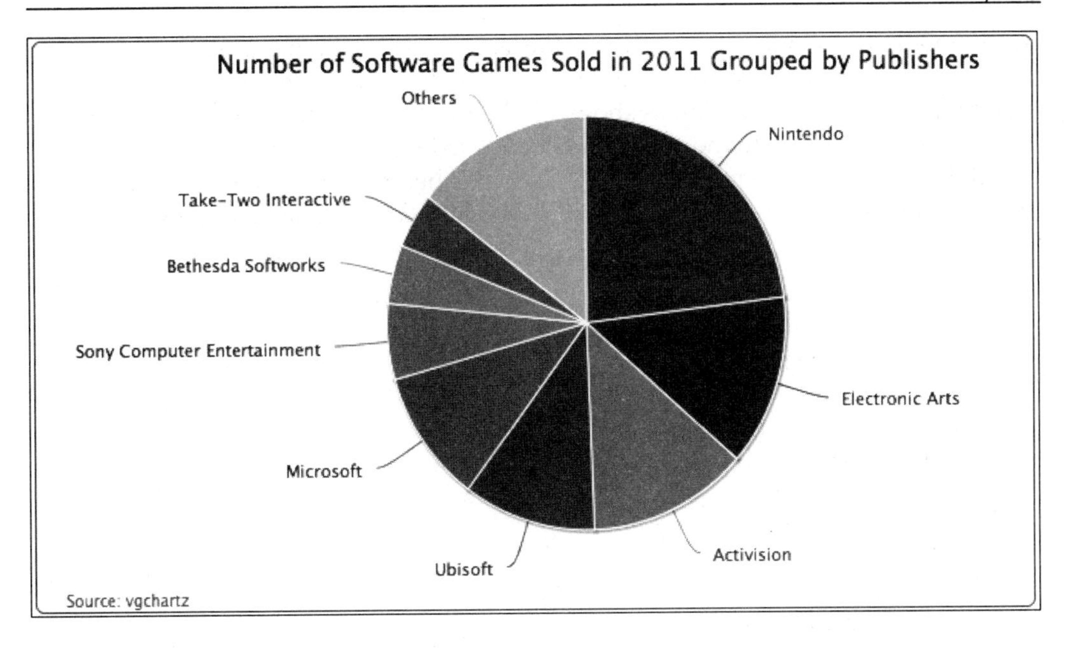

Configuring the pie with sliced off sections

We can improve the previous pie chart to include values in the labels and word wrap some of the long names of the publishers. The following is the configuration code for the pie series. The `allowPointSelect` option allows the users to interact with the chart by clicking on the data points. As for the pie series, this is used for slicing off a section of the pie chart (see the following screenshot). The `slicedOffset` option is used to adjust how far the section is sliced off from the pie chart. For word wrap labels, we set the labels style, `dataLabels.style.width`, to 140 pixels wide:

```
plotOptions: {
  pie: {
    slicedOffset: 20,
    allowPointSelect: true,
    dataLabels: {
      style: {
           width: '140px'
      },
      formatter: function() {
        var str = this.point.name + ': ' +
          Highcharts.numberFormat(this.y, 0);
        return str;
      }
    }
  }
},
```

Additionally, we would like to slice off the largest section in the initial display; its label is shown in bold type font. To do that, we will need to change the largest data point into object configuration as shown in the following screenshot. Then we put the `sliced` property into the object and change from the default, `false`, to `true`, which forces the slice to part from the center. Furthermore, we set `dataLabels` with the assignment of the `fontWeight` option to overwrite the default settings:

```
series: [{
  data: [ {
    name: 'Nintendo',
    y: 54030288,
    sliced: true,
    dataLabels: {
      style: {
      fontWeight: 'bold'
      }
    }
   }, [ 'Electronic Arts', 31367739 ],
    [ 'Activision', 30230170 ], .... ]
}]
```

The following is the chart with the refined labels:

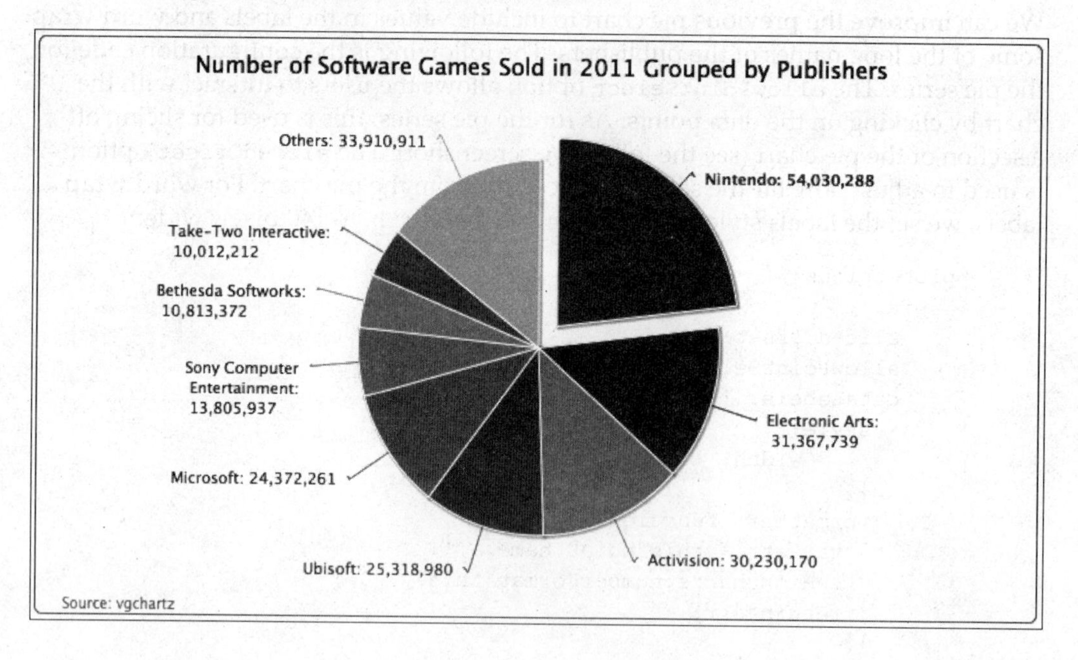

As mentioned earlier, the `slicedOffset` option has also pushed the sliced off section further than the default distance, which is 10 pixels. The `slicedOffset` option applies to all the sliced off sections, which means that we cannot control the distance of individually parted sections. It is also worth noticing that the connectors (the lines between the slice and the data label) become crooked as a result of that. In the next example, we demonstrate that the `sliced` property can be applied to as many data points as we want, and remove the `slicedOffset` option to resume the default settings to show the difference. The following chart illustrates this with three parted slices, by repeating the data object settings (Nintendo) for two other points:

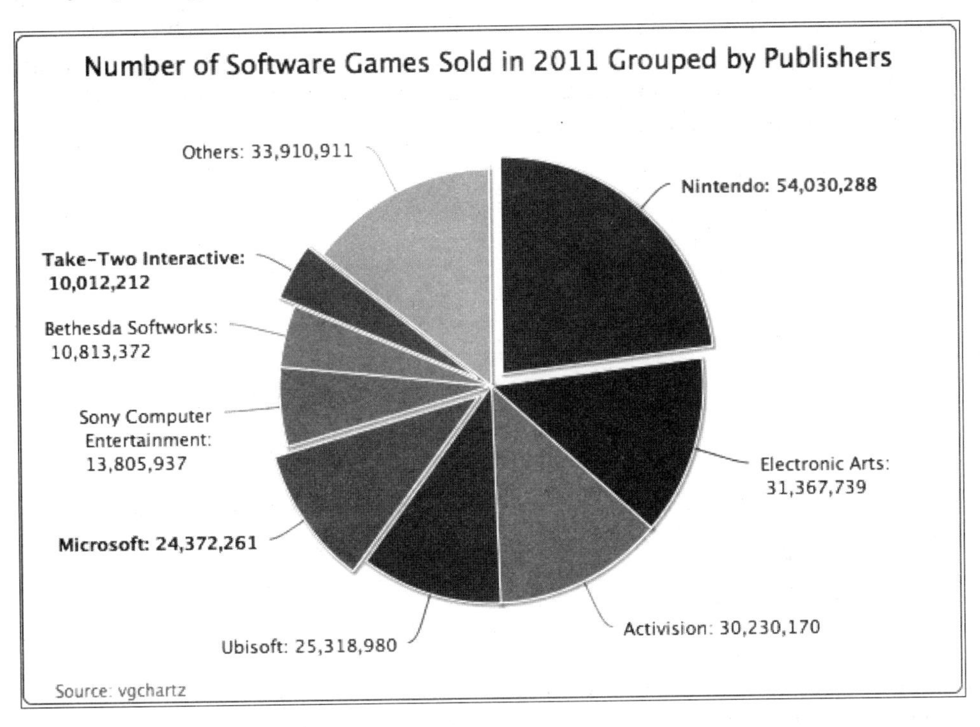

Notice that the connectors go back to being smooth lines. However, there is another interesting behavior for the `sliced` option. For those slices with `sliced` as the default setting (`false`), only one of them can be sliced off. For instance, the user clicks on the **Others** section and it moves away from the chart. Then, clicking on **Activision** will slice off the section and the **Others** section moves back towards the center, whereas the three configured `sliced: true` sections maintain their parted positions. In other words, with the `sliced` option set to `true`, this enables its state to be independent of others with the `false` setting.

Applying a legend to a pie chart

So far, the chart contains large numbers, and it is difficult to really comprehend how much larger one section is than the other. We can print all the labels in percentages. Let's put all the publisher names inside a legend box and print the percentage values inside each slice.

The plotting configuration is redefined as follows. To enable the legend box, we set `showInLegend` to `true`. Then, we set the data labels' font color and style to bold and white respectively, and change the `formatter` function slightly to use the `this.percentage` variable that is only available for the pie series. The `distance` option is the distance between the data label and the outer edge of the pie. A positive value will shift the data label outside of the edge and a negative value will do the same in the opposite direction:

```
plotOptions: {
  pie: {
    showInLegend: true,
    dataLabels: {
      distance: -24,
      color: 'white',
      style: {
        fontWeight: 'bold'
      },
      formatter: function() {
        return Highcharts.numberFormat(this.percentage) + '%';
      }
    }
  }
},
```

Then, for the legend box, we add in some padding as there are more than a few legend items, and set the legend box closer to the pie, as follows:

```
legend: {
  align: 'right',
  layout: 'vertical',
  verticalAlign: 'middle',
  itemMarginBottom: 4,
  itemMarginTop: 4,
  x: -40
},
```

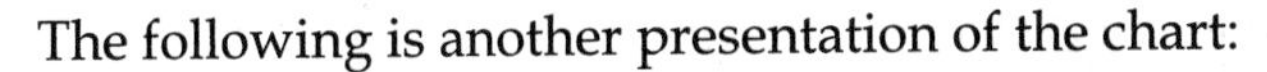

The following is another presentation of the chart:

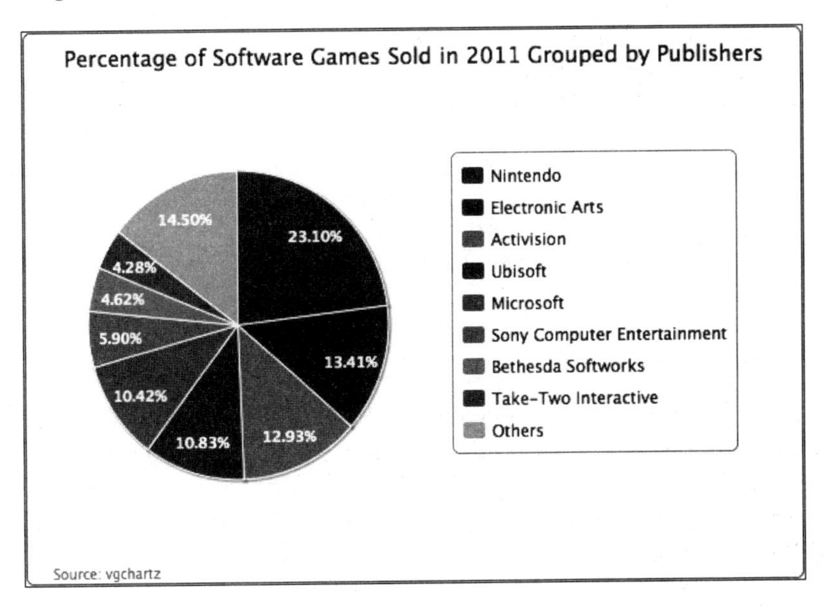

Plotting multiple pies in a chart – multiple series

With pie charts, we can do something more informative by displaying another pie chart side by side to compare data. This can be done by simply specifying two series configurations in the series array.

We can continue to use the previous example for the chart on the left-hand side and we create a new category series from the same dataset, but this time grouped by platforms. The following is the series configuration for doing so:

```
plotOptions:{
  pie: {
     ....,
     size: '75%'
  }
},
series: [{
  center: [ '25%', '50%' ],
  data: [ [ 'Nintendo', 54030288 ],
          [ 'Electronic Arts', 31367739 ],
          .... ]
}, {
```

```
            center: [ '75%', '50%' ],
            dataLabels: {
              formatter: function() {
                var str = this.point.name + ': ' +
    Highcharts.numberFormat(this.percentage, 0) + '%';
                return str;
              }
            },
            data: [ [ 'Xbox', 80627548 ],
                    [ 'PS3', 64788830 ],
                      ....   ]
          }]
```

As we can see, we use a new option, `center`, to position the pie chart. The option contains an array of two percentage values—the first is the ratio of the x position to the whole container width, whereas the second percentage value is the y ratio. The default value is `['50%', '50%']`, which is in the middle of the container. In this example, we specify the first percentage values as `'25%'` and `'75%'`, which are in the middle of the left- and the right-hand halves respectively. We set the size for the pie series to 75 percent with `plotOptions.pie.size`. This ensures that both pies are the same size, otherwise the left hand pie will appear smaller.

In the second series, we will choose to display the pie chart with percentage data labels instead of unit values. The following is the screenshot of a chart with double pies:

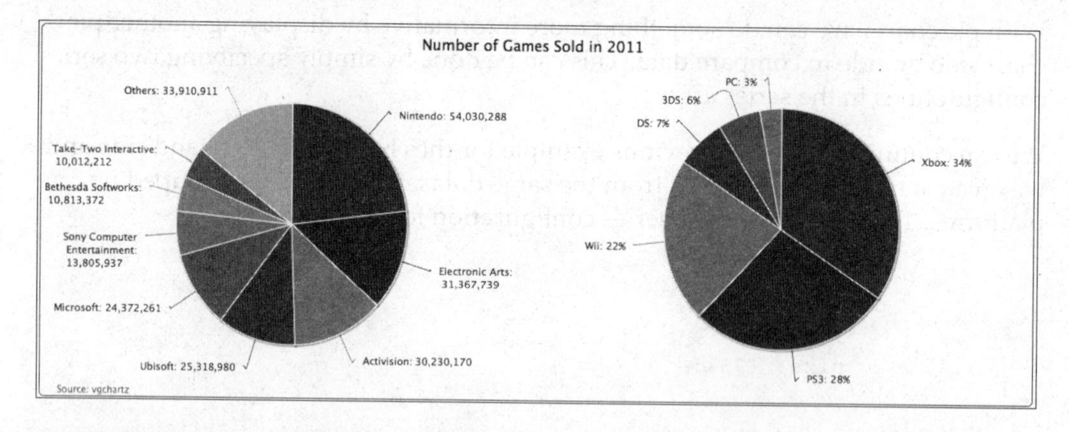

On the surface, this is not much different to plotting two separate pie charts in an individual `<div>` tag, apart from sharing the same title. The main benefit is that we can combine different series type presentations under the same chart. For instance, let's say we want to present the distribution in ratio in pie series directly above each group of multiple column series. We will learn how to do this later in the chapter.

Preparing a donut chart – multiple series

Highcharts offers another type of pie chart, a **donut chart**. It has the effect of drilling down on a category to create subcategories, and is a convenient way of viewing data in greater detail. This drill-down effect can be applied on multiple levels. In this section, we will create a simple donut chart that has an outer ring of subcategories (game titles) that align with the inner categories (publishers).

For the sake of simplicity, we will only use the top three games publishers for the inner pie chart. The following is the series array configuration for the donut chart:

```
      series: [{
        name: 'Publishers',
        dataLabels : {
          distance: -70,
          color: 'white',
          formatter: function() {
             return this.point.name + ':<br/> ' +
  Highcharts.numberFormat(this.y / 1000000, 2);
          },
          style: {
             fontWeight: 'bold'
          }
        },
        data: [ [ 'Nintendo', 54030288 ],
                [ 'Electronic Arts', 31367739 ],
                [ 'Activision', 30230170 ] ]
      }, {
        name: 'Titles',
        innerSize: '60%',
        dataLabels: {
          formatter: function() {
            var str = '<b>' + this.point.name + '</b>: ' +
                 Highcharts.numberFormat(this.y / 1000000, 2);
            return str;
          }
        },
        data: [ // Nintendo
            { name: 'Pokemon B&W', y: 8541422,
             color: colorBrightness("#4572A7",
                       0.05) },
            { name: 'Mario Kart', y: 5349103,
             color: colorBrightness('#4572A7',
                       0.1) },
            ....
```

```
            // EA
            { name: 'Battlefield 3', y: 11178806,
              color: colorBrightness('#AA4643',
                          0.05) },
            ....

            // Activision
            { name: 'COD: Modern Warfare 3',
              y: 23981182,
              color: colorBrightness('#89A54E',
                          0.1) },

            ....
        }]
    }]
```

First, we have two series—the inner pie series, or the **Publishers**, and the outer ring series, or the **Titles**. The **Titles** series has all the data for the subcategories together, and it aligns with the **Publisher** series. The order is such that the values of the subcategories for the **Nintendo** category are before the subcategory data for **Electronic Arts,** and so on (see the order of data arrays in the **Title** series).

Each data point in the subcategories series is declared as a data point object for assigning the color in a similar range to their main category. This can be achieved by following the Highcharts demo to fiddle with the color brightness:

```
color: Highcharts.Color(color).brighten(brightness).get()
```

Basically, what this does is to use the main category color value to create a `Color` object and then adjust the color code with the brightness parameter. This parameter is derived from the ratio of the subcategory value. We rewrite this example into a function known as `colorBrightness`, and call it in the chart configuration:

```
function colorBrightness(color, brightness) {
  return
    Highcharts.Color(color).brighten(brightness).get();
}
```

The next part is to specify which series goes to the inner pie and which goes to the outer ring. The `innerSize` option is used by the outer series, **Title**, to create an inner circle. As a result, the **Title** series forms a donut/concentric ring. The value for the `innerSize` option can be either in pixels or percentage values of the plot area size.

The final part is to decorate the chart with data labels. Obviously, we want to position the data labels of the inner charts to be over the inner pie, so we assign a negative value to the `dataLabels.distance` option. Instead of printing long values, we define `formatter` to convert them into units of millions.

The following is the display of the donut chart:

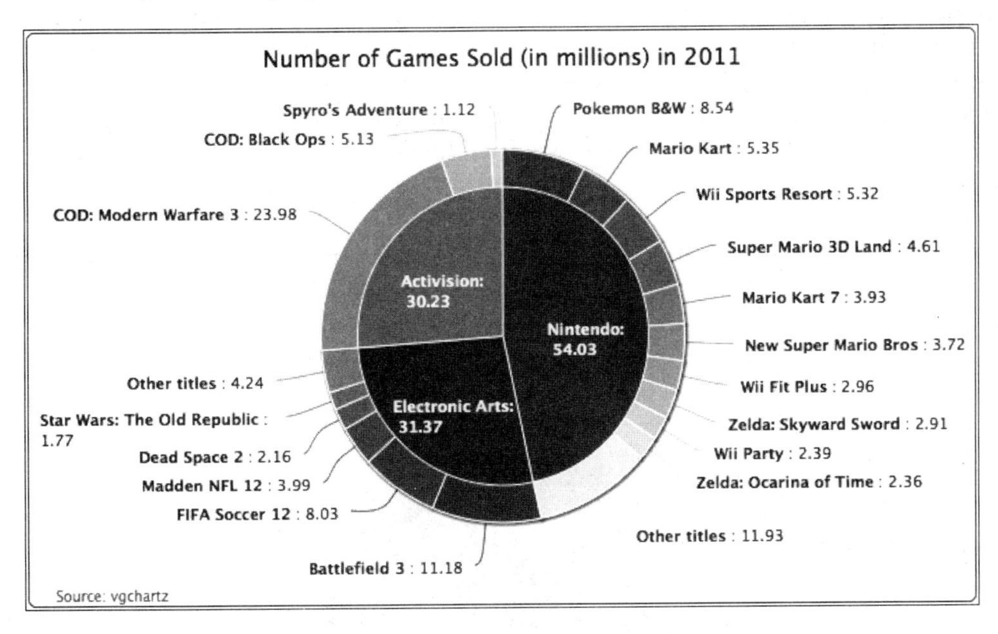

Note that it is not mandatory to put a pie chart in the center of a donut chart. It is just the presentation style of this example. We can have multiple concentric rings instead. The following chart is exactly the same example as mentioned earlier, with the addition of the `innerSize` option to the inner series of **Publishers**:

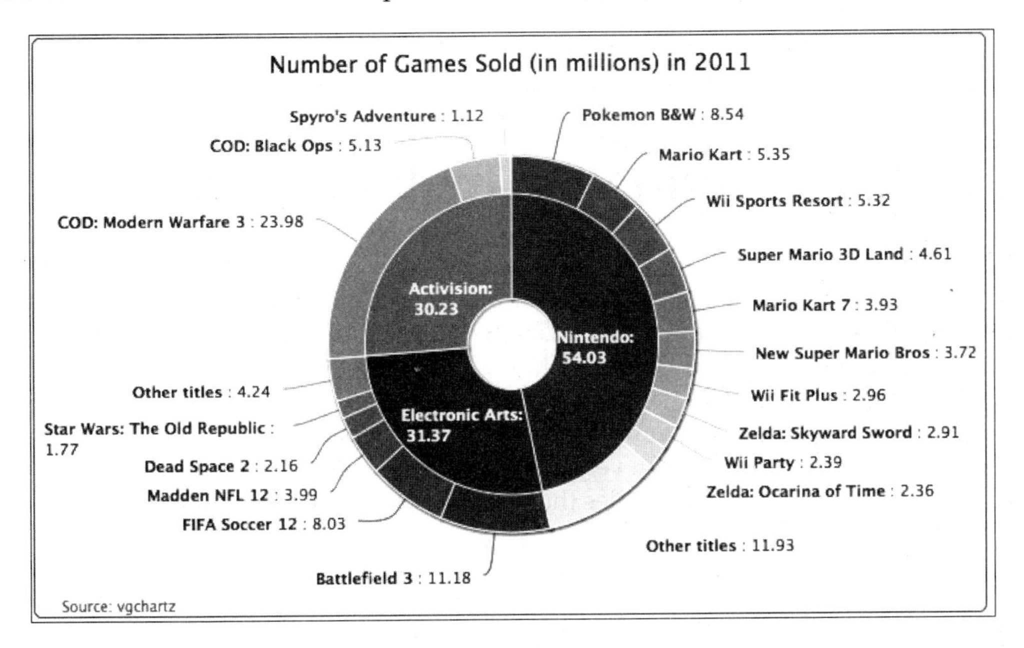

We can even further complicate the donut chart by introducing a third series. We plot the following chart in three layers. The code is simply extended from the example with another series and includes more data. The source code and the demo are available at `http://joekuan.org/Learning_Highcharts/`. The two outer series use the `innerSize` option. As the inner pie will become even smaller and will not have enough space for the labels, we therefore enable the legend box for the innermost series with the `showInLegend` option.

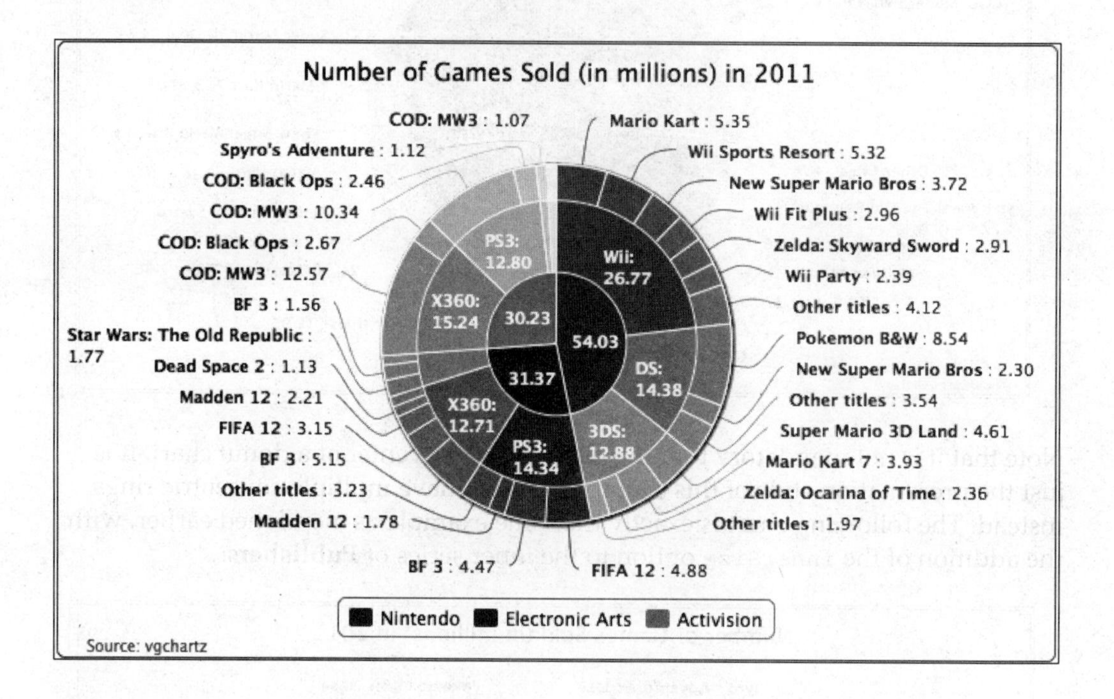

Building a chart with multiple series types

So far, we have learned about the line, column, and pie series types. It's time to bring all these different series presentations together in a single chart. In this section, we will use the annual data from 2008 through 2011 to plot three different kinds of series type: column, line, and pie. The column type represents the yearly number of games sold for each type of gaming console. The pie series shows the annual number of gaming consoles sold for each vendor. The last one is the spline series type that discloses how many new game titles there are released in total for all the consoles each year.

In order to ensure the whole graph uses the same color scheme for each type of gaming console, we have to manually assign a color code for each data point in the pie charts and the columns series:

```
var wiiColor = '#BBBBBB',
    x360Color = '#89A54E',
    ps3Color = '#4572A7',
    splineColor = '#FF66CC';
```

We then decorate the chart in a more funky way. First, we give the chart a dark background with a color gradient:

```
var chart = new Highcharts.Chart({
  chart: {
    renderTo: 'container',
    borderWidth: 1,
    spacingTop: 40,
    backgroundColor: {
      linearGradient: { x1: 0, y1: 0,
                        x2: 0, y2: 1 },
      stops: [ [ 0, '#0A0A0A' ],
               [ 1, '#303030' ] ]
    }
  },
  subtitle: {
      floating: true,
      text: 'Number of consoles sold (in millions)',
      y: -5
  },
```

Then, we need to shift the columns to the right-hand side, so that we have enough room for an image (described later) that we are going to put in the top left-hand corner:

```
  xAxis: {
    minPadding: 0.2,
    tickInterval: 1,
    labels: {
      formatter: function() {
        return this.value;
        },
      style: {
       color: '#CFCFCF'
      }
    }
  },
```

The next task is to make enough space for the pie charts to locate them above the columns. This can be accomplished by introducing the `maxPadding` option on both y axes:

```
yAxis: [{
  title: {
    text: 'Number of games sold',
    align: 'low',
    style: {
      color: '#CFCFCF'
    }
  },
  labels: {
    style: {
     color: '#CFCFCF'
    }
  },
  maxPadding: 0.5
}, {
  title: {
    text: 'Number of games released',
    style: {
     color: splineColor
    }
  },
  labels: {
    style: {
     color: splineColor
    }
  },
  maxPadding: 0.5,
  opposite: true
}],
```

Each pie series is displayed separately and aligned to the top of the columns, as well as with the year category. This is done by adjusting the pie chart's `center` option in the series array. We also want to reduce the display size for the pie series, as there are other types of series to share within the chart. We will use the `size` option and set the value in percentages. The percentage value is the diameter of the pie series compared to the size of the plot area:

```
series: [{
  type: 'pie',
  name: 'Hardware 2011',
```

```
size: '25%',
center: [ '88%', '5%' ],
data: [{ name: 'PS3', y: 14128407,
    color: ps3Color },
   { name: 'X360', y: 13808365,
    color: x360Color },
   { name: 'Wii', y: 11567105,
    color: wiiColor } ],
.....
```

The spline series is defined to correspond to the opposite y axis. To make the series clearly associated with the second axis, we apply the same color scheme for the line, axis title, and labels:

```
{  name: "Game released",
  type: 'spline',
  showInLegend: false,
  lineWidth: 3,
  yAxis: 1,
  color: splineColor,
  pointStart: 2008,
  pointInterval: 1,
  data: [ 1170, 2076, 1551, 1378 ]
},
```

We use the `renderer.image` method to insert the image into the chart and make sure that the image has a higher `zIndex`, so that the axis line does not lie at the top of the image. Instead of including a PNG image, we use an SVG image. This way the image stays sharp and avoids the pixelation effect when the chart is resized:

```
chart.renderer.image('./pacman.svg', 0,
           0, 200, 200).attr({
    'zIndex': 10
}).add();
```

The following is the final look of the graph with a Pac-Man SVG image to give a gaming theme to the chart:

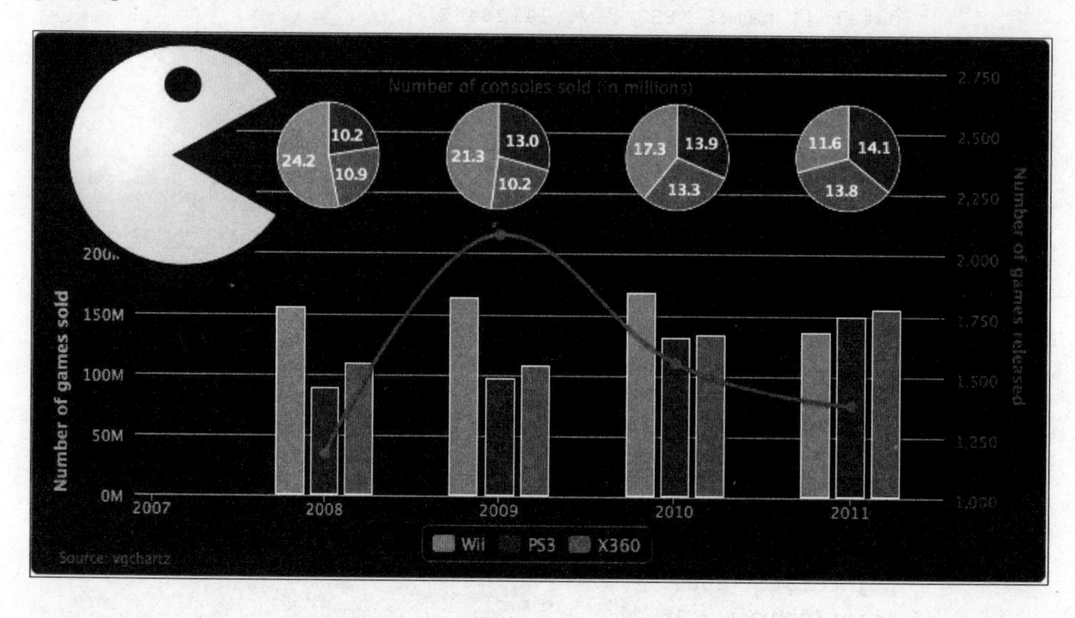

Creating a stock picking wheel

A stock picking wheel chart is a designer financial chart created by Investors Intelligence (`http://www.investorsintelligence.co.uk/wheel/`), and the chart provides an interactive way to visualize both the overall and detailed performance of major shares from a market index. The following is a screenshot of the chart:

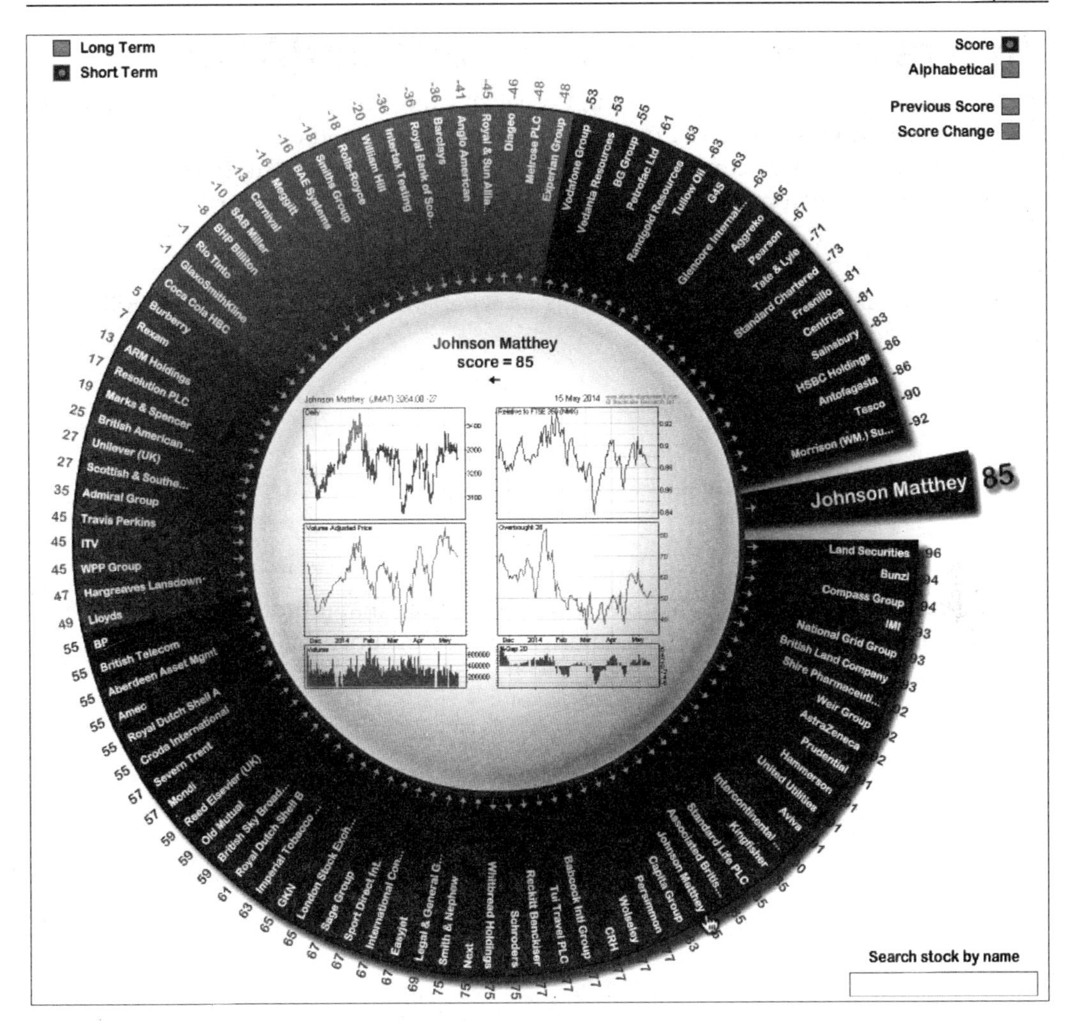

Basically, it is an open-ended donut chart. Each slice represents a blue chip company and all the slices have equal width. The slices are divided into different color bands (from red, to orange, green, and blue) based on the share performance score. When the user hovers the mouse over a company name, a bigger slice appears at the gap of the donut chart with the company name and detailed performance information. Although this impressive chart is implemented in Adobe Flash, we are going to push our luck to see whether Highcharts can produce a lookalike cousin in this section. We will focus on the donut chart and leave the middle charts as an exercise for you later.

Understanding startAngle and endAngle

Before we start plotting a Stock Picking Wheel chart, we need to know how to create an open-ended donut chart. The options `startAngle` and `endAngle` can make a pie chart split open in any size and direction. However, we need to familiarize ourselves with these options before plotting the actual chart. By default, the starting position for the first slice is at 12 o'clock, which is regarded as 0 degrees for the `startAngle` option. The `startAngle` option starts from the origin and moves in a clockwise direction to `endAngle`. Let's make a simple pie chart with `startAngle` and `endAngle` assigned to `30` and `270` respectively:

```
plotOptions: {
   pie: {
        startAngle: 30,
        endAngle: 270
   }
},
```

The preceding code produces the following chart. Note that although the pie area for the series data becomes smaller, the data will still be proportionally adjusted:

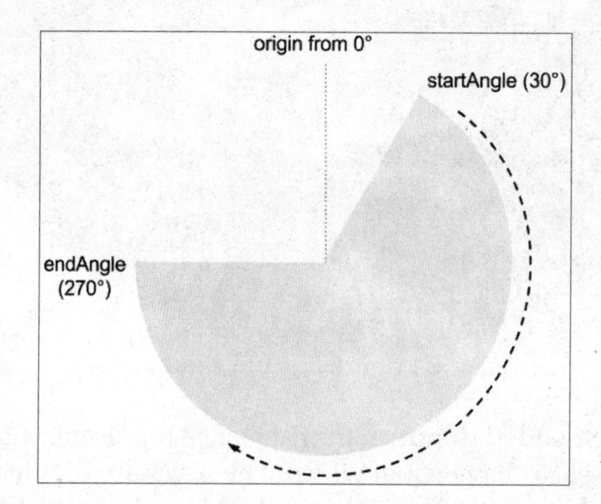

In order to display a gap at the right-hand side of the pie, the `endAngle` option needs to be handled slightly differently. For `endAngle` to end up above the `startAngle` option, that is, the `endAngle` option goes round and passes the origin at 0 degrees, the `endAngle` value has to exceed 360, not start from 0 degrees again. Let's set the `startAngle` and `endAngle` values to 120 and 420 respectively and increase the `innerSize` value to form a donut chart. Here is the new configuration:

```
plotOptions: {
    pie: {
         startAngle: 120,
```

```
                endAngle: 420,
                innerSize: 110,
                ....
            }
        }
```

We can see the `endAngle` option goes past the origin and ends above the `startAngle` option. See the following screenshot:

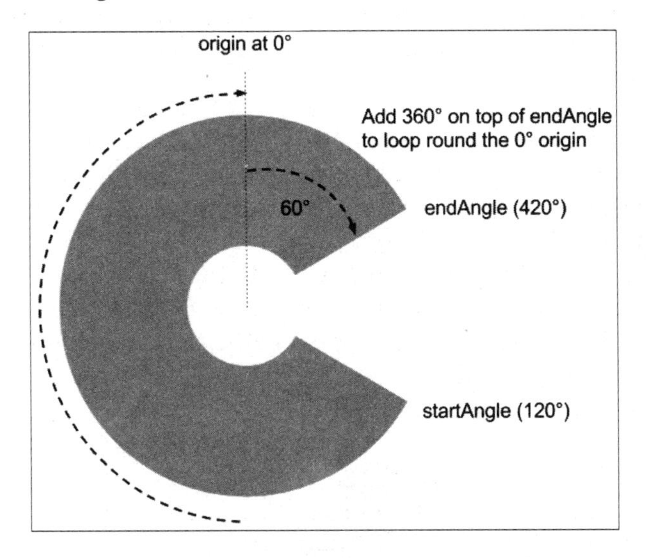

Creating slices for share symbols

So far, we have successfully created the general shape of the stock picking wheel chart. The next task is to populate financial data labels on each slice with equal size. Instead of having a hundred shares from the FTSE index like the original chart, we will use the Dow Jones Industrial Average instead, which is composed of 30 major shares. First, we generate a series data array with share symbols in descending order of percentage change:

```
var data = [{
     name: 'MCD 0.96%',
     y: 1,
     dataLabels: { rotation: 35 },
     color: colorBrightness('#365D97', ratio1)
}, {
    name: 'HD 0.86%',
    y: 1,
    dataLabels: { rotation: 45 },
```

```
                color: colorBrightness('#365D97', ratio2)
            }, {
                ....
            }]
```

We set the y-axis value to 1 for each share symbol in order to make it the same size for all the slices. Then, we evaluate the positive and negative gradual color change based on the ratio between the percentage change of those shares and the overall maximum and minimum values respectively. We then apply this ratio value to the `colorBrightness` method (which we have discussed previously) to achieve the gradual change effect.

As for the label rotation, since we know where to put each share symbol and what rotation to apply in advance, we can compute the label rotation based on the company's share symbol position from the `startAngle`. The orientation for the `dataLabel` is different to the `startAngle`, which starts with the 3 o'clock position as 0 degrees (because we read text horizontally). With the `startAngle` as 120 degrees, we can easily calculate the `dataLabel` rotation for the first slice as 30 degrees (we add another 5 degrees for the `dataLabel` to appear in the middle of the slice) and each slice is another 10 degrees of rotation more than the previous slice. The following illustrates the logic of this computation:

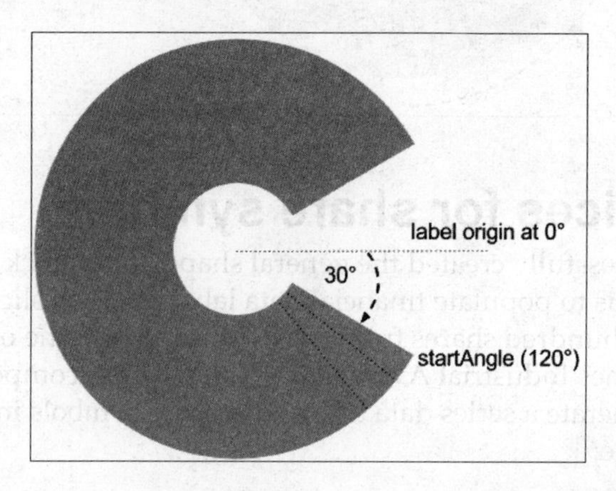

Our next task is to remove the connector and move the default `dataLabel` position (which is outside the slice) towards the center of the pie chart. We achieve that by setting the `connectorWidth` value to zero and assigning the `distance` option to a negative value to drag the `dataLabel` towards the center of the chart, such that the label is placed inside the slice. Here is the `plotOptions.pie` config:

```
    plotOptions: {
        pie: {
            size: '100%',
```

```
                startAngle: 120,
                endAngle: 420,
                innerSize: 110,
                dataLabels: {
                    connectorWidth: 0,
                    distance: -40,
                    color: 'white',
                    y: 15
                }
    },
```

The following screenshot shows the result of rotating labels along the slice's orientation for a stock picking wheel effect:

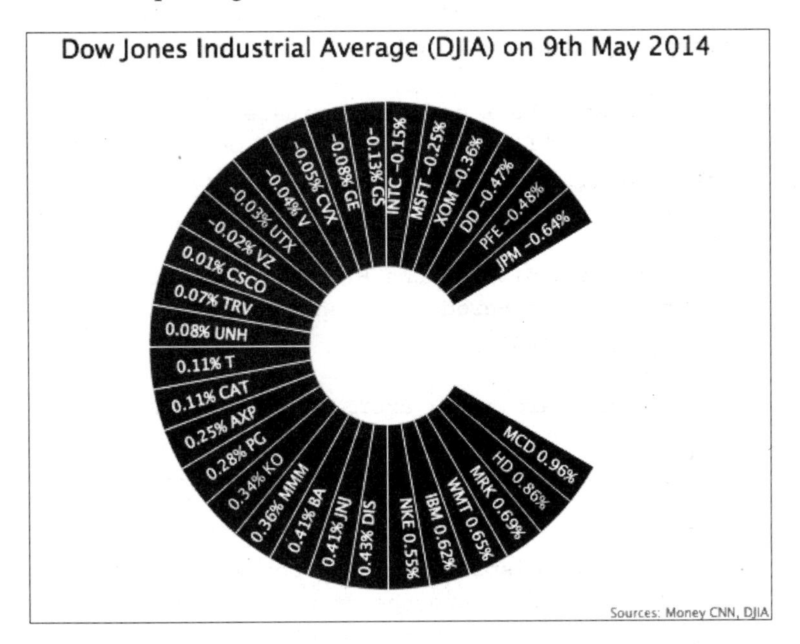

Creating shapes with Highcharts' renderer

The final missing piece for this experiment is to display an arc showing share symbol details when users hover the pointer over a slice. To do that, we make use of the Highcharts renderer engine, which provides a set of APIs to draw various shapes and text in the chart. In this exercise, we call the method `arc` to create a dynamic slice and position it inside the gap.

Since this is triggered by a user action, the process has to be performed inside an event handler. The following is the event handler code:

```
var mouseOver = function(event) {
        // 'this' keyword in this handler represents
        // the triggered object, i.e. the data point
        // being hovered over
        var point = this;

        // retrieve the renderer object from the
        // chart hierarchy
        var renderer = point.series.chart.renderer;

        // Initialise the SVG elements group
        group = renderer.g().add();

        // Calculate the x and y coordinate of the
        // pie center from the plot dimension
        var arcX = c.plotWidth/2 - c.plotLeft;
        var arcY = c.plotHeight/2 + c.plotTop;

        // Create and display the SVG arc with the
        // same color hovered slice and add the arc
        // into the group
        detailArc =
            renderer.arc(arcX, arcY, 230, 70, -0.3, 0.3)
                      .css({
                            fill: point.options.color
                       }).add(group);

        // Create and display the text box with stock
        // detail and add into the group
        detailBox = renderer.text(sharePrice,
                                     arcX + 80, arcY - 5)
                             .css({
                                     color: 'white',
                                     fontWeight: 'bold',
                                     fontSize: '15px'
                             }).add(group);
};

plotOptions: {
    pie: {
        ....
```

```
point: {
    events: {
        mouseOver: mouseOver,
        mouseOut: function(event) {
            group && (group = group.destroy());
        }
}
```

In the mouse over event, we first extract the triggered data point object from the `'this'` keyword inside the handler according to the Highcharts API documentation. From that, we can propagate down to restore the `renderer` object. Then, we create a group object with the renderer call `g( )` and then call the object method `add` to insert it into the chart. This group handling object enables us later to gather multiple SVG elements into one group for easier handling.

We then call the arc method to generate a slice with the chart's x and y center location evaluated from the chart plot dimension. We also specify the `innerRadius` parameter with a nonzero value (70) to make the arc emerge as a donut chart slice. Then, we chain the call with `css` for the same color as the point object and chain the `add` method to the group. Then, we create a textbox with stock details inside the new arc; the call is chained in the same manner with `css` and `add` to the same group.

For the mouse out event, we can take advantage of the group object. Instead of removing each element individually, we can simply call the `destroy` method on the group object. The following screenshot shows the new slice in display:

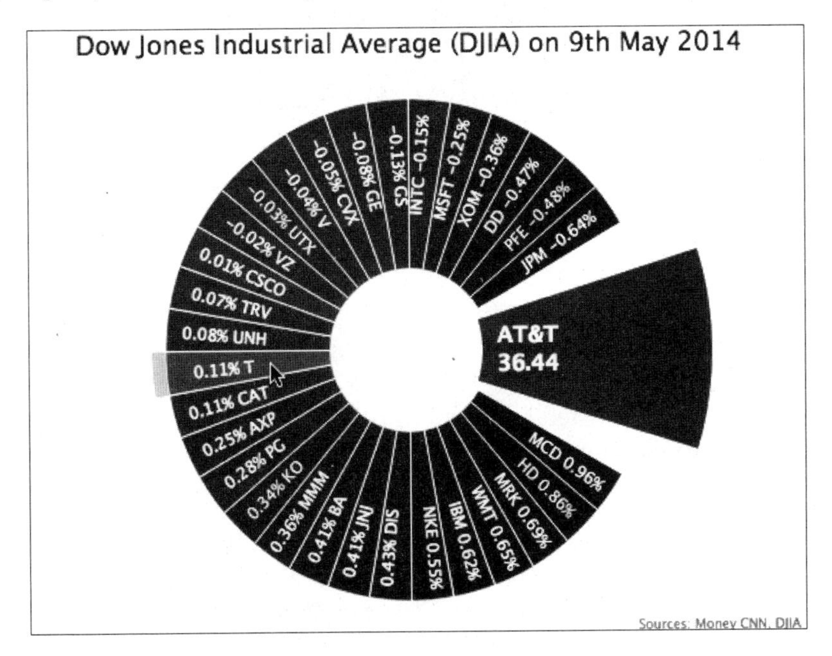

So we have created a Highcharts look-alike of the stock picking wheel, and this proves how versatile Highcharts can be. However, for the sake of simplicity, this exercise omits some complexity, which compromises the finished look. In fact, if we push Highcharts hard enough, we can finish with a chart that looks even closer to the original. The following is such an example, with several color bands and color gradients on each slice:

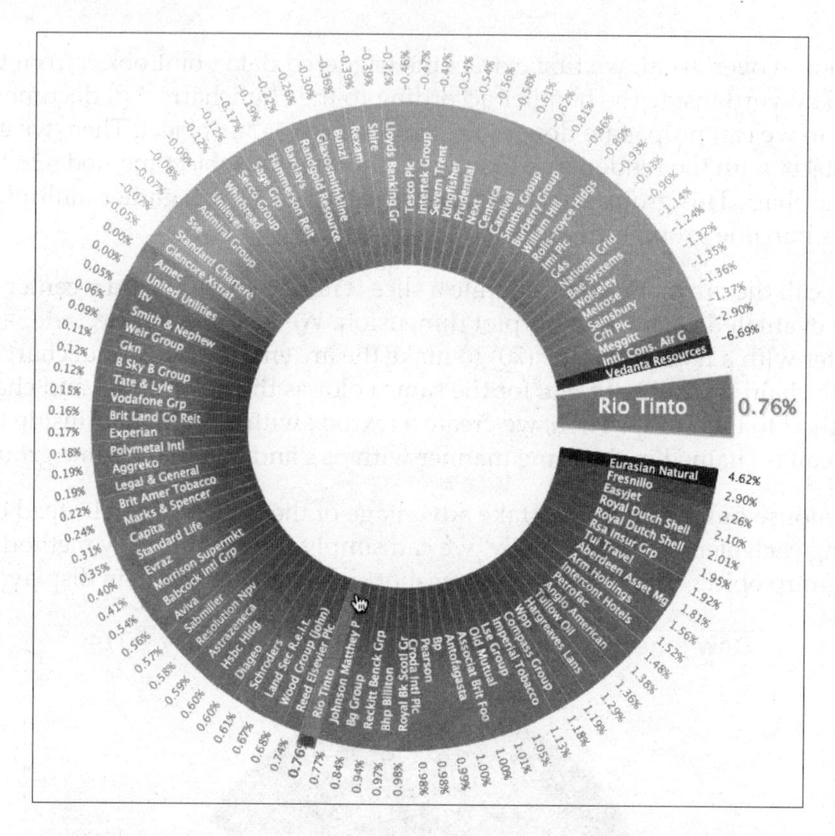

The online demo can be found at `http://joekuan.org/demos/ftse_donut`.

Summary

In this chapter, we have learned how to outline a pie chart and its variant, the donut chart. We also sketched a chart that included all the series types we have learned so far.

In the next chapter, we will explore the Highcharts APIs that are responsible for making a dynamic chart, such as using AJAX queries to update the chart content, accessing components in Highcharts objects, exporting charts to SVG, and so on.

6
Gauge, Polar, and Range Charts

In this chapter, we will learn how to create a gauge chart step by step. A gauge chart is very different from other Highcharts graphs. We will explore the new settings by plotting something similar to a twin-dials Fiat 500 speedometer. After that, we will review the structure of the polar chart and its similarity with other charts. Then, we will move on to examine how to create a range chart by using examples from the previous chapter. Finally, we will use a gauge chart to tweak the radial gradient in stages to achieve the desired effect. In this chapter, we will cover the following topics:

- Plotting a speedometer gauge chart
- Demonstrating a simple solid gauge chart
- Converting a spline chart to a polar/radar chart
- Plotting range charts with market index data
- Using a radial gradient on a gauge chart

Loading gauge, polar, and range charts

In order to use any gauge, polar, and range type charts, first we need to include an additional file, `highcharts-more.js`, provided in the package:

```
<script type="text/javascript"
    src="http://code.highcharts.com/highcharts.js"></script>
<script type="text/javascript"
    src="http://code.highcharts.com/highcharts-more.js"></script>
```

Plotting a speedometer gauge chart

Gauge charts have a very different structure compared to other Highcharts graphs. For instance, the backbone of a gauge chart is made up of a pane. A **pane** is a circular plot area for laying out chart content. We can adjust the size, position, and background of the pane. Once a pane is laid out, we can then put the axis on top of it. Highcharts supports multiple panes within a chart, so we can display multiple gauges (that is, multiple gauge series) within a chart. A gauge series is composed of two specific components—a pivot and a dial.

Another distinct difference in gauge charts is that the series is actually one dimensional data: a single value. Therefore, there is only one axis, the y axis, used in this type of chart. The `yAxis` properties are used the same way as other series type charts which can be on a `linear`, `datetime`, or `logarithmic` scale, and it also responds to the `tickInterval` option and so on.

Plotting a twin dials chart – a Fiat 500 speedometer

So far, we have mentioned all the parts that make up a gauge chart, and their relationships. There are many selectable options that can be used with gauge charts. In order to fully utilize them, we are going to learn in stages how to construct a sophisticated gauge chart by following the design of a Fiat 500 speedometer, as follows:

The speedometer is assembled with two dials on top of each other. The outer dial has two axes—mph and km/h. The inner dial is the rpm meter, which has a different scale and style. Another uncommon feature is that the body parts of both dials are hidden underneath: only the top needle parts are displayed. In the center of the gauge is an LED screen showing journey information. Despite all these unique characteristics, Highcharts provides enough flexibility to assemble a chart that looks very similar.

Plotting a gauge chart pane

First, let's see what a pane does in Highcharts. In order to do that, we should start by building a single dial speedometer. The following is the chart configuration code for a gauge chart with a single pane and a single axis:

```
chart: {
        renderTo: 'container'
},
title: {  text: 'Fiat 500 Speedometer' },
pane: [{
         startAngle: -120,
         endAngle: 120,
         size: 300,
         backgroundColor: '#E4E3DF'
}],
yAxis: [{
         min: 0,
         max: 140,
         labels: {
                rotation: 'auto'
         }
}],
series: [{
         type: 'gauge',
         data: [ 0 ]
}]
```

The preceding code snippet produces the following gauge chart:

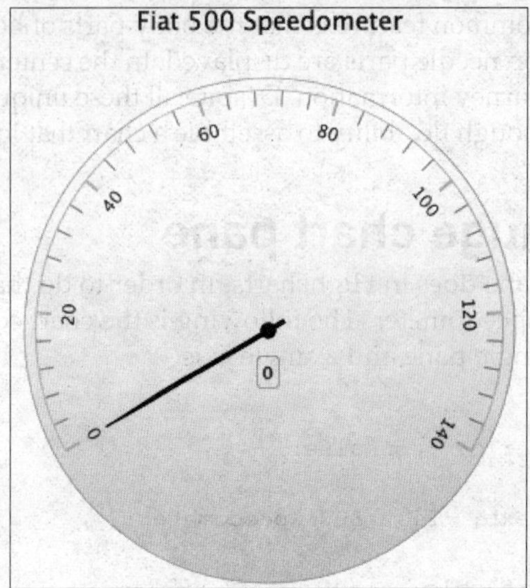

At the moment, this looks nothing like a Fiat 500 speedometer, but we will see the chart evolving gradually. The configuration declares a pane with a circular plot area from -120 to 120 degrees, with the y axis laying horizontally, whereas 0 degrees is at the 12 o'clock position. The `rotation` option generally takes a numerical degree value; `'auto'` is the special keyword to enable the y axis labels to automatically rotate so that they are aligned with the pane angle. The little box below the dial is the default data label showing the current value in the series.

Setting pane backgrounds

Gauge charts support more advanced background settings than just a single background color, as we saw in the previous example. Instead, we can specify another property, `background`, inside the `pane` option that accepts an array of different background settings. Each setting can be declared as an inner ring with both the `innerRadius` and `outerRadius` values defined, or as a circular background with only the `outerRadius` option. Both options are assigned percentage values with respect to the size of the pane. Here we set multiple backgrounds to the pane, as follows:

```
chart: {
    type: 'gauge',
    ....
},
title: { .... },
series: [{
```

```
        name: 'Speed',
        data: [ 0 ],
        dial: { backgroundColor: '#FA3421' }
    }],
    pane: [{
        startAngle: -120,
        endAngle: 120,
        size: 300,
        background: [{
            backgroundColor: {
                radialGradient: {
                    cx: 0.5,
                    cy: 0.6,
                    r: 1.0
                },
                stops: [
                    [0.3, '#A7A9A4'],
                    [0.45, '#DDD'],
                    [0.7, '#EBEDEA'],
                ]
            },
            innerRadius: '72%',
            outerRadius: '105%'
        }, {
            // BG color in between speed and rpm
            backgroundColor: '#38392F',
            outerRadius: '72%',
            innerRadius: '67%'
        }, {
            // BG color for rpm
            .....
        }]
```

As we can see, several backgrounds are defined, which include the backgrounds for the inner gauge: the rpm dial. Some of these backgrounds are rings and the last one is a circular background. Moreover, the needle color, `series.dial.backgroundColor`, is set to black by default. We are setting the dial color to red initially, so that we can still see the needle with the black background, which also more closely resembles the real-life example. Later in this section, we will explore the details of shaping and coloring the dial and pivot. As for the backgrounds with the `radialGradient` feature, we will examine them later in this chapter.

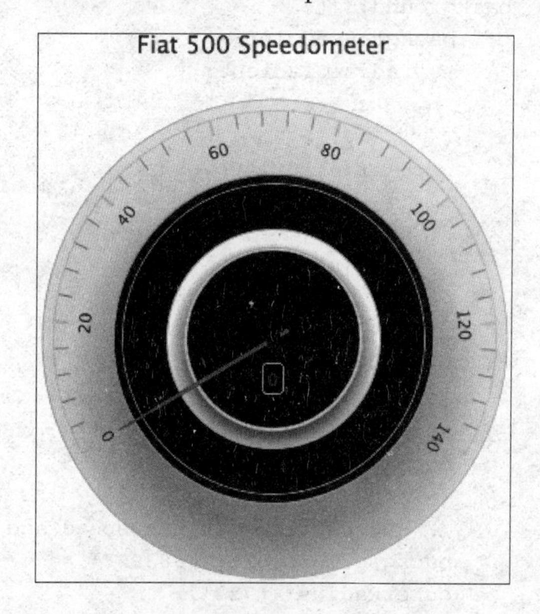

Managing axes with different scales

The next task is to lay a secondary *y* axis for the km/h scale, and we will set the new axis below the current display axis. We insert the new axis configuration as follows:

```
yAxis: [{
    min: 0,
    max: 140,
    labels: {
       rotation: 'auto'
    }
}, {
    min: 0,
    max: 220,
    tickPosition: 'outside',
```

```
        minorTickPosition: 'outside',
        offset: -40,
        labels: {
            distance: 5,
            rotation: 'auto'
        }
    }],
```

The new axis has a scale from `0` to `220` and we use the `offset` option with a negative value, which pushes the axes' line towards the center of the pane. In addition, both `tickPosition` and `minorTickPosition` are set to `'outside'`. This changes the interval ticks' direction to the opposite of the default settings. Both axes are now facing each other, which makes them similar to the the following figure:

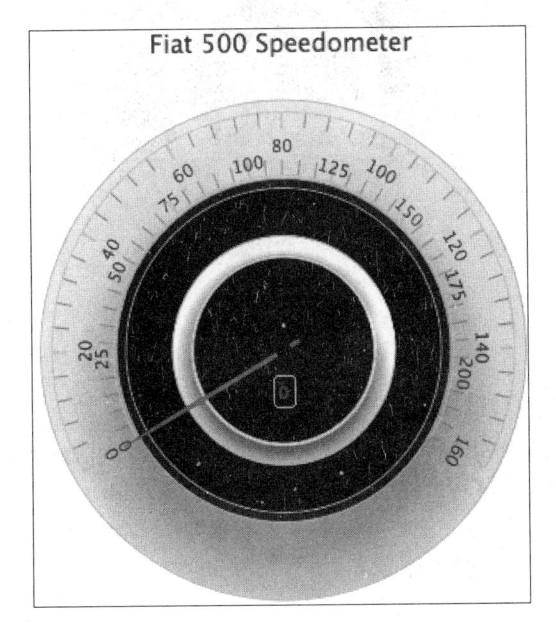

However, there is an issue that has arisen, as the scale of the top axis has been disturbed: it is no longer between 0 and 140. This is because the default action for a secondary axis is to align the intervals between multiple axes. To resolve this issue, we must set the `chart.alignTicks` option to `false`. After that, the issue is resolved and both axes are laid out as expected:

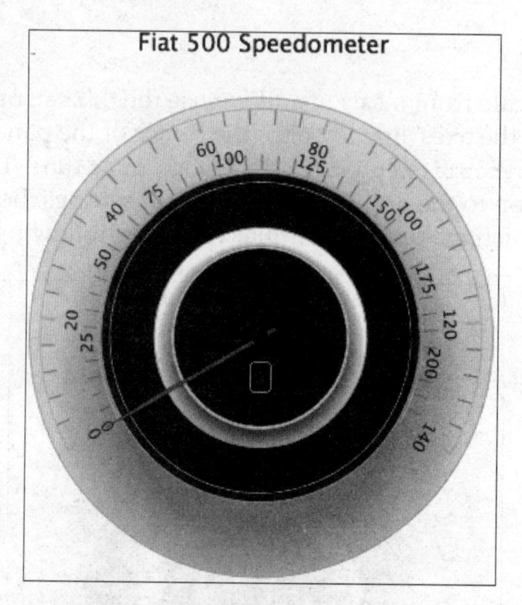

Extending to multiple panes

Since the gauge consists of two dials, we need to add an extra pane for the second dial. The following is the pane configuration:

```
pane: [{
    // First pane for speed dial
    startAngle: -120,
    endAngle: 120,
    size: 300,
    background: [{
        backgroundColor: {
            radialGradient: {
            .....
    }]
}, {
    // Second pane for rpm dial
    startAngle: -120,
    endAngle: 120,
    size: 200
}]
```

The second pane's plot area starts and ends at the same angles as the first pane, but has a smaller size. Since we haven't used the `center` option to position any panes within the chart, the inner pane is automatically placed at the center of the outer pane. The next step is to create another axis, rpm, which has a red region marked between the values of 4.5 and 6. Then, we bind all the axes to their panes, as follows:

```
yAxis: [{
    // axis for rpm - pane 1
    min: 0,
    max: 6,
    labels: {
       rotation: 'auto',
       formatter: function() {
           if (this.value >= 4.5) {
              return '<span style="color:' +
              '#A41E09">' + this.value +
              "</span>";
           }
           return this.value;
       }
    },
    plotBands: [{
        from: 4.5,
        to: 6,
        color: '#A41E09',
        innerRadius: '94%'
    }],
    pane: 1
}, {
    // axis for mph - pane 0
    min: 0,
    max: 140,
    .....
    pane: 0
}, {
    // axis for km/h - pane 0
    min: 0,
    max: 220,
    ....
    pane: 0
}]
```

For the rpm axis, we use `labels.formatter` to mark up the font color in the high-revolution region and also create a plot band for the axis. The `innerRadius` option is to control how thick the red area appears to be. The next task is to create a new gauge series, that is, a second dial for the new pane. Since the chart contains two different dials, we need to make the dial movement relative to an axis, therefore we assign the `yAxis` option to bind the series to an axis. Also, we set the initial value for the new series to `4`, just to demonstrate how the two dials are constructed, not superimposed on each other, as follows:

```
series: [{
    type: 'gauge',
    name: 'Speed',
    data: [ 0 ],
    yAxis: 0
}, {
    type: 'gauge',
    name: 'RPM',
    data: [ 4 ],
    yAxis: 2
}]
```

With all these additional changes, the following is the new look of the internal dial:

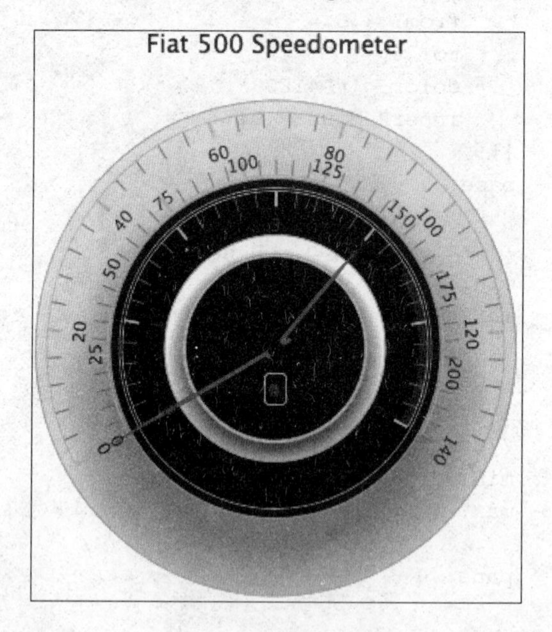

In the next part, we will learn how to set up the look and feel of the dial needles.

Gauge series – dial and pivot

There are a couple of properties specific to the gauge series, which are `plotOptions.gauge.dial` and `plotOptions.gauge.pivot`. The `dial` option controls the look and feel of the needle itself, whereas `pivot` is the tiny circle object at the center of the gauge attached to the dial.

First of all, we want to change the color and the thickness of the dials, as follows:

```
series: [{
    type: 'gauge',
    name: 'Speed',
    ....
    dial: {
        backgroundColor: '#FA3421',
        baseLength: '90%',
        baseWidth: 7,
        topWidth: 3,
        borderColor: '#B17964',
        borderWidth: 1
    }
}, {
    type: 'gauge',
    name: 'RPM',
    ....
    dial: {
        backgroundColor: '#FA3421',
        baseLength: '90%',
        baseWidth: 7,
        topWidth: 3,
        borderColor: '#631210',
        borderWidth: 1
    }
}]
```

The preceding code snippet results in the following output:

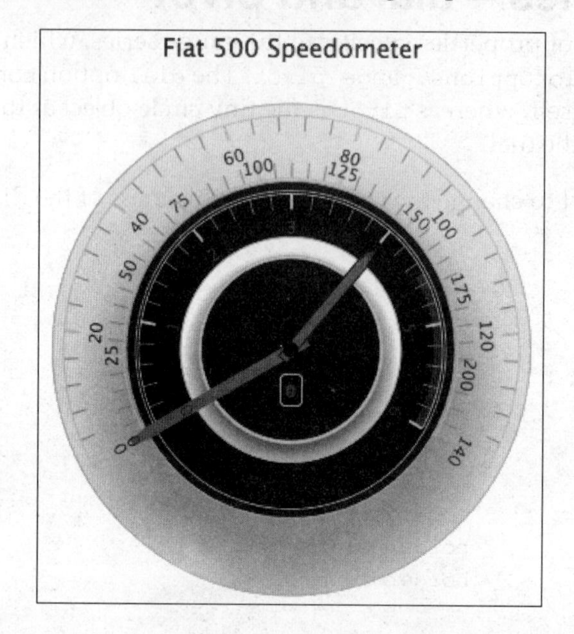

First, we widen the needle by setting the `baseWidth` option to `7` pixels across and `3` pixels at the end of the needle. Then, instead of having the needle narrowing down gradually to the end of the tip, we set the `baseLength` option to `'90%'`, which is the position on the dial where the needle starts to narrow down to a point.

As we can see, the dials are still not quite right in that they are not long enough to reach to their axis lines, as shown in the preceding screenshot. Secondly, the rest of the dial bodies are not covered up. We can resolve this issue by fiddling with the `rearLength` option. The following are the amended series settings:

```
series: [{
    type: 'gauge',
    name: 'Speed',
    .....
    dial: {
        .....
        radius: '100%',
        rearLength: '-74%'
    },
    pivot: { radius: 0 }
}, {
    type: 'gauge',
    name: 'RPM',
    .....
```

```
        dial: {
            .....
            radius: '100%',
            rearLength: '-74%'
        },
        pivot: { radius: 0 }
    }]
```

The trick is that instead of having a positive value like most of the gauge charts would have, we input a negative value that creates the covered-up effect. Finally, we remove the pivot by specifying the `radius` value as `0`. The following is the final adjustment of the dials:

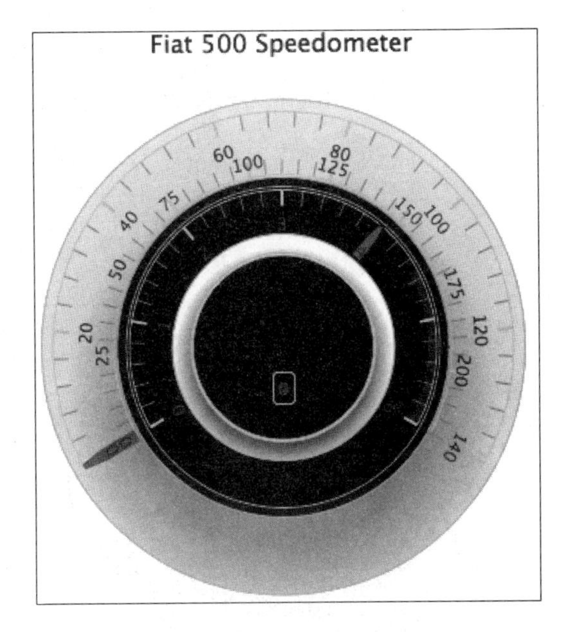

Polishing the chart with fonts and colors

The next step is to apply the axis options to tweak the tick intervals' color and size. The axis labels use fonts from the Google web fonts service (refer to `http://www.google.com/fonts` for Google web fonts). Then, we adjust the font size and color to that shown in the screenshot. There are a myriad of fonts to choose from with Google web fonts and they come with easy instructions to apply them. The following is an example of embedding the Squada One font into the `<head>` section of an HTML file:

```
<link href='http://fonts.googleapis.com/css?family=Squada One'
 rel='stylesheet' type='text/css'>
```

We apply the new imported font to the y axis titles and labels as shown in the following example:

```
yAxis: [{
      // axis for rpm - pane 1
      min: 0,
      max: 6,
      labels: {
         style: {
            fontFamily: 'Squada One',
            fontSize: 22
            ....
         }
      }
```

This significantly improves the look of the gauge, as follows:

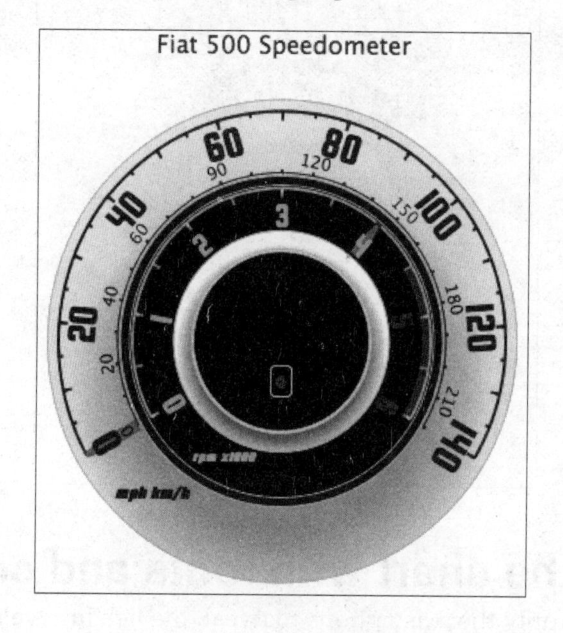

The final part is to transform the series data labels to resemble an LED screen. We will change the data labels' font size, style, and color and remove the border of the label box. The rpm data label has a smaller font size and moves above the mph data label. To make it look more realistic, we will also set the background for the data labels to a pale orange color. All the details of the tunings can be found in the online example at `http://joekuan.org/Learning_Highcharts/`. The following is the final look of the polished gauge chart:

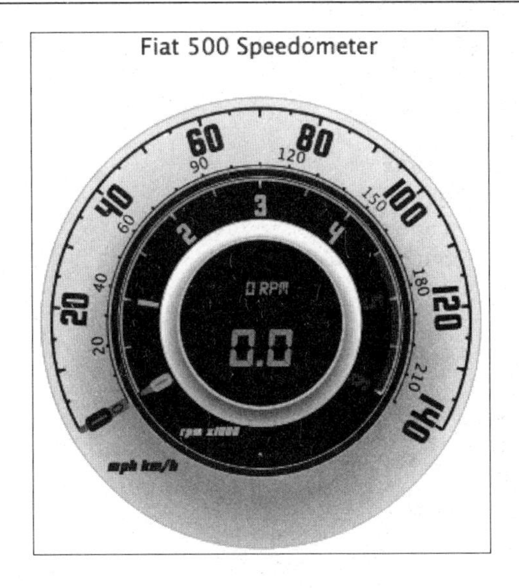

Plotting the solid gauge chart

Highcharts provides another type of gauge chart, solid gauge, which has a different presentation. Instead of having a dial, the chart background is filled with another color to indicate the level. The following is an example derived from the Highcharts online demo:

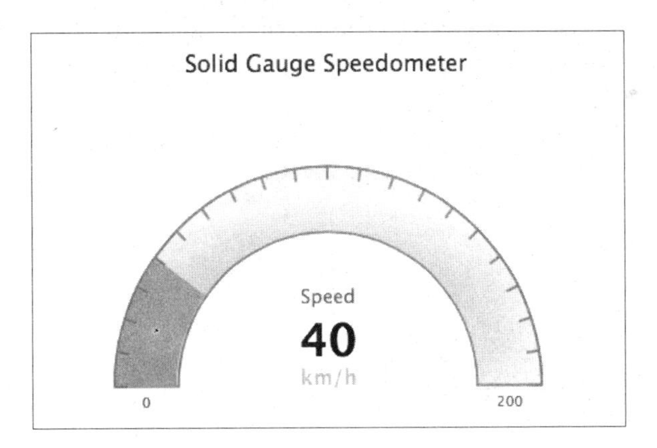

The principle of making a solid gauge chart is the same as a gauge chart, including the pane, y axis and series data, but without the dial setup. The following is our first attempt:

```
chart: {
    renderTo: 'container',
```

```
        type: 'solidgauge'
    },
    title: ....,
    pane: {
       size: '90%',
        background: {
            innerRadius: "70%",
            outerRadius: "100%"
        }
    },
    // the value axis
    yAxis: {
        min: 0,
        max: 200,
        title: {
            text: 'Speed'
        }
    },
    series: [{
       data: [40],
       dataLabels: {
           enabled: false
       }
    }]
```

We start with `pane` in a donut shape by specifying the `innerRadius` and `outerRadius` options. Then, we assign the y axis range along the background pane and set the initial value to 40 for a visible gauge level. The following is the result of our first solid gauge:

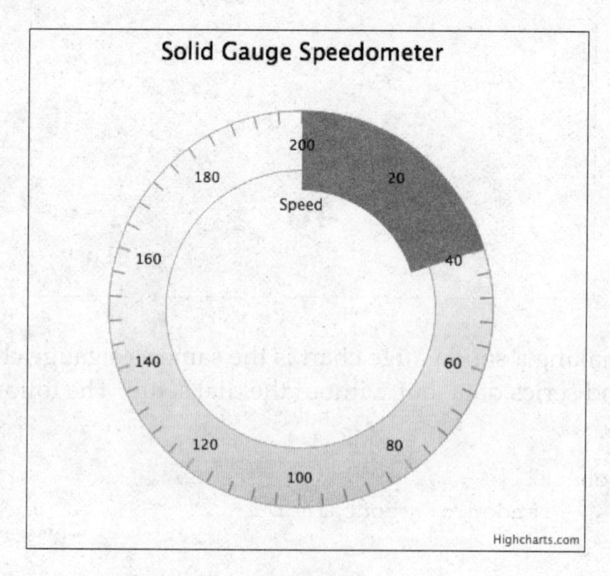

As we can see, the pane starts and ends at the 12 o'clock position with a section of the pane filled up with blue as the solid gauge dial. Let's make the axis progress clockwise from 9 o'clock to 3 o'clock and configure the intervals to become visible over the color gauge. In addition, we will increase the thickness of the intervals and only allow the first and last interval labels to be displayed. Here is the modified configuration:

```
chart: .... ,
title: .... ,
pane: {
    startAngle: -90,
    endAngle: 90,
    ....
},
yAxis: {
    lineColor: '#8D8D8D',
    tickColor: '#8D8D8D',
    intervalWidth: 2,
    zIndex: 4,
    minorTickInterval: null,
    tickInterval: 10,
    labels: {
        step: 20,
        y: 16
    },
    ....
},
series: ....
```

The `startAngle` and `endAngle` options are the effective start and end positions for labelling on the pane, so -90 and 90 degrees are the 9 o'clock and 3 o'clock positions respectively. Next, both `lineColor` (the axis line color at the outer boundary) and `tickColor` are applied with a darker color, so that by combining them with the `zIndex` option, the intervals become visible with the different levels of gauge colors.

The following is the output of the modified chart:

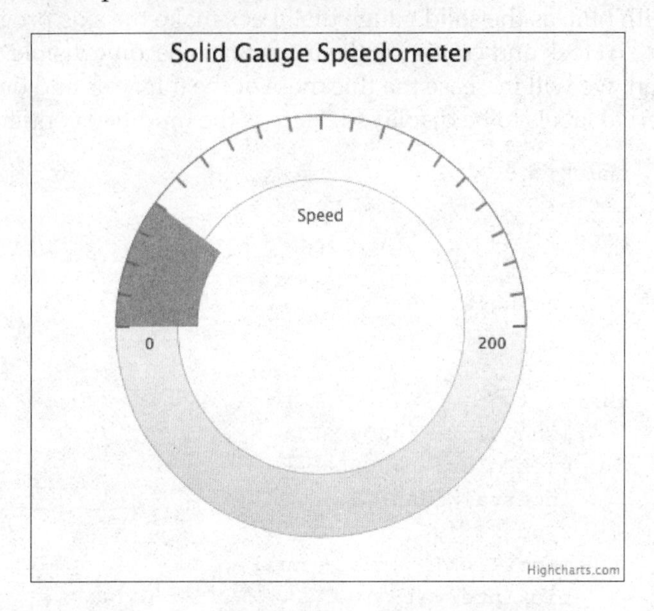

Next, we remove the bottom half of the background pane to leave a semi-circle shape and configure the background with shading (described in detail later in this chapter). Then, we adjust the size of the color gauge to be the same width as the background pane. We provide a list of color bands for different speed values, so that Highcharts will automatically change the gauge level from one color band to another according to the value:

```
chart: ...
            title: ...
            pane: {

                background: {

                    shape: 'arc',
                    borderWidth: 2,
                    borderColor: '#8D8D8D',
                    backgroundColor: {
                        radialGradient: {
                            cx: 0.5,
                            cy: 0.7,
                            r: 0.9
                        },
                        stops: [
                            [ 0.3, '#CCC' ],
                            [ 0.6, '#E8E8E8' ]
                        ]
```

```
            },
        }
    },
    yAxis: {
                ... ,
  stops: [
            [ 0, '#4673ac' ], // blue
            [ 0.2, '#79c04f' ], // green
            [ 0.4, '#ffcc00'], // yellow
            [ 0.6, '#ff6600'], // orange
            [ 0.8, '#ff5050' ], // red
            [ 1.0, '#cc0000' ]
        ],
    },
    plotOptions: {
              solidgauge: {
                           innerRadius: '71%'
                     }
            },
          series: [{
                  data: [100],
                  }]
```

The shape option, 'arc', turns the pane into an arc shape. The color bands along the y axis are provided with a stops option, which is an array of ratios (the ratio of y axis range) and the color values. In order to align the gauge level with the pane, we assign the innerRadius value of plotOptions.solidgauge to be slightly smaller than the pane innerRadius value, so that the movement of the gauge level doesn't cover the pane's inner border. We set the series value to 100 to show that the gauge level displays a different color as follows:

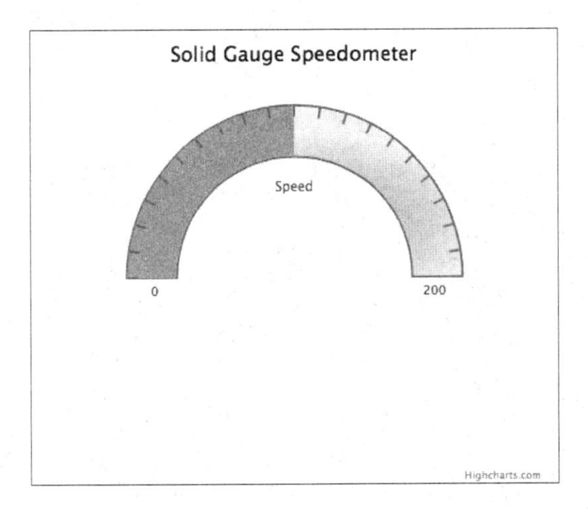

Now we have extra space in the bottom half of the chart, we can simply move the chart lower and decorate it with a data label as follows:

```
pane: {
    .... ,
    center: [ '50%', '65%' ]
},
plotOptions: {
    solidgauge: {
       innerRadius: '71%',
        dataLabels: {
                y: 5,
                borderWidth: 0,
                useHTML: true
        }
    }
},
series: [{
      data: [100],
      dataLabels: {
            y: -60,
            format: '<div style="text-align:center">' +
   '<span style="font-size:35px;color:black">{y}</span><br/>' +
  '<span style="font-size:16px;color:silver">km/h</span></div>'
      }
}]
```

We set the chart's vertical position to 65 percent of the plot area height with the `center` option. Finally, we enable `dataLabels` to be shown as in HTML with textual decoration:

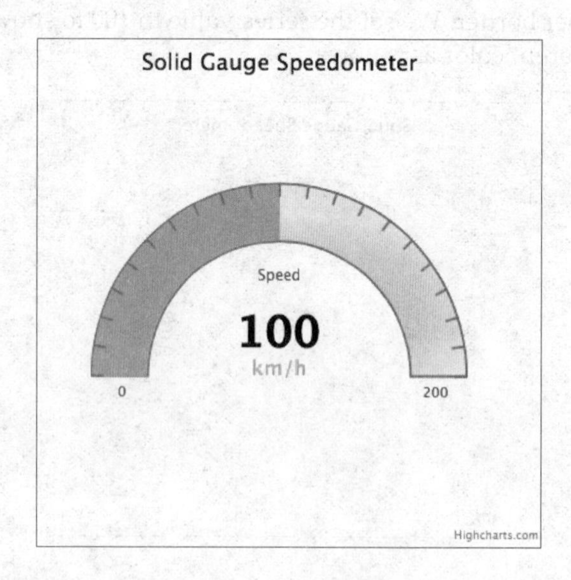

Converting a spline chart to a polar/radar chart

Polar (or radar) charts are generally used to spot data trends. They have a few differences from line and column type charts. Even though they may look like pie charts, they have nothing in common with them. In fact, a polar chart is a round representation of the conventional two-dimensional charts. To visualize it another way, it is a folded line or a column chart placed in a circular way with both ends of the x axis meeting.

Again, we need to include `highcharts-more.js` in order to plot a polar chart. The following screenshot illustrates the structure of a polar chart:

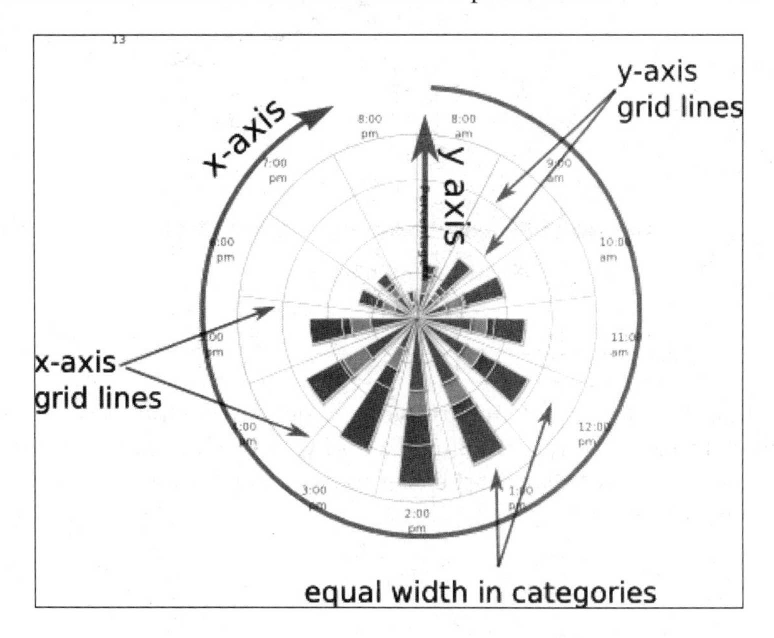

There are very few differences in principle, and the same also applies to the Highcharts configuration. Let's use our very first example of a browser usage chart in *Chapter 1, Web Charts*, and turn it into a radar chart. Recalling the browser line chart, we made the following:

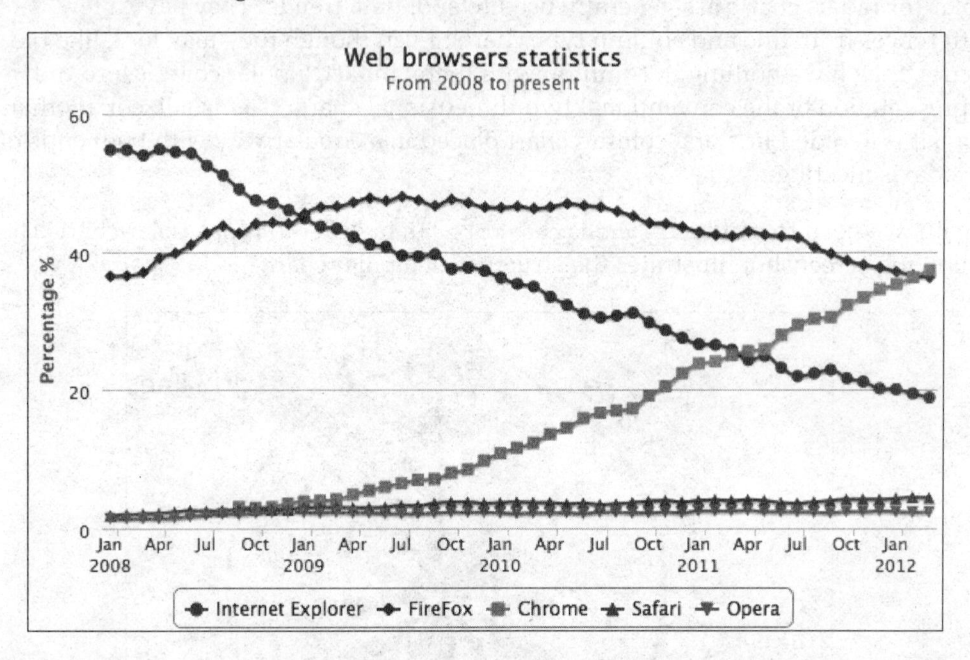

To turn the line chart into a polar chart, we only need to set the `chart.polar` option to `true`, which transforms the orthogonal x and y coordinates into a polar coordinate system. To make the new polar chart easier to read, we set the *x* axis labels' `rotation` value to `'auto'`, as follows:

```
chart: {
    ....,
    polar: true
},
....,
xAxis: {
    ....,
    labels: { rotation:  'auto' }
},
```

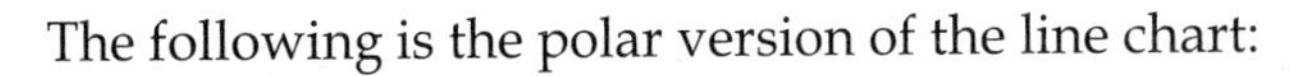

The following is the polar version of the line chart:

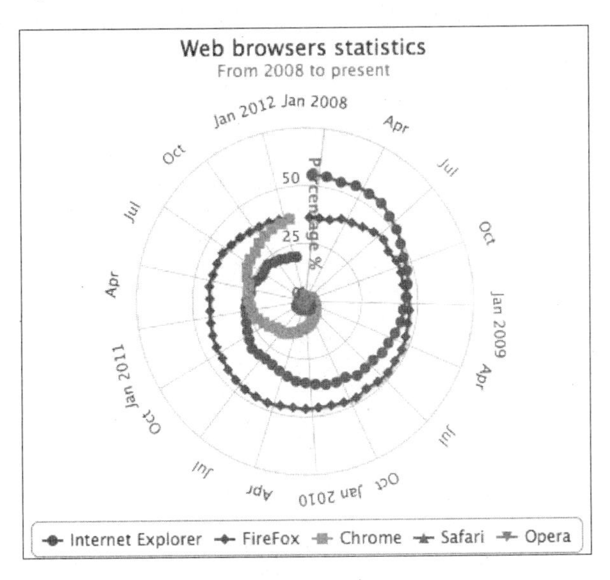

As we can see, one characteristic of a polar chart is that it reveals data trends differently compared to conventional charts. From a clockwise direction, we see the data line "spirals up" for an upward trend (Chrome) and "spirals down" for a downward trend (Internet Explorer), whereas the data line for Firefox doesn't show much movement. As for Safari and Opera, these series are essentially lost as they are completely invisible unless we enlarge the chart container, which is impractical. Another characteristic is that the last and first data points in the series are connected together. As a result, the Firefox series shows a closed loop and there is a sudden jump in the Internet Explorer series (the Chrome series is not connected because it has null values at the beginning of the series as the Chrome browser was not released until late 2008).

To correct this behavior, we can simply add a null value at the end of each series data array to break the continuity, which is demonstrated in the following screenshot:

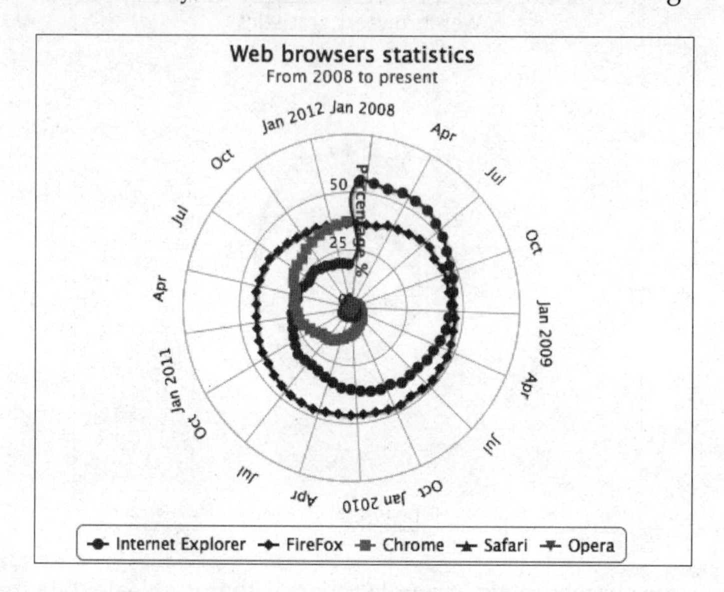

Instead of having a round polar chart, Highcharts supports polygon interpolation along the *y* axis grid lines. This means that the grid lines are straightened and the whole chart becomes like a spider web.

For the sake of illustration, we set the width of *x* and *y* axis lines to 0, which removes the round outline from the chart. Then, we set a special option on the *y* axis, `gridLineInterpolation`, to `polygon`. Finally, we change the `tickmarkPlacement` option of the x axis to `'on'` instead of the default value, `between`. This gets the interval ticks on the x axis to align with the start of each category. The following code snippet summarizes the changes that we need to make:

```
xAxis: {
    categories: [ ..... ],
    tickmarkPlacement: 'on',
    labels: {
        rotation: 'auto',
    },
    lineWidth: 0,
    plotBands: [{
        from: 10,
        to: 11,
        color: '#FF0000'
    }]
},
```

```
yAxis: {
    .....,
    gridLineInterpolation: 'polygon',
    lineWidth: 0,
    min: 0
},
```

In order to demonstrate a spider web shape, we will remove most of the data samples from the previous chart (alternatively, you can enlarge the chart container to keep all the data). We will also add a couple of grid line decorations and an x-axis plot band (Nov - Dec) just to show that other axis options can still be applied to a polar chart:

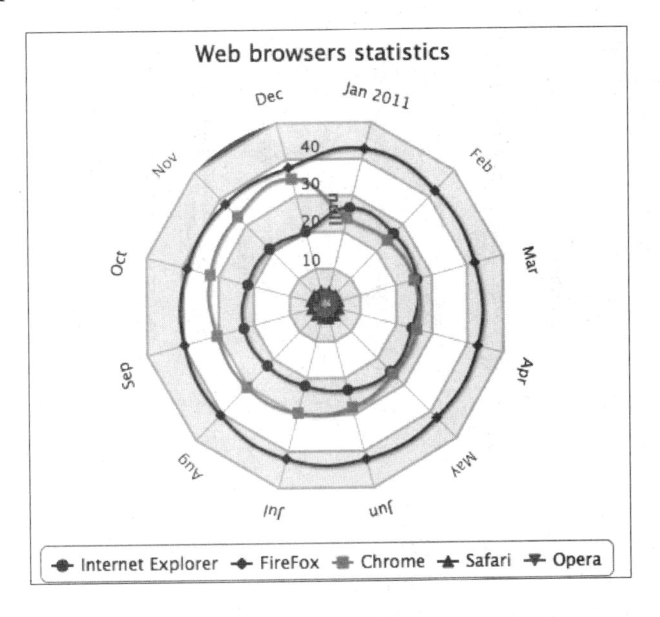

Plotting range charts with market index data

Range charts are really line and column type charts presenting a series of data in range. The set of range type series can be `arearange`, `areasplinerange`, and `columnrange`. These series expect an array of three data points, x, y min, y max, in the data option or array of y min, y max if `xAxis.cateogries` has already been specified.

Let's use our past examples to see whether we can make an improvement to the range charts. In *Chapter 2, Highcharts Configurations*, we have a five-series graph showing the monthly data of Nasdaq 100: open, close, high, low, and volume, as shown in the following screenshot:

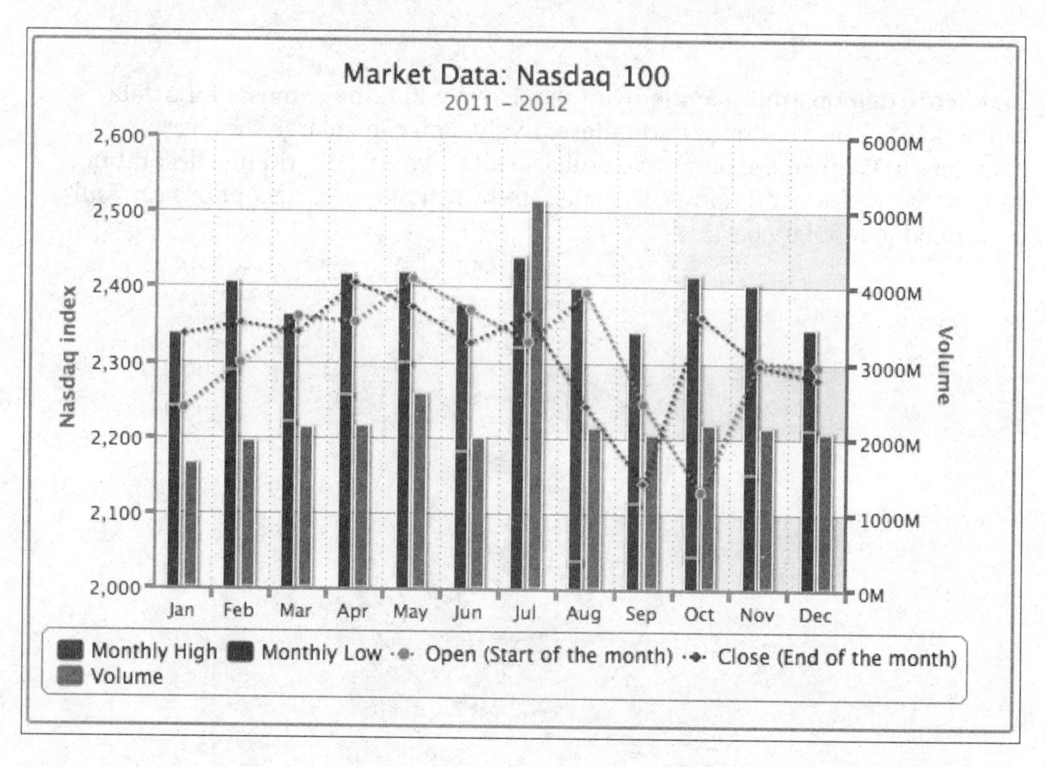

With the new range series, we sort the series data and merge the **Monthly High** and **Monthly Low** columns into a column range series, and the **Open** and **Low** columns into an area spline range series, as follows:

```
series: [{
  type: 'columnrange',
  name: 'High & Low',
  data: [ [ 2237.73, 2336.04 ],
          [ 2285.44, 2403.52 ],
          [ 2217.43, 2359.98 ],  ...... ]
}, {
  type: 'areasplinerange',
  name: 'Open & Close',
  // This array of data is pre-sorted,
  // not in Open, Close order.
  data: [ [ 2238.66, 2336.04 ],
```

```
                    [ 2298.37, 2350.99 ],
                    [ 2338.99, 2359.78 ], ...... ]
        }, {
          name: 'Volume',
          ......
```

The following screenshot shows the range chart version:

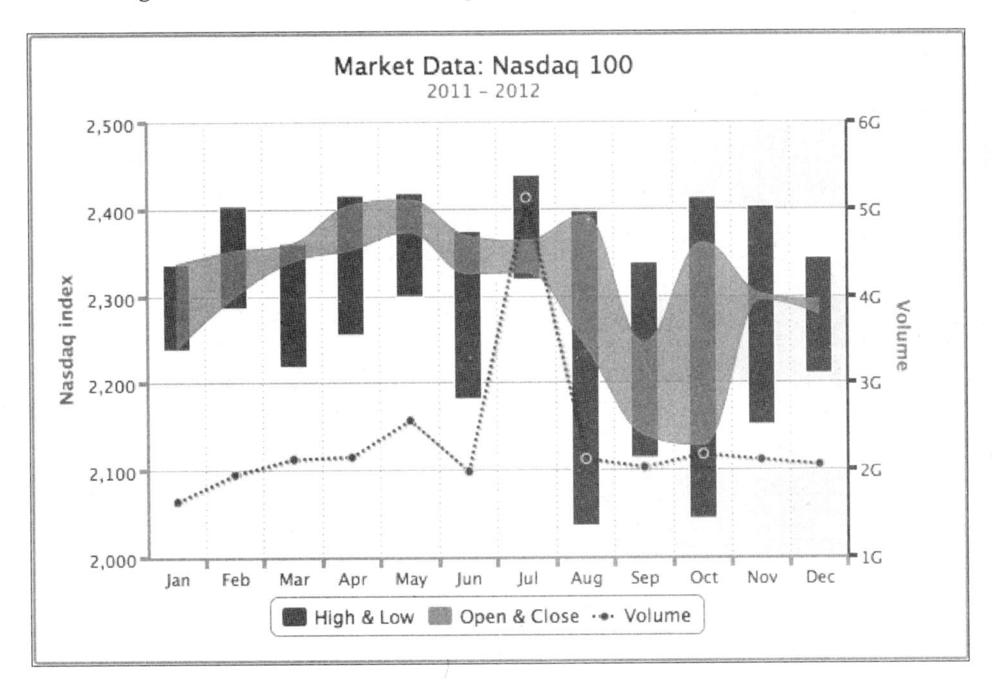

The new chart looks much simpler to read and the graph is less packed. It is worth noting that for the column range series, it is mandatory to keep the range in min to max order. As for the area spline and area range series types, we can still plot the range series even without sorting them beforehand.

For instance, the **High & Low** range series have to be in min and max order, according to the natural meaning of the name of the series. However, this is not the same for the **Open & Close** range series as we wouldn't know which way is open or close. If we plot the **Open & Close** area range series by keeping keeping the range in open to close order, instead of y min to y max, the area range is displayed differently, as shown in the following screenshot:

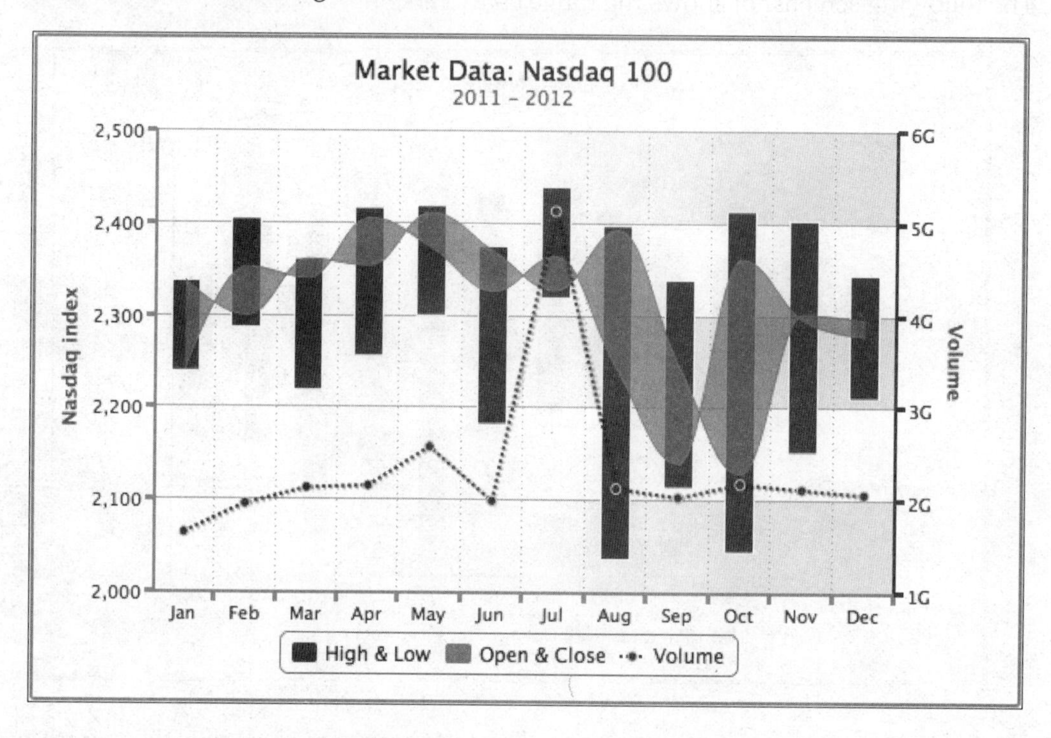

As we can see, there are twisted bits in the area range series; these crossovers are caused by the reverse order in the data pairs. Nonetheless, we won't know whether open is higher than close or vice versa. If we only want to know how wide the range between the **Open & Close** series is, then the preceding area range chart achieves the goal. By keeping them as separate series, there will be no such issue. In a nutshell, this is the subtle difference when plotting range series data with ambiguous meanings.

Using a radial gradient on a gauge chart

The radial gradient setting is based on SVG. As its name implies, a **radial gradient** is color shading radiating outwards in a circular direction. Therefore, it requires three properties to define the gradient circle—`cx`, `cy`, and `r`. The gradient circle is the outermost circle for shading, such that no shading can go outside of this.

All the gradient positions are defined in ratio values between zero and one with respect to their containing elements. The `cx` and `cy` options are at the x, y center position of the outermost circle, whereas `r` is the radius of the outmost circle. If `r` is 0.5, it means the gradient radius is half the diameter of its element, the same size as the containing pane. In other words, the gradient starts from the center and goes all the way to the edge of the gauge. The `stop` offsets option works the same way as the linear gradient: the first parameter is the ratio position in the gradient circle to stop the shading. This controls the intensity of shading between the colors. The shorter the gap, the higher the contrast between the colors.

Let's explore how to set up the color gradient. The following is a mood swing detector without any color gradient:

We will apply a radial gradient to the preceding chart with the following settings:

```
background: [{
    backgroundColor: {
        radialGradient: {
            cx: 0.5,
            cy: 0.5,
             r: 0.5
        },
        stops: [
            [ 0, '#CCD5DE' ],
            [ 1, '#002E59' ]
        ]
    }
}]
```

We have set cx, cy, and r to 0.5 for the gradient to start shading from the center position all the way towards the edge of the circle, as follows:

As we can see, the preceding chart shows white shading evenly radiating from the center. Let's change some of the parameters and see the effect:

```
backgroundColor: {
    radialGradient: {
        cx: 0.5,
        cy: 0.7,
        r: 0.25
    },
    stops: [
        [ 0.15, '#CCD5DE' ],
        [ 0.85, '#002E59' ]
    ]
}
```

Here, we have changed the size of the gradient circle to half the size of the gauge and moved the circle down. The bright color doesn't start shading until it reaches 15 percent of the size of the gradient circle, so there is a distinct white blob in the middle, and the shading stops at 85 percent of the circle:

In the SVG radiant gradient, there are two other options, `fx` and `fy`, which are used to set the focal point position for the shading; they are also referred to as the inner circle settings.

The '`fx`'and '`fy`' options may not work properly with Chrome and Safari browsers. At the time of writing this, only Firefox and Internet Explorer browsers are working properly with the images shown here.

Let's experiment with how the focal point can affect the shading:

```
backgroundColor: {
    radialGradient: {
        cx: 0.5,
        cy: 0.7,
         r: 0.25,
        fx: 0.6,
        fy: 1.0
    },
    stops: [
        [ 0.15, '#CCD5DE' ],
        [ 0.85, '#002E59' ]
    ]
}
```

The preceding code snippet produces the following output:

We can observe that the `fx` and `fy` options move the bright color starting from the bottom of the gradient circle and slightly to the right-hand side. This makes the shading much more directional. Let's make our final change by moving the bright spot position towards the center of the gauge chart. In addition, we align the focal directional along the **Good** label:

```
background: [{
    backgroundColor: {
        radialGradient: {
            cx: 0.32,
            cy: 0.38,
             r: 0.25,
            fx: 1.3,
            fy: 0.95
        },
        stops: [
            [ 0.1, '#CCD5DE' ],
            [ 0.9, '#002E59' ]
        ]
    }]
```

Finally, we can finish the chart by moving the bright side to where we want it to be, as follows:

The `fx` and `fy` options are only for SVG, which older versions of Internet Explorer (8.0 or earlier) using VML won't support.

Summary

In this chapter, we learned about gauge, polar, and range charts. An extensive step-by-step demonstration showed how to plot a complex speedometer by utilizing most of the gauge options. We also demonstrated the differences between polar, column, and line charts with respect to principle and configuration. We used range charts to improve past chapter examples and study the subtle differences they insert into the chart. Finally, we explored how to define radial gradients by tweaking the options in stages.

In the next chapter, we will learn the structure of a bubble chart by recreating it from an online sport chart. Then, we will briefly discuss the properties of a box plot chart by transforming data from a spider chart, following with presenting the F1 race data with an error bar chart.

7
Bubble, Box Plot, and Error Bar Charts

In this chapter, we will explore bubble charts by first studying how the bubble size is determined from various series options, and then familiarizing ourselves with it by replicating a real-life chart. After that, we will study the structure of the box plot chart and discuss an article converting an over-populated spider chart into a box plot. We will use that as an exercise to familiarize ourselves with box plot charts. Finally, we will move on to the error bar series to understand its structure and apply the series to some statistical data.

This chapter assumes that you have some basic knowledge of statistics, such as mean, percentile, standard error, and standard deviation. For readers needing to revise these topics, there are plenty of online materials covering them. Alternatively, the book *Statistics For Dummies* by Deborah J. Rumsey provides great explanations and covers some fundamental charts such as the box plot.

In this chapter, we will cover the following topics:

- How bubble size is determined
- Bubble chart options when reproducing a real-life chart in a step-by-step approach
- Box plot structure and the series options when replotting data from a spider chart
- Error bar charts with real-life data

The bubble chart

The bubble series is an extension of the scatter series where each individual data point has a variable size. It is generally used for showing 3-dimensional data on a 2-dimensional chart, and the bubble size reflects the scale between z-values.

Understanding how the bubble size is determined

In Highcharts, bubble size is decided by associating the smallest z-value to the `plotOptions.bubble.minSize` option and the largest z-value to `plotOptions.bubble.maxSize`. By default, `minSize` is set to 8 pixels in diameter, whereas `maxSize` is set to 20 percent of the plot area size, which is the minimum value of width and height.

There is another option, `sizeBy`, which also affects the bubble size. The `sizeBy` option accepts string values: `'width'` or `'area'`. The width value means that the bubble width is decided by its z-value ratio in the series, in proportion to the `minSize` and `maxSize` range. As for `'area'`, the size of the bubble is scaled by taking the square root of the z-value ratio (see `http://en.wikipedia.org/wiki/Bubble_chart` for more description on `'area'` implementation). This option is for viewers who have different perceptions when comparing the size of circles. To demonstrate this concept, let's take the `sizeBy` example (`http://jsfiddle.net/ZqTTQ/`) from the Highcharts online API documentation. Here is the snippet of code from the example:

```
plotOptions: {
    series: { minSize: 8, maxSize: 40 }
},
series: [{
    data: [ [1, 1, 1], [2, 2, 2],
            [3, 3, 3], [4, 4, 4], [5, 5, 5] ],
    sizeBy: 'area',
    name: 'Size by area'
}, {
    data: [ [1, 1, 1], [2, 2, 2],
            [3, 3, 3], [4, 4, 4], [5, 5, 5] ],
   sizeBy: 'width',
    name: 'Size by width'
}]
```

Two sets of bubbles are set with different `sizeBy` schemes. The minimum and maximum bubble sizes are set to 8 and 40 pixels wide respectively. The following is the screen output of the example:

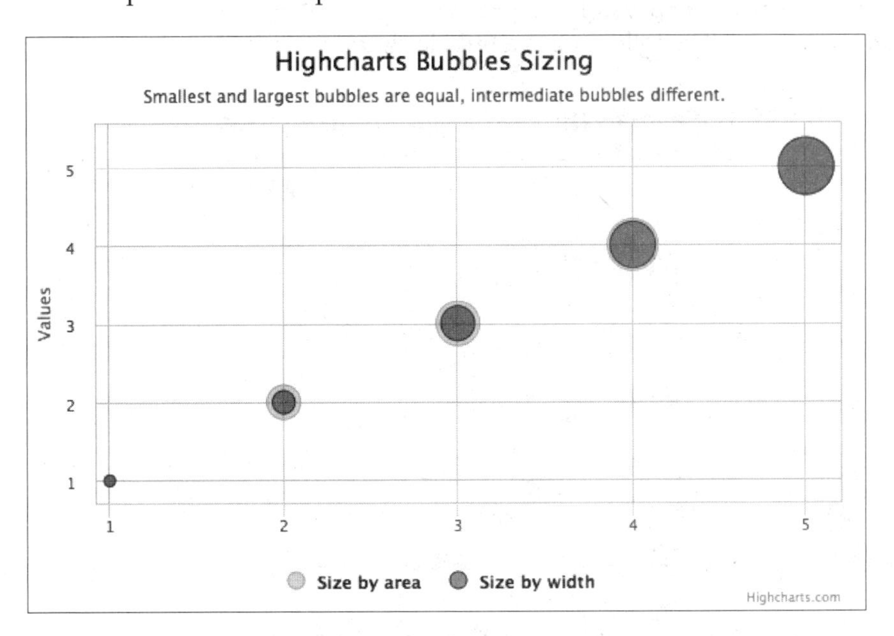

Both series have the exact same x, y, and z values, so they have the same bubble sizes at the extremes (1 and 5). With the middle values 2 to 4, the **Size by area** bubbles start off with a larger area than the **Size by width**, and gradually both schemes narrow to the same size. The following is a table showing the final size values in different z-values for each method, whereas the associating value inside the bracket is the z-value ratio in the series:

Z-Value	1	2	3	4	5
Size by width (Ratio)	8 (0)	16 (0.25)	24 (0.5)	32 (0.75)	40 (1)
Size by area (Ratio)	8 (0)	24 (0.5)	31 (0.71)	36 (0.87)	40 (1)

Let's see how the bubble sizes are computed in both approaches. The ratio in **Size by width** is calculated as $(Z - Zmin) / (Zmax - Zmin)$. So, for z-value 3, the ratio is computed as $(3 - 1) / (5 - 1) = 0.5$. To evaluate the ratio for the **Size by area** scheme, simply take the square root of the **Size by width** ratio. In this case, for z-value 3, it works out as $\sqrt{0.5} \approx 0.71$. We then convert the ratio value into the number of pixels based on the `minSize` and `maxSize` range. **Size by width** with z-value 3 is calculated as:

$$ratio * (maxSize - minSize) + minSize = 0.5 * (40 - 8) + 8 = 24$$

Reproducing a real-life chart

In this section, we will examine bubble series options by replicating a real-life example (MLB Players Chart: `http://fivethirtyeight.com/datalab/has-mike-trout-already-peaked/`). The following is a bubble chart of baseball players' milestones:

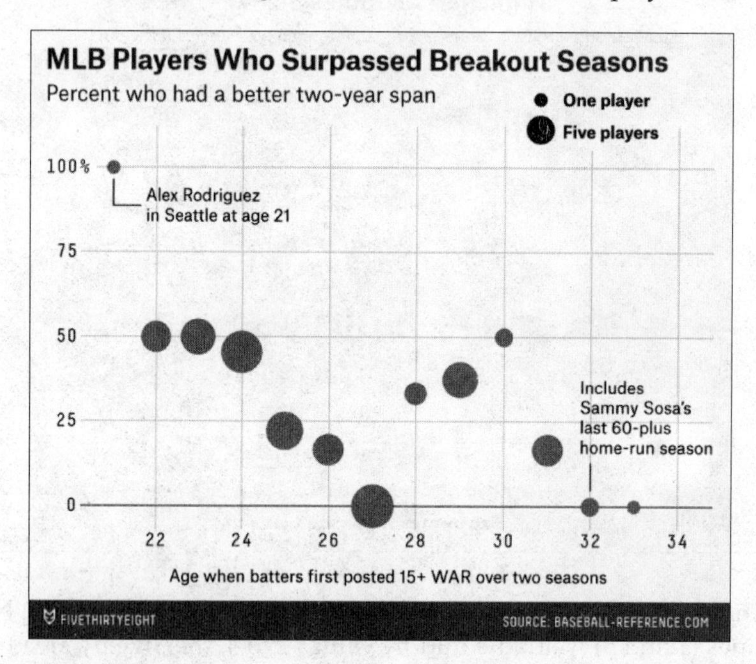

First, there are two ways that we can list the data points (the values are derived from best estimations of the graph) in the series. The conventional way is an array of x, y, and z values where x is the age value starting from 21 in this example:

```
series: [{
    type: 'bubble',
    data: [ [ 21, 100, 1 ],
            [ 22, 50, 5 ],
            ....
```

Alternatively, we can simply use the `pointStart` option as the initial age value and miss out the rest:

```
series: [{
   type: 'bubble',
   pointStart: 21,
   data: [ [ 100, 1 ],
           [ 50, 5 ],
           ....
```

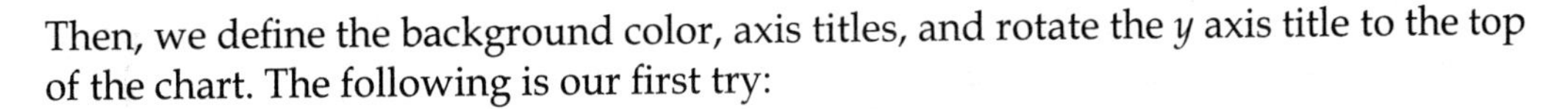

Then, we define the background color, axis titles, and rotate the *y* axis title to the top of the chart. The following is our first try:

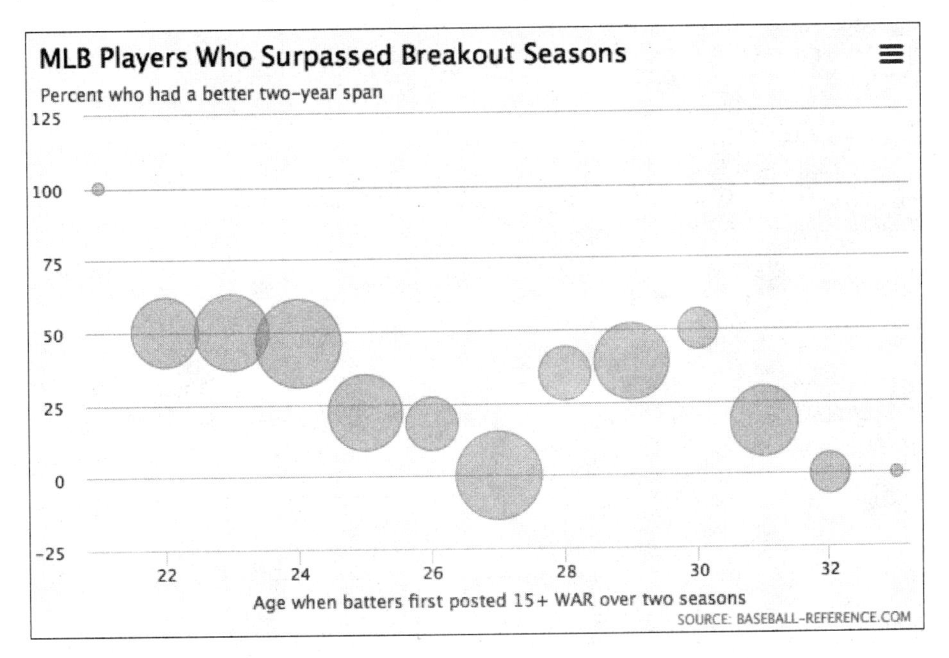

As we can see, there are a number of areas that are not quite right. Let's fix the bubble size and color first. Compared to the original chart, the preceding chart has a larger bubble size for the upper value and the bubbles should be solid red. We update the `plotOptions.bubble` series as follows:

```
plotOptions: {
     bubble: {
          minSize: 9,
          maxSize: 30,
          color: 'red'
     }
},
```

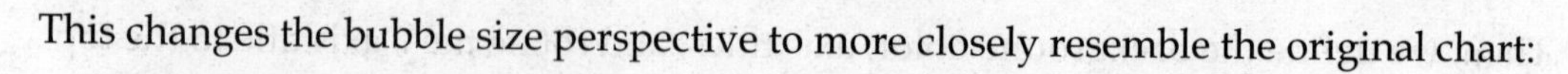

This changes the bubble size perspective to more closely resemble the original chart:

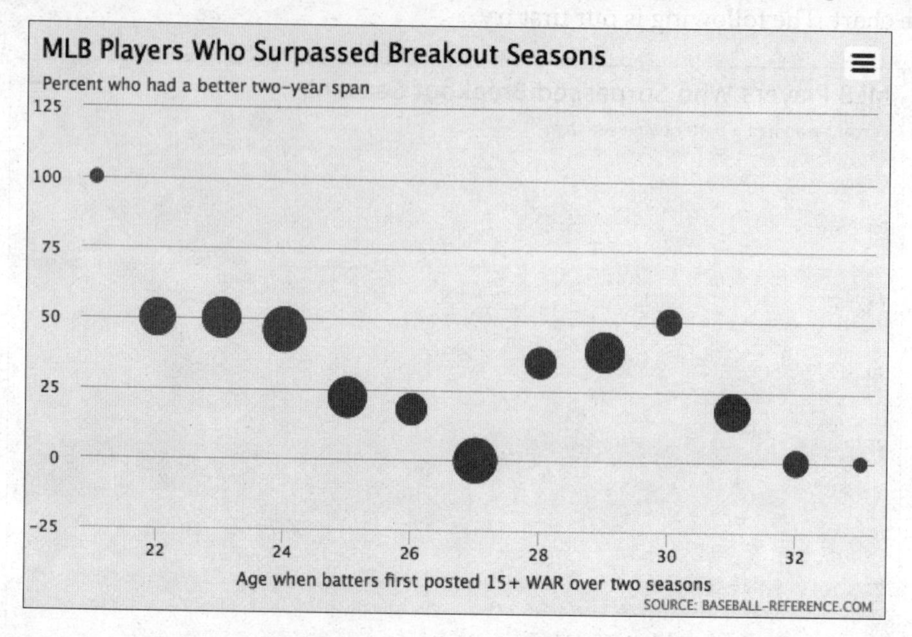

The next step is to fix the y-axis range as we want it to be between 0 and 100 only. So, we apply the following config to the `yAxis` option:

```
yAxis: {
    endOnTick: false,
    startOnTick: false,
    labels: {
        formatter: function() {
            return (this.value === 100) ?
                this.value + ' %' : this.value;
        }
    }
},
```

By setting the options `endOnTick` and `startOnTick` to `false`, we remove the extra interval at both ends. The label formatter only prints the `%` sign at the `100` interval. The following chart shows the improvement on the *y* axis:

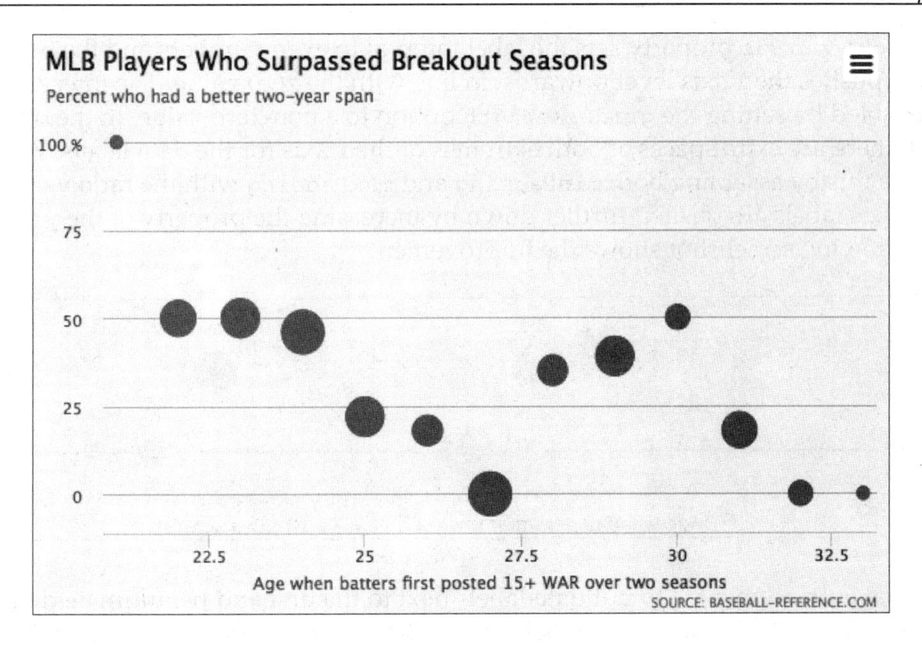

The next enhancement is to move the *x* axis up to the zero value level and refine the *x* axis into even number intervals. We also enable the grid lines on each major interval and increase the width of the axis line to resemble the original chart:

```
xAxis: {
    tickInterval: 2,
    offset: -27,
    gridLineColor: '#d1d1d1',
    gridLineWidth: 1,
    lineWidth: 2,
    lineColor: '#7E7F7E',
    labels: {
        y: 28
    },
    minPadding: 0.04,
    maxPadding: 0.15,
    title: ....
},
```

The `tickInterval` property sets the label interval to even numbers and the `offset` option pushes the *x* axis level upwards, in line with the zero value. The interval lines are enabled by setting the `gridLineWidth` option to a non-zero value. In the original chart, there are extra spaces at both extremes of the *x* axis for the data labels. We can achieve this by assigning both `minPadding` and `maxPadding` with the ratio values. The *x* axis labels are pushed further down by increasing the property of the `y` value. The following screenshot shows the improvement:

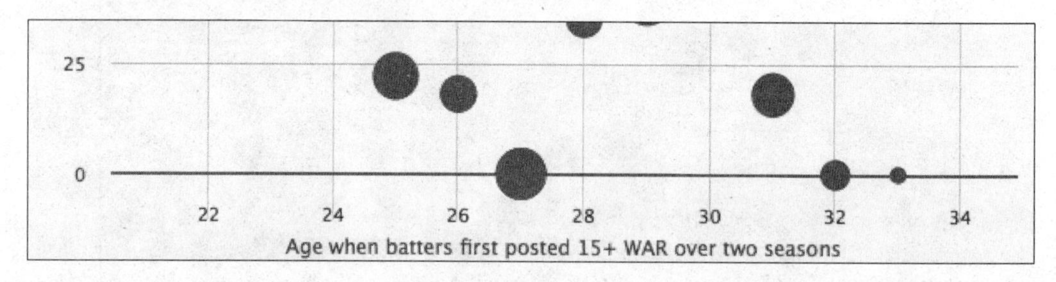

The final enhancement is to put data labels next to the first and penultimate data points. In order to enable a data label for a particular point, we turn the specific point value into an object configuration with the `dataLabels` option as follows:

```
series: [{
    pointStart: 21,
    data: [{
        y: 100,
        z: 1,
        name: 'Alex Rodriguez <br>in Seattle at age 21',
        dataLabels: {
            enabled: true,
            align: 'right',
            verticalAlign: 'middle',
            format: '{point.name}',
            color: 'black',
            x: 15
        }
    }, ....
```

We use the `name` property for the data label content and set the `format` option pointing to the `name` property. We also position the label on the right-hand side of the data point and assign the label color.

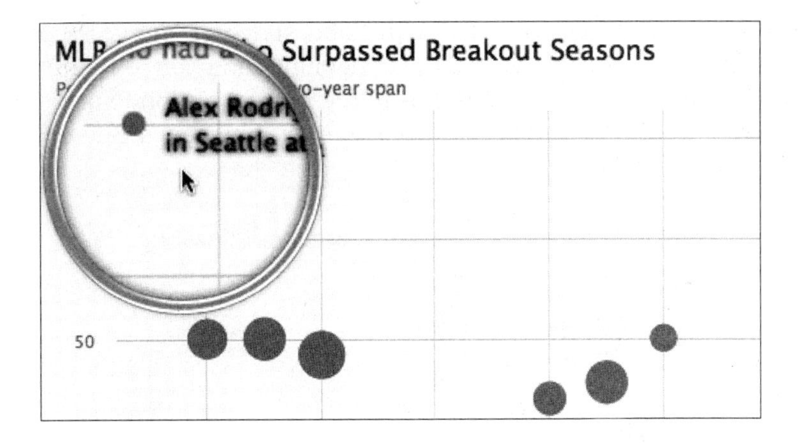

From the preceding observation, we notice that the font seems quite blurred. Actually, this is the default setting for `dataLabels` in the bubble series. (The default label color is white and the position inside the bubble is filled with the series color. As a result, the data label actually looks clear even when the text shadow effect is applied) Also, there is a connector between the bubble and data label in the original chart. Here is our second attempt to enhance the chart:

```
{ y: 100,
  z: 1,
  name: 'Alex Rodriguez <br>in Seattle at age 21',
  dataLabels: {
      enabled: true,
      align: 'right',
      verticalAlign: 'middle',
      format: '<div style="float:left">' +
              '<font size="5">⊠</font></div>' +
              '</span><div>{point.name}</div>',
      color: 'black',
      shadow: false,
      useHTML: true,
      x: -2,
      y: 18,
      style: {
          fontSize: '13px',
          textShadow: 'none'
      }
  }
},
```

To remove the blurred effect, we redefine the label style without CSS `textShadow`. As for the L-shaped connector, we use the trick of alt-code (alt-code: 28) with a larger font size. Then, we put the inline CSS style in the `format` option to make the two `DIV` boxes connector and text label adjacent to each other. The new arrangement looks considerably more polished:

We apply the same trick to the other label; here is the final draft of our bubble chart:

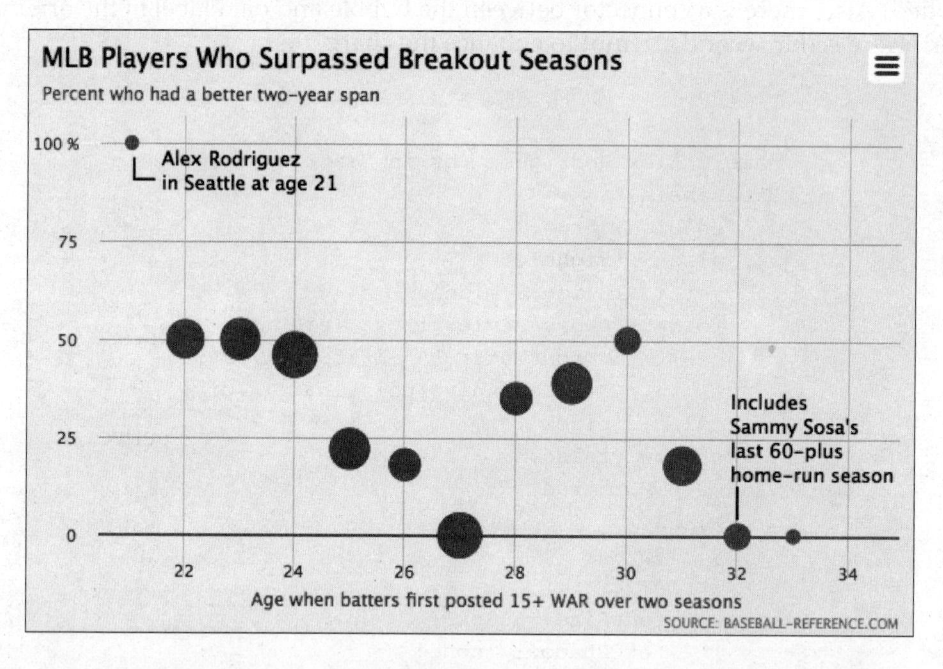

The only part left undone is the bubble size legend. Unfortunately, Highcharts currently doesn't offer such a feature. However, this can be accomplished by using the chart's Renderer engine to create a circle and text label. We will leave this as an exercise for readers.

Technically speaking, we can create the same chart with a scatter series instead, with each data point specified in object configuration, and assign a precomputed z-value ratio to the `data.marker.radius` option.

Understanding the box plot chart

A box plot is a technical chart that shows data samples in terms of the shape of distribution. Before we can create a box plot chart, we need to understand the basic structure and concept. The following diagram illustrates the structure of a box plot:

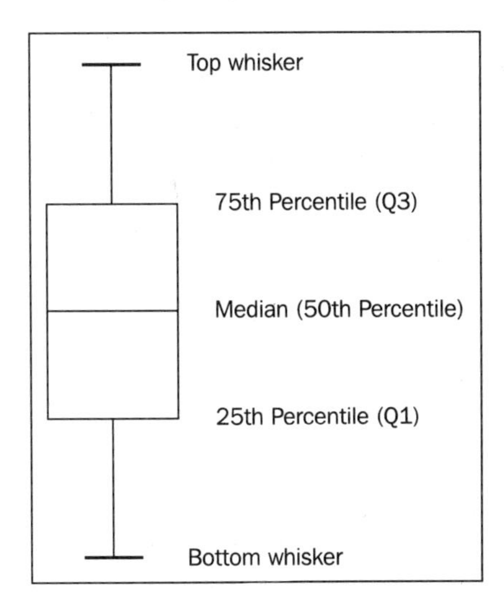

In order to find out the percentile values, the entire data sample needs to be sorted first. Basically, a box plot is composed of top and bottom whisker values, first (Q1) and third (Q3) quartile values, and the median. The quartile Q1 represents the median value between the 50th percentile and the minimum data. Quartile Q3 works in a similar fashion but with maximum data. For data with a perfectly normal distribution, the box plot will have an equal distance between each section.

Strictly speaking, there are other types of box plot that differ in how much the percentiles of both whiskers cover. Some use the definition of 1.5 times the inter-quartile range, that is, *1.5 * (Q3 - Q1)*, or standard deviation. The purpose is to isolate the outlier data and plot them as separate points which can be put into scatter data points along with the box plot. Here, we use the simplest form of box plot: the maximum and minimum data points are regarded as the top and bottom whiskers respectively.

Plotting the box plot chart

In order to create a box plot chart, we need to load an additional library, `highcharts-more.js`:

```
<script src="http://code.highcharts.com/highcharts-more.js"></script>
```

Highcharts offers a set of options to shape and style the box plot series, such as the line width, style, and color, which are shown in the following code snippet:

```
plotOptions: {
    boxplot: {
        lineWidth: 2,
        fillColor: '#808080',
        medianColor: '#FFFFFF',
        medianWidth: 2,
        stemColor: "#808080",
        stemDashStyle: 'dashdot',
        stemWidth: 1,
        whiskerColor: '#808080',
        whiskerWidth: 2,
        whiskerLength: '120%'
    }
},
```

The `lineWidth` option is the overall line width of the boxplot, and `fillColor` is for the color inside the box. The `median` options refer to the horizontal median line inside the box whereas the `stem` options are for the line between the quartile and whisker. The `whiskerLength` option is the ratio that corresponds to the width of the quartile box. In this example, we will enlarge the `whiskerLength` option for ease of visualization, as there are a number of box plots packed into the graph.

The series data values for a box plot are listed in array form in ascending order, so from the bottom to top whisker. The following shows a sample of series data:

```
series: [{
    type: 'boxplot',
    data: [
            [16.855, 19.287, 26.537, 31.368, 33.035 ],
            [16.139, 18.668, 25.33, 30.632, 32.385 ],
            [12.589, 15.536, 23.5495, 28.960, 30.848 ],
            [13.395, 16.399, 22.078, 27.013, 29.146 ],
            ....
    ]
}]
```

Making sense with the box plot data

Before we dive into an example with real-life data, it is worth looking at an excellent article (`http://junkcharts.typepad.com/junk_charts/2014/04/an-overused-chart-why-it-fails-and-how-to-fix-it.html`) by Kaiser Fund, a marketing analytics and data visualization expert who also authored a couple of books on big data crunching. In the article, Kaiser raises an observation of a spider chart from a video *Arctic Death Spiral* (`http://youtu.be/20pjigmWwiw`), as follows:

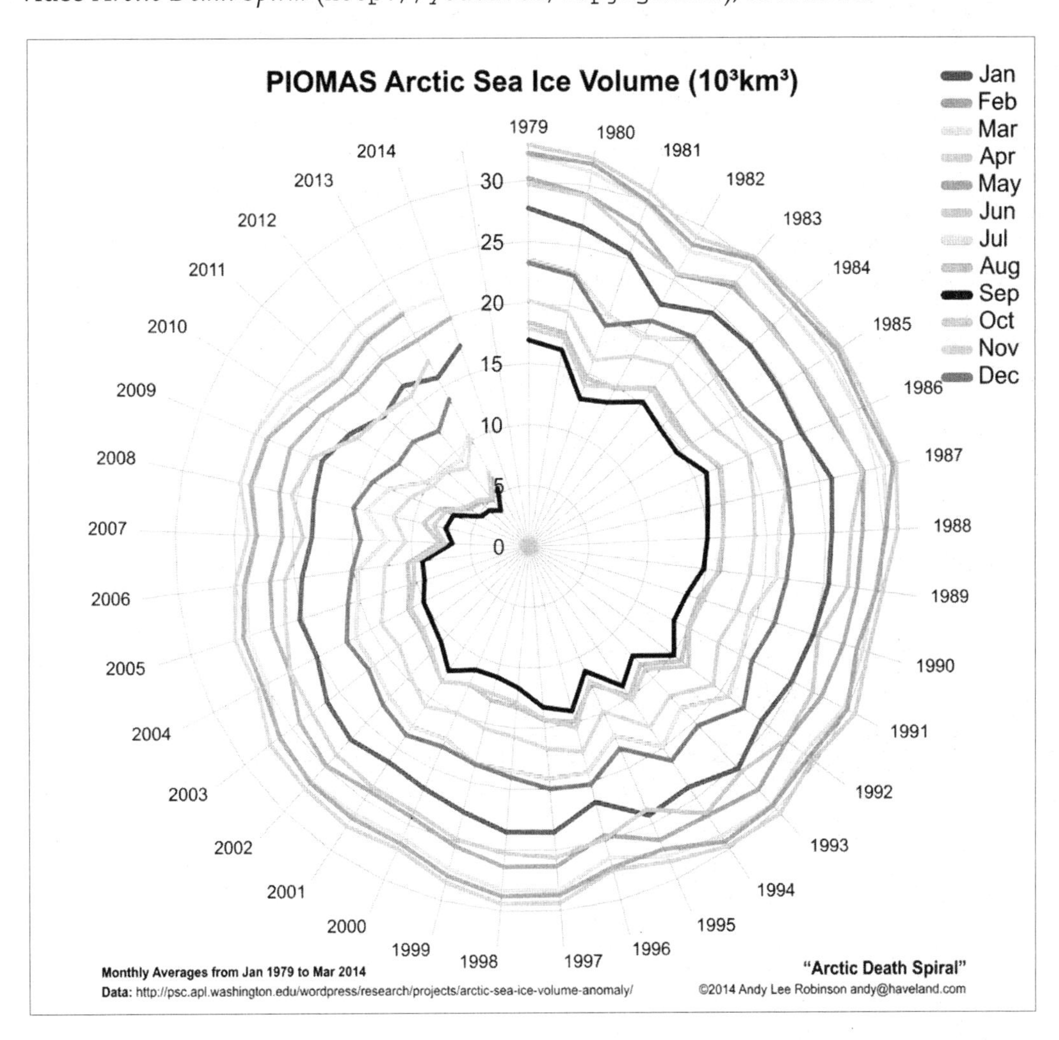

The video demonstrates how the arctic sea ice volume (each month per series over the years) spirals towards the center at an alarming rate. He argues that using a spider chart doesn't do justice to the important message in the data. To summarize his arguments:

- It is difficult for readers to comprehend the real downward trend scale in a circular chart.
- Humans perceive time series data more naturally in a horizontal progression than in a circular motion.
- If the movement of monthly data within a year fluctuates more than other years, we will have multiple line series crossing each other. As a result, we have a plateful of spaghetti instead of a comprehensible chart.

In order to fix this, Kaiser suggests that a box plot is the best candidate. Instead of having 12 multiple series lines crammed together, he uses a box plot to represent the annual data distribution. The 12 months' data for each year are sorted and only the median, quartiles, and extreme values are substituted into the box plot. Although small details are lost due to less data, the range and scale of the downward trend over time are better represented in this case.

The following is the final box plot presentation in Highcharts:

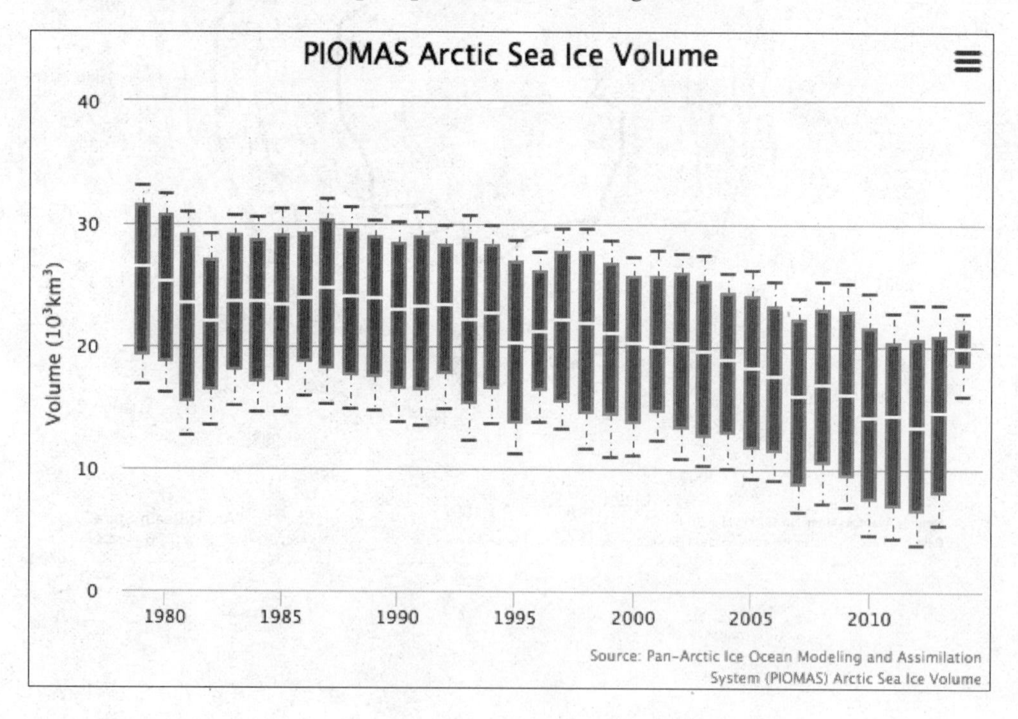

The box plot tooltip

Since the box plot series holds various values, the series has different property names—`low`, `q1`, `median`, `q3`, `high`—to refer to them. The following illustrates an example of `tooltip.formatter`:

```
chart: {
    ....
},
....,
tooltip: {
    formatter: function() {
        return "In year: " + this.x + ", <br>" +
               "low: " + this.point.low + ", <br>" +
               "Q1: " + this.point.q1 + ", <br>" +
               "median: " + this.point.median + ", <br>" +
               "Q3: " + this.point.q3 + ", <br>" +
               "high: " + this.point.high;
        }
    },
series: [{
    ....
}]
```

Note that `formatter` should be added to the `tooltip` property of the main options object, and not in the `series` object. Here is what the box plot tooltip looks like:

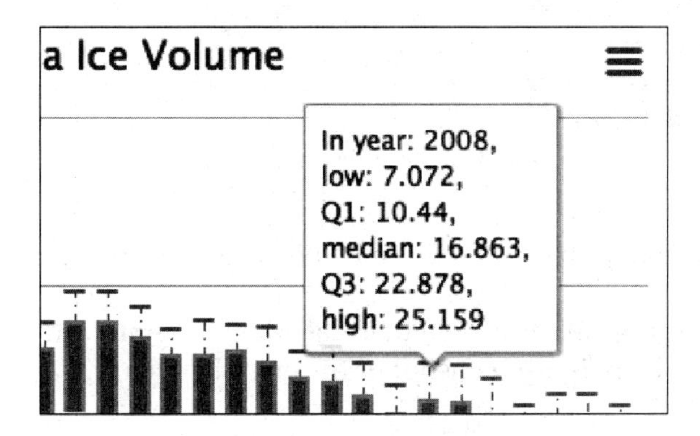

The error bar chart

An error bar is another technical chart that shows the standard error, which is the standard deviation divided by the square root of the sample size. It means that as the sample size increases, the variation from the sample mean diminishes. The error bar series has similar color and style options as the box plot but only applies to whisker and stem:

```
plotOptions: {
    errorbar: {
        stemColor: "#808080",
        stemDashStyle: 'dashdot',
        stemWidth: 2,
        whiskerColor: '#808080',
        whiskerWidth: 2,
        whiskerLength: '20%'
    }
},
```

The same also applies to the tooltip formatter, in which `low` and `high` refer to both ends of the error bar. As for the series data option, it takes an array of tuples of lower and upper values:

```
series: [{
    type: 'column',
    data: ....
}, {
    name: 'error range',
    type: 'errorbar',
    data: [
            [ 22.76, 23.404 ],
            [ 25.316, 29.976 ],
```

To demonstrate the error bar, we use the F1 pit stop times from all the teams in each circuit from `http://www.formula1.com/results/season/2013`. We plot the mean of each circuit in a column series. We then calculate the standard error and apply the result to the mean. Here is a screenshot of the error bar chart:

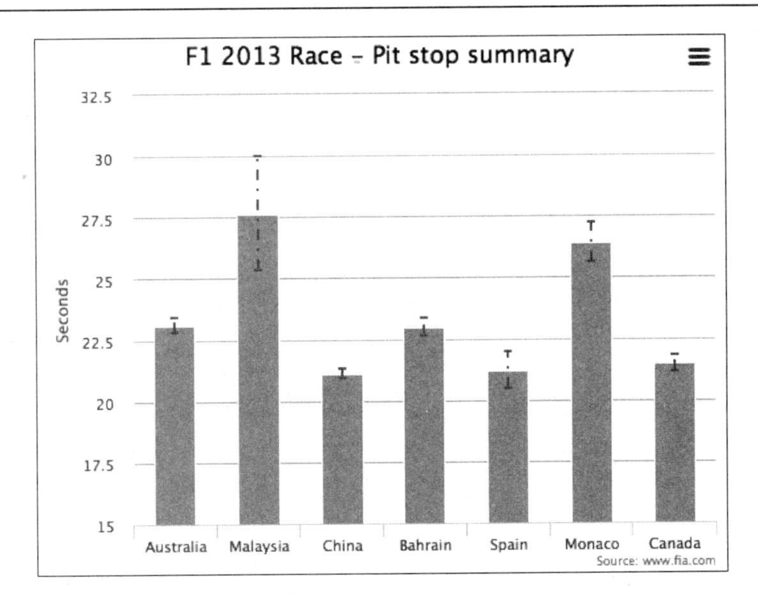

Note that when displaying an error bar series with a column series, the column series has to be specified before the error bar series in the series array. Otherwise, half of the error bar is blocked by the column, as in the following example:

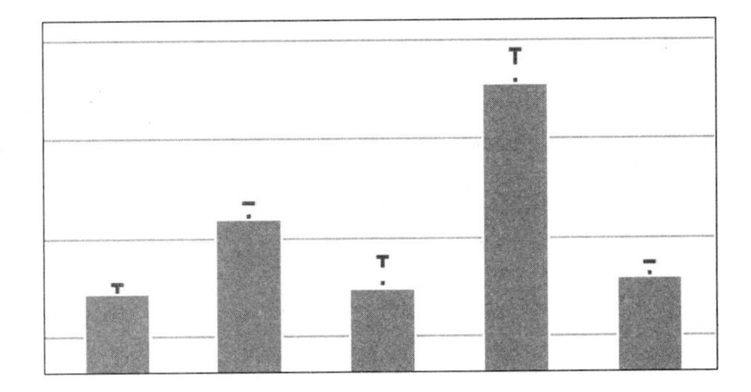

Summary

In this chapter, we have learned about bubble charts and how bubble sizes are determined. We also tested the bubble chart series by recreating a real-life chart. We examined the box plot principle and structure, and practiced how to turn data samples into percentile data and layout into a box plot chart. Finally, we studied error bar charts by using some statistical data.

In the next chapter, we will investigate the properties of waterfall, funnel, pyramid, and heatmap charts and experiment with them.

8

Waterfall, Funnel, Pyramid, and Heatmap Charts

In this chapter, we explore some of the more unusual charts, such as waterfall and funnel charts, by using simple sales and accounting figures. We experiment with how we can link both the charts together with related information. Then, we move on to investigate how to plot a pyramid chart and assess the flexibility of Highcharts by imitating a financial pyramid chart from a proper commercial report. Finally, we study the heatmap chart and tweak various specific options with real-life statistics in a step-by-step approach. In this chapter, we will cover the following topics:

- Constructing a waterfall chart
- Making a funnel chart
- Joining both waterfall and funnel charts
- Plotting a commercial pyramid chart
- Exploring a heatmap chart with inflation data
- Experimenting with the dataClasses and nullColor options of a heatmap chart

Constructing a waterfall chart

A waterfall chart is a type of column chart where the positive and negative values are accumulated along the *x* axis categories. The start and end of adjacent columns are aligned on the same level, with columns going up or down depending on the values. They are mainly used to present cash flow at a higher level.

To use a waterfall chart, the `highcharts-more.js` library must be included:

```
<script type="text/javascript"
 src="http://code.highcharts.com/highcharts-more.js"></script>
```

Let's start with a simple waterfall, with a couple of values for income and expenditure. The following is the series configuration:

```
series: [{
    type: 'waterfall',
    upColor: Highcharts.getOptions().colors[0],
    color: '#E64545',
    data: [{
        name: 'Product Sales',
        y: 63700
    }, {
        name: 'Renew Contracts',
        y: 27000
    }, {
        name: 'Total Revenue',
        isIntermediateSum: true,
        color: '#4F5F70'
    }, {
        name: 'Expenses',
        y: -43000
    }, {
        name: 'Net Profit',
        isSum: true,
        color: Highcharts.getOptions().colors[1]
    }]
}]
```

First, we specify the series `type` as `waterfall`, then assign a color value (blue) to the `upColor` option, which sets the default color for any columns with positive values. In the next line, we declare another color value (red) which is for columns with negative values. The data point entries are the same as a normal column chart, except those columns representing the total sum so far. These columns are specified with the `isIntermediateSum` and `isSum` Boolean options without *y* axis values. The waterfall chart will display those as accumulated columns. We specify the accumulated columns with another color since they are neither income nor expenditure.

Finally, we need to enable the data labels to show the value flow, which is the sole purpose of a waterfall chart. The following is the `plotOptions` setting with the data label styling and formatting:

```
        plotOptions: {
            series: {
                borderWidth: 0,
                dataLabels: {
                    enabled: true,
                    style: {
                        fontWeight: 'bold'
                    },
                    color: 'white',
                    formatter: function() {
                        return Highcharts.numberFormat(this.y / 1000, 1,
    '.') + ' k';
                    }
                }
            }
        },
```

Here is a screenshot of the waterfall chart:

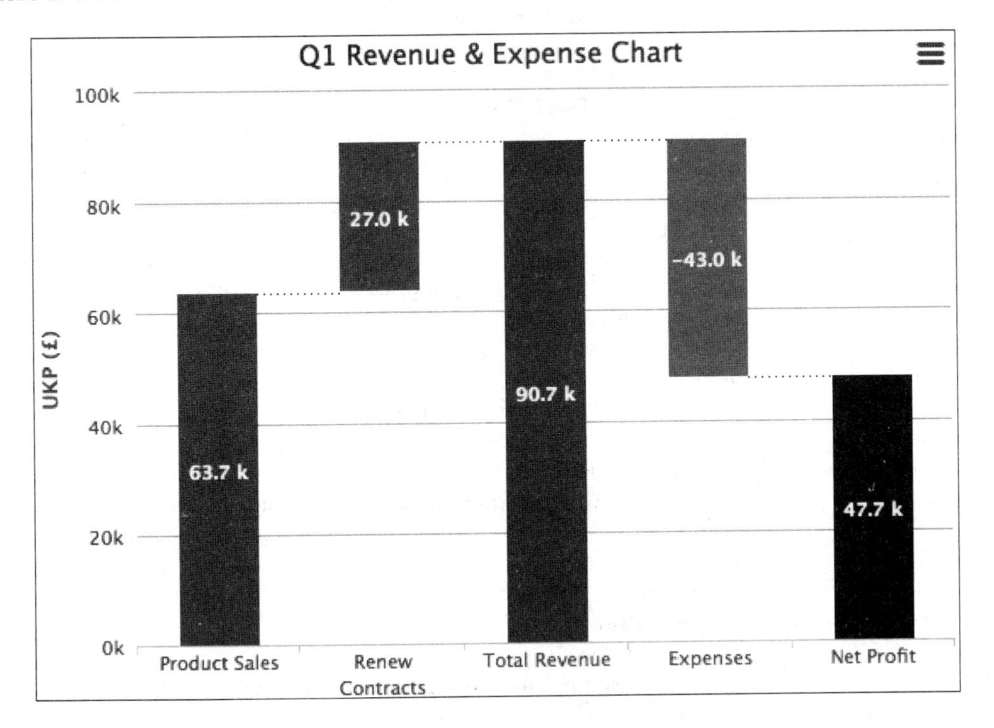

Making a horizontal waterfall chart

Let's assume that we need to plot a slightly more complicated waterfall chart, as we have a few more figures for revenue, expenses, and expenditure plans. Moreover, we need to display multiple stages of accumulated income. Instead of having a traditional waterfall chart with vertical columns, we declare the `inverted` option to switch the columns to horizontal. Since the columns are more compacted, the data labels may not fit inside the columns. Therefore, we put the data labels at the end of the columns. Here is a summary of the changes:

```
chart: {
    renderTo: 'container',
    inverted: true
},
....,
plotOptions: {
    series: {
        borderWidth: 0,
        dataLabels: {
            enabled: true,
            inside: false,
            style: {
                fontWeight: 'bold'
            },
            color: '#E64545',
            formatter: ....
        }
    }
},
series: [{
    type: 'waterfall',
    upColor: Highcharts.getOptions().colors[0],
    color: '#E64545',
    name: 'Q1 Revenue',
    data: [{
        name: 'Product A Sales',
        y: 63700,
        dataLabels: {
            color: Highcharts.getOptions().colors[0]
        }
    }, {
        name: 'Product B Sales',
        y: 33800,
        dataLabels: {
            color: Highcharts.getOptions().colors[0]
        }
    }, {
```

```
            // Product C & Renew Maintenance
            ....
        }, {
            name: 'Total Revenue',
            isIntermediateSum: true,
            color: '#4F5F70',
            dataLabels: {
                color: '#4F5F70'
            }
        }, {
            name: 'Staff Cost',
            y: -83000
        }, {
            // Staff Cost, Office Rental and Tax
            ....
        }, {
            name: 'Net Profit',
            isSum: true,
            color: Highcharts.getOptions().colors[1],
            dataLabels: {
                color: Highcharts.getOptions().colors[1]
            }
    }]
```

The following is the new look of the waterfall chart:

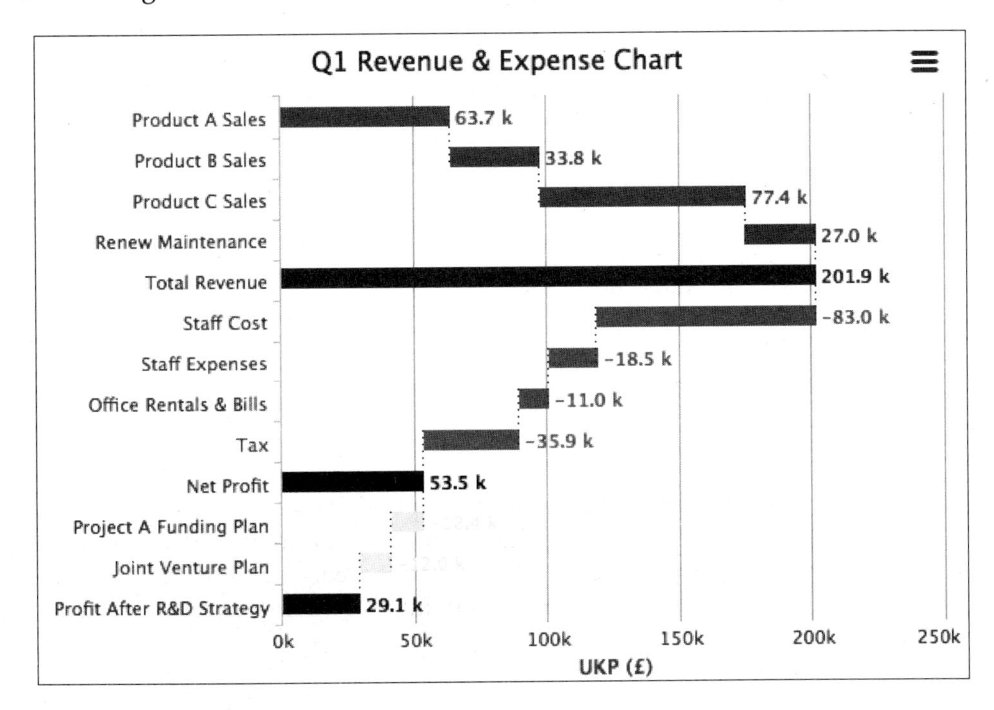

Constructing a funnel chart

As the name implies, the shape of a funnel chart is that of a funnel. A funnel chart is another way of showing how data flows in a diminished fashion. For example, it can be used to show the stages from the number of sale leads, demos, and meetings to actual sales, or the number of job applicants that pass through exams, interviews, and an offer to actual acceptance.

The funnel chart is packaged as a separate module which needs to include the JavaScript file as follows:

```
<script type="text/javascript" src="http://code.highcharts.com/
modules/funnel.js"></script>
```

In this funnel chart example, we will plot a product starting from web visits and leading to product sales. Here is the series configuration:

```
series: [{
      type: 'funnel',
      width: '60%',
      height: '80%',
      neckWidth: '20%',
      dataLabels: {
          format: '{point.name} - {y}',
          color: '#222'
      },
      title: {
          text: "Product from Web Visits to Sales"
      },
      data: [{
          name: 'Website visits',
          y: 29844
      }, {
          name: 'Product downloads',
          y: 9891
      }, {
          ....
      }]
}]
```

Apart from the series type option, Highcharts simply draws the set of data points in a funnel shape configuration. We use the `neckWidth` option to control the width of the narrow part of the funnel as a ration of the plot area. The same applies to the `width` and `height` options. Here is a screenshot illustrating this:

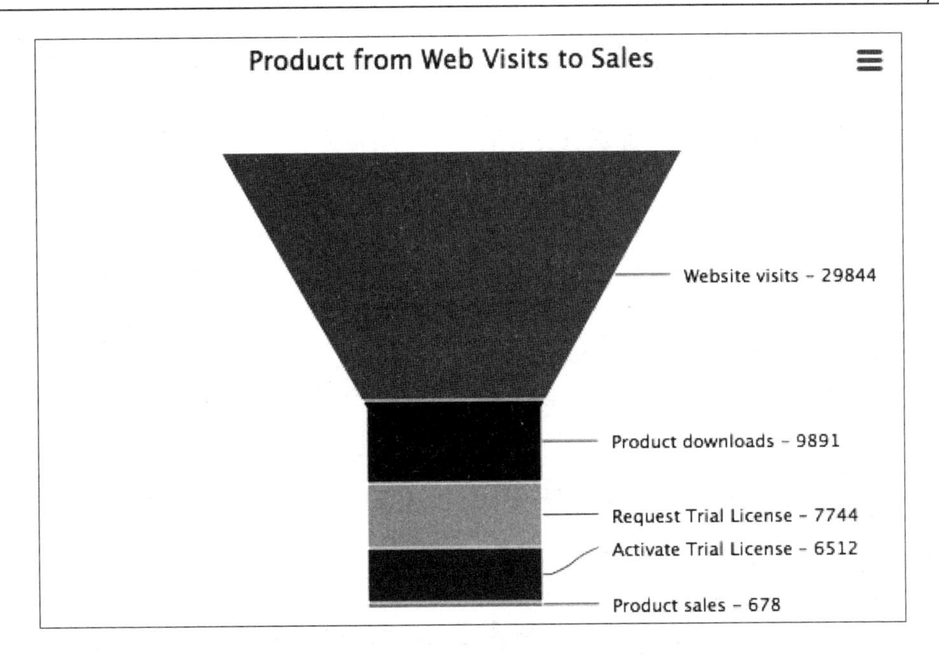

Joining both waterfall and funnel charts

So far, we have explored both waterfall and funnel charts. Both types of chart have been used in sales data. In this section, we are going to join the previous waterfall and funnel examples together, so that when the client clicks on a horizontal bar (let's say **Product B sales**), the waterfall chart is zoomed into the funnel chart to show the prospect ratio of the product. To achieve that, we use the drilldown feature (described and demonstrated in *Chapter 2, Highcharts Configurations*).

First, we set the data point with a drilldown identifier, like `productB`. Then, we import the whole funnel series configuration from the previous example into the drilldown option with the matching `id` value. Finally, we relocate the drill up back button to the bottom of the screen. The final change should be as follows:

```
series: [{
   type: 'waterfall',
   ....
   data: [{
        ....
   }, {
        name: 'Product B Sales',
        y: 33800,
        drilldown: 'productB',
        ....
```

```
            }]
        }],
        drilldown: {
            drillUpButton: {
                position: {
                    verticalAlign: 'bottom',
                    align: 'center',
                    y: -20
                }
            },
            series: [{
                type: 'funnel',
                id: 'productB',
```

Now, we have a fully interactive chart, from a revenue breakdown waterfall chart to the product prospect funnel chart. The following is the waterfall chart with the drilldown:

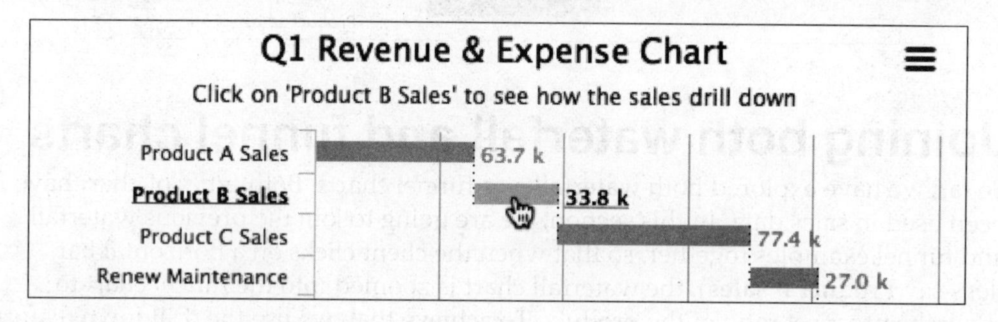

Clicking on the **product B sales** column, the chart transforms to a funnel to dissect the sales from product activities:

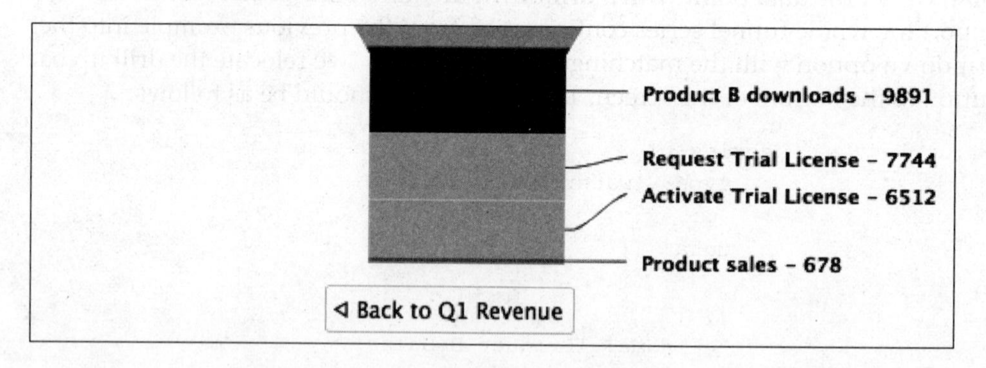

Plotting a commercial pyramid chart

A pyramid chart is the inverse of a funnel chart, but in a pyramid shape, generally used for representing a top-down hierarchical ordering of data. Since this is a part of the funnel chart module, `funnel.js` is required.

The default order for data entries is that the last entry in the series data array is shown at the top of the pyramid. This can be corrected by switching the `reverse` option to false. Let's take a real-life example for this exercise. The following is a picture of the Global Wealth Pyramid chart taken from the Credit Suisse Global Wealth Databook 2013:

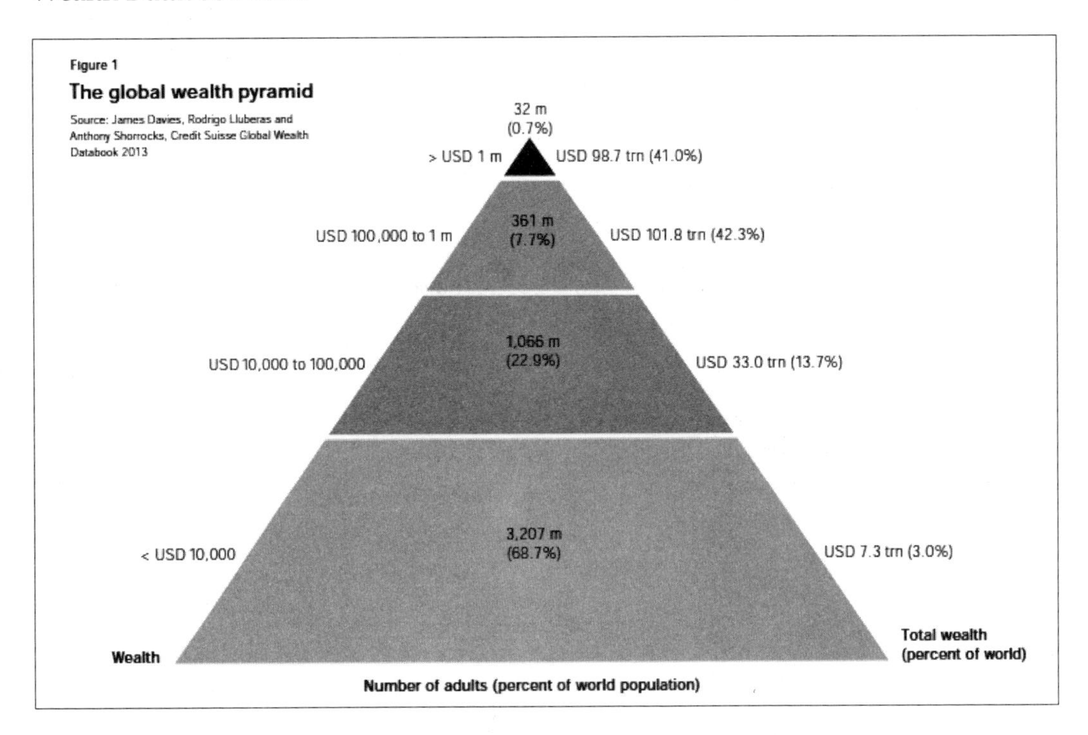

As we can see, it looks stylish with data labels along each side as well as in the middle. Let's make an initial attempt to reproduce the chart:

```
title: {
    text: "The global wealth pyramid",
    align: 'left',
    x: 0,
    style: {
        fontSize: '13px',
        fontWeight: 'bold'
    }
```

```
    },
    credits: {
        text: 'Source: James Davies, ....',
        position: {
            align: 'left',
            verticalAlign: 'top',
            y: 40,
            x: 10
        },
        style: {
            fontSize: '9px'
        }
    },
    series: [{
        type: 'pyramid',
        width: '70%',
        reversed: true,
            dataLabels: {
                enabled: true,
                format: '<b>{point.name}</b> ({point.y:,.1f}%)',
            },
        data: [{
            name: '3,207 m',
            y: 68.7,
            color: 'rgb(159, 192, 190)'
        }, {
            name: '1,066 m',
            y: 22.9,
            color: 'rgb(140, 161, 191)'
        }, {
            name: '361 m',
            y: 7.7,
            color: 'rgb(159, 165, 157)'
        }, {
            name: '32 m',
            y: 0.7,
            color: 'rgb(24, 52, 101)'
        }]
    }]
```

First, we move the title and credits to the top-left corner. Then, we copy the color and percentage values into the series, which gives us the following chart:

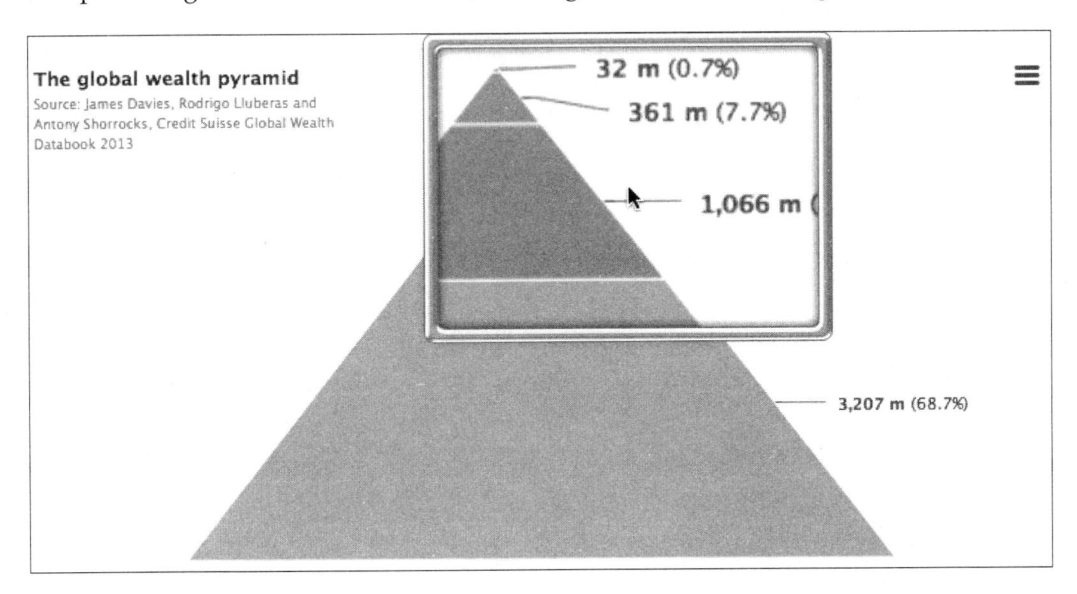

There are two main problems when compared to the original: the scale of each layer is wrong and the data labels are either missing or in the wrong place. As we can see, the reality is that the rich population is actually much smaller than the original chart showed, amusingly a single pixel. In the next section, we are going to see how far we can push Highcharts to make the financial chart.

Plotting an advanced pyramid chart

Let face it. The layers in the original chart do not reflect their true percentage, proved in the previous chart. So we need to rig the ratio values to make them similar to the original chart. Secondly, the repositioning of data labels is limited in a pyramid chart. The only way to move the data label into the center of each layer of a pyramid chart is to disable the connectors and gradually adjust the `distance` option in `plotOptions.pyramid.dataLabels`. However, this only allows us a single label per data layer and we can only apply the same positional settings to all the labels.

So how can we put extra data labels along each side of the layer and the titles along the bottom of the pyramid? The answer (after hours of trial and error) is to use multiple y axes and `plotLines`. The idea is to have three y axes to put data labels on the left, center, and right side of the pyramid. Then, we hide all the y axes' interval lines and their labels. We place the y axis title at the bottom of the axis line with no rotation, we only rotate for the left (**Wealth**) and right (**Total Wealth**) axes.

We disable the remaining y axis title. The trick is to use the x axis title instead for the center label (**Number of adults**), because it sits on top of the pyramid. Here is the code snippet for the configuration of these axes:

```
            chart: {
                renderTo: 'container',
                // Force axes to show the title
                showAxes: true,
                ....
            },
            xAxis: {
                title: {
                    text: "Number of adults (percentage of world
    population)"
                },
                // Only want the title, not the line
                lineWidth: 0
            },
            yAxis: [{
                // Left Y-axis
                title: {
                    align: 'low',
                    text: 'Wealth',
                    rotation: 0,
                    x: 30
                },
                // Don't show the numbers
                labels: {
                    enabled: false
                },
                gridLineWidth: 0,
                min: 0,
                max: 100,
                reversed: true,
                ....
            }, {
                // Center Y-axis
                title: {
                    text: ''
                },
                // Same setting to hide the labels and lines
                ....
            }, {
```

```
    // Right Y-axis
    opposite: true,
    title: {
        align: 'low',
        text: 'Total Wealth <br>(percent of world)',
        rotation: 0,
        x: -70
    },
    // Same setting to hide the labels and lines
    ....
```

The following screenshot demonstrates the titles aligned at the base of pyramid chart:

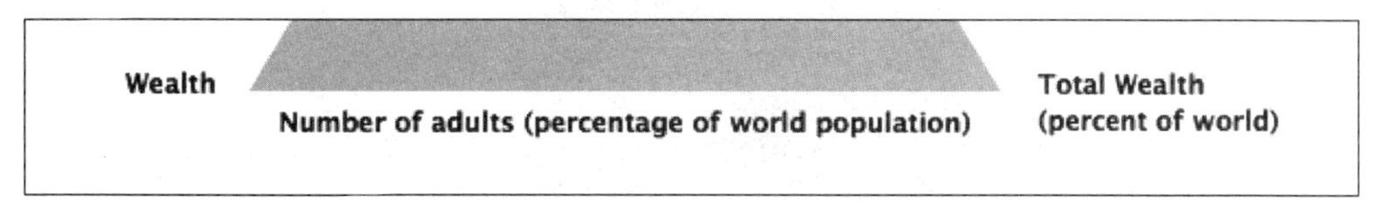

The next part is to simply create a `plotLine` for each y-axis in each layer of the pyramid and set the label for these `plotLines`. Since we know the data value range of each pyramid layer, we can set the same `y` value for `plotLines` across all the y axes to display the labels on the same level. Here is an example of a `plotLines` option for the left y-axis:

```
plotLines: [{
    value: 40,
    label: {
        text: '< USD 10,000',
        x: 20
    },
    width: 1,
}, {
    value: 30,
    label: {
        text: 'USD 10,000 - 100,000',
        x: 35
    },
    width: 1
}, {
```

We do the same for the center labels instead of using the points' data labels. Note that the x position here caters for the data in this chart. If the graph has dynamic data, a new x position must be calculated to fit the new data. Here is the final look in Highcharts:

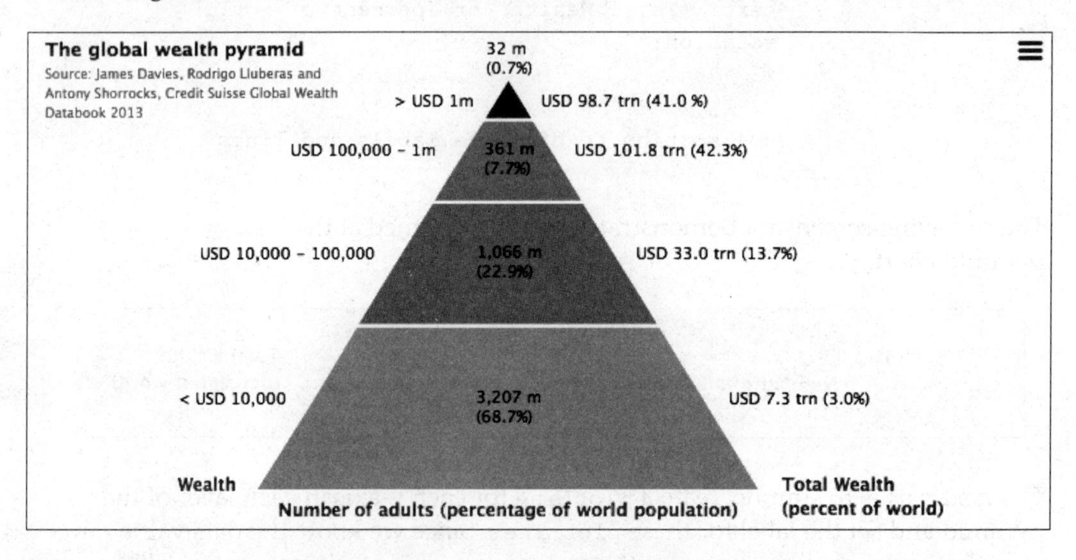

Exploring a heatmap chart with inflation data

Heatmap is the latest chart addition in Highcharts 4. As its name implies, the data values are represented in temperature colors, where the data is arranged in a grid-based layout, so the chart expresses three-dimensional data. The heatmap appeals to some readers because of the instant awareness of the data trend it generates, as temperature is one of the most natural things to sense. The chart is generally used for demonstrating climate data, but it has been used for other purposes. Let's explore the heatmap chart with some real-life data. Here, we use it to show the inflation of various countries, as in this example in the Wall Street Journal (`http://blogs.wsj.com/economics/2011/01/27/feeling-the-heat-comparing-global-inflation/`):

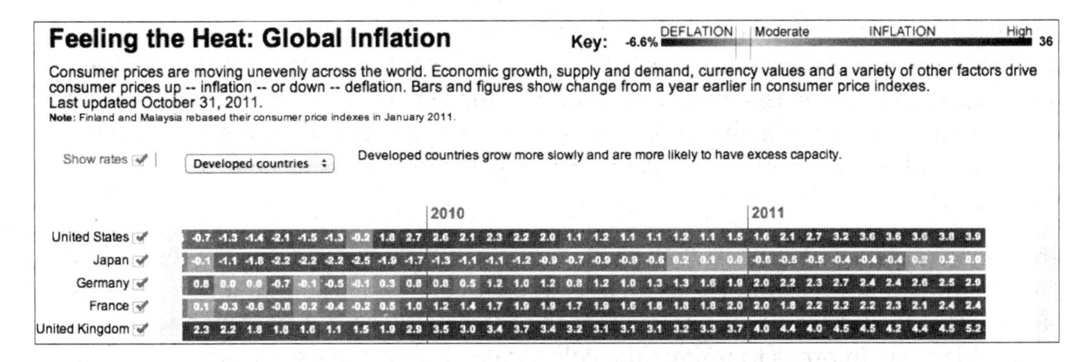

In Highcharts, the heatmap module was released as part of the Highmaps extension, which can be used as part of Highmaps or as a module of Highcharts. In order to load a heatmap as a Highcharts module, it includes the following library:

```
<script src="http://code.highcharts.com/modules/heatmap.js"></script>
```

First, the months are sketched along the y axis, whereas the x axis holds the country names. So we need to invert the chart:

```
chart: {
    renderTo: 'container',
    type: 'heatmap',
    inverted: true
},
```

The next step is to set both x and y axes as categories with specific labels. Then, we set the x axis without interval lines. As for the y axis, we position the labels at the top of the cells by assigning the opposite option to true and including an offset distance to make sure the axis lines are not too close:

```
xAxis: {
    tickWidth: 0,
    // Only show a subset of countries
    categories: [
        'United States', 'Japan', 'United Kingdom',
        'Venezuela', 'Singapore', 'Switzerland', 'China'
    ]
},
yAxis: {
    title: { text: null },
    opposite: true,
    offset: 8,
    categories: [ 'Aug 2010', 'Sept', 'Oct', 'Nov', 'Dec', ...
```

There is one more axis that we need to configure, `colorAxis` (see `http://api.highcharts.com/highmaps`), which is specific to heatmap charts and Highmaps. The `colorAxis` option is similar to `x/yAxis`, which shares a number of options with it. The major difference is that `colorAxis` is the definition in mapping between the color and value. There are two ways to define color mapping: discrete and linear color range.

In this example, we demonstrate how to define multiple linear color ranges. As we can see, the inflation example has an asymmetric color scale, so that it ranges from -6.6 percent to 36 percent, but notice that the color between -0.1 percent and 0.1 percent is gray. In order to imitate the color scale closely, we use the `stops` option to define fragments of discrete spectrums. The `stops` option takes an array of tuples of ratio range and color, and we transform the inflation and color values from the example into a number of ratios (we take the range from -1 percent to 30 percent instead because of the subset samples):

```
colorAxis: {
    min: -0.9,
    max: 30,
    stops: [
        [0, '#1E579F'],
        // -6.6
        [0.085, '#467CBA'],
        // -6
        [0.1, '#487EBB'],
        // -2
        [0.2, '#618EC4'],
        // -1
        [0.225, '#7199CA'],
        // -0.2
        [0.245, '#9CB4D9'],
        // Around 0
        [0.25, '#C1C1C1'],
        // Around 0.2
        [0.256, '#ECACA8'],
        // Around 10
        [0.5, '#D02335'],
        // Around 20
        [0.75, '#972531'],
        [1.0, '#93212E']
    ],
    labels: {
        enabled: true
    }
}
```

Additionally, we replicate the title (top-left), color scale legend (top-right), and credits (bottom-right) as in the original chart, with the following configurations:

```
title: {
    text: "Feeling the Heat: Global Inflation",
    align: 'left',
    style: {
        fontSize: '14px'
    }
},
subtitle: {
    text: "From Aug 2010 - Aug 2011",
    align: 'left',
    style: {
        fontSize: '12px'
    }
},
legend: {
    align: 'right',
    verticalAlign: 'top',
    floating: true,
    x: -60,
    y: -5
},
credits: {
    text: 'Sources: CEIC Data; national statistical ....',
    position: {
        y: -30
    }
},
```

The final step is to define the three-dimensional data and switch the `dataLabels` option on:

```
series: [{
    dataLabels: {
        enabled: true,
        color: 'white'
    },
    // Country Category Index, Month/Year Index, Inflation
    data: [
        // US
        [ 0, 0, 1.1 ],
        [ 0, 1, 1.1 ],
        ....,
```

```
            // Japan
            [ 1, 0, -0.9 ],
            [ 1, 1, -0.6 ],
```

Here is the display:

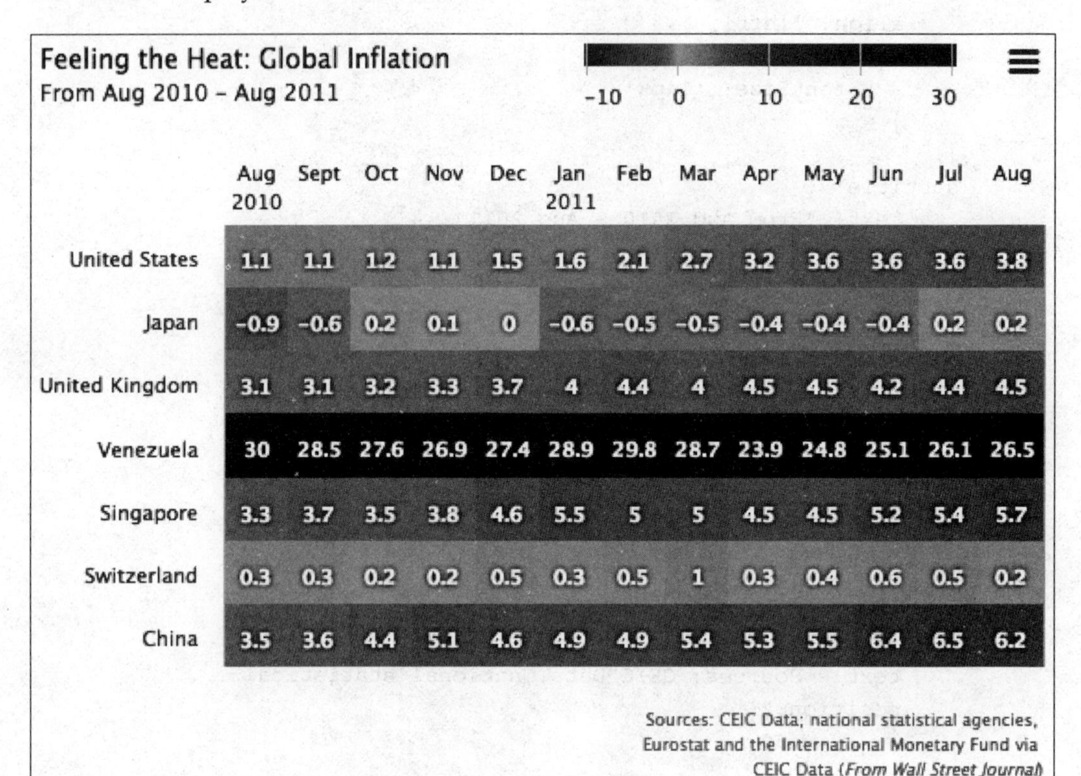

Experimenting with dataClasses and nullColor options in a heatmap

An alternative way to define the color axis is to have a specific range of values associated with a color. Let's plot another heatmap chart. In this example, we reconstruct a graph taken from `http://kindofnormal.com/truthfacts`, shown here:

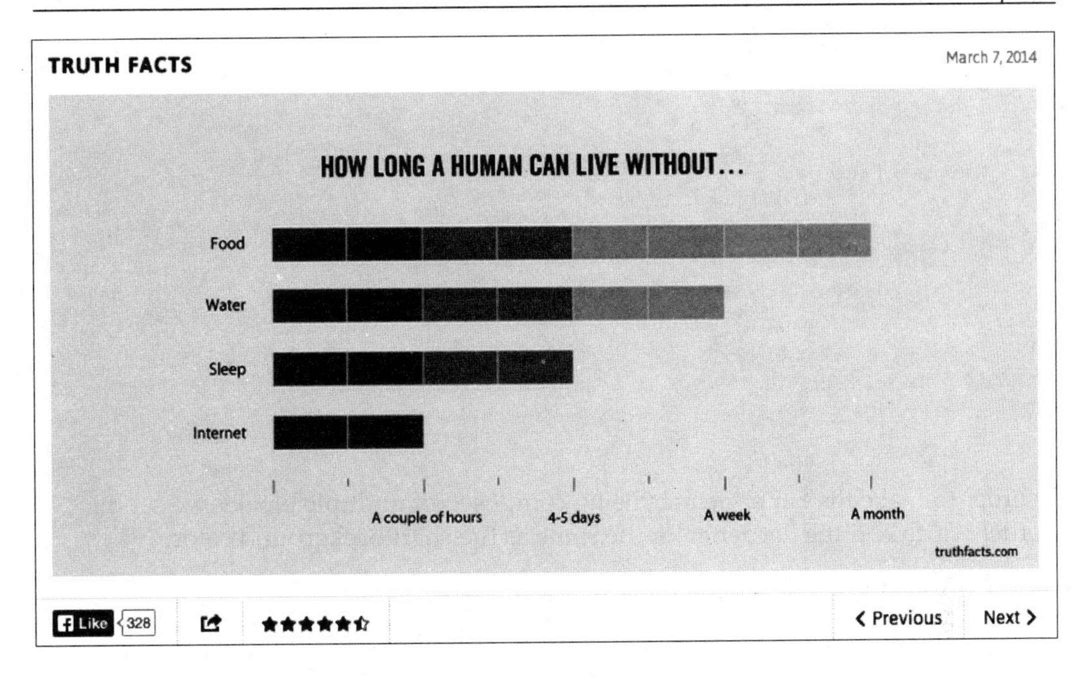

To recreate the preceding chart, we first use an inverted heatmap to emulate it as a bar chart, but the bars itself are composed of cells with a gradual change of color. We treat each block as a unit of y axis value and every two intervals associates with a color value. Hence, the range along the y-axis is between 0 and 8. Here is the trimmed configuration:

```
yAxis: {
    title: { text: null },
    gridLineWidth: 0,
    minorTickWidth: 1,
    max: 8,
    min: 0,
    offset: 8,
    labels: {
        style: { .... },
        formatter: ....,
```

Then, we specify the `colorAxis` with `dataClasses` options which divide the value range into four groups of color:

```
colorAxis: {
    dataClasses: [{
        color: '#2D5C18',
        from: 0,
        to: 2
```

```
        }, {
            color: '#3B761D',
            from: 2,
            to: 4
        }, {
            color: '#70AD28',
            from: 4,
            to: 6
        }, {
            color: '#81C02E',
            from: 6,
            to: 8
        }]
    },
```

In order to make the bar appear as being composed of multiple blocks, we set the border width and the border color the same as the chart background color:

```
plotOptions: {
    heatmap: {
        nullColor: '#D2E4B4',
        borderWidth: 1,
        borderColor: '#D2E4B4',
    },
},
```

Notice that there is a `nullColor` option; this is to set the color for a data point with null value. We assign the null data point the same color as the background. We will see later what this null color can do in a heatmap.

In heatmaps, unlike column charts, we can specify the distance and grouping between columns. The only way to have a gap between categories is to fake it, hence a category with an empty title:

```
xAxis: {
        tickWidth: 0,
        categories: [ 'Food', '', 'Water', '', 'Sleep',
                      '', 'Internet' ],
        lineWidth: 0,
        ....,
},
```

Since we are emulating a bar chart in which the change of color values correlate to y-axis values, the z-value is the same as the y-value. Here is the series data configuration for **Food** and **Water** categories:

```
series: [{
    data: [  [ 0, 7, 7 ], [ 0, 6, 6 ], [ 0, 5, 5 ], [ 0, 4, 4 ],
```

```
[ 0, 3, 3 ], [ 0, 2, 2 ], [ 0, 1, 1 ], [ 0, 0, 0 ],
[ 2, 0, 0 ], [ 2, 1, 1 ], [ 2, 2, 2 ], [ 2, 3, 3 ],
[ 2, 4, 4 ], [ 2, 5, 5 ],
```

The first group of data points have zero values on the inverted x axis which is the index to the categories—**Food**. The group has a full range of eight data points. The next has a group of six data points (value two on the x axis because of the dummy category, see `xAxis.categories` in preceding code) which corresponds to six blocks in the **Water** category.

For the sake of demonstrating how `nullColor` works, instead of displaying a block of cells as in the original chart, let's change it slightly to have a fraction unit of cells. In the **Sleep** and **Internet** categories, we are going to change the values to 1.5 and 0.25 respectively. The trick to displaying a heatmap cell that is not in a full block is to use the `nullColor` option, that is, a fraction of the unit value is assigned to null and the color for the null value is the same as the chart background color to "hide" the rest of the unit:

```
[ 4, 0, 0 ],
[ 4, 1, 1 ],
[ 4, 1.5, null ],

[ 6, 0, 0 ],
[ 6, 0.25, null ]
```

Here is a screenshot of the reproduction in Highcharts:

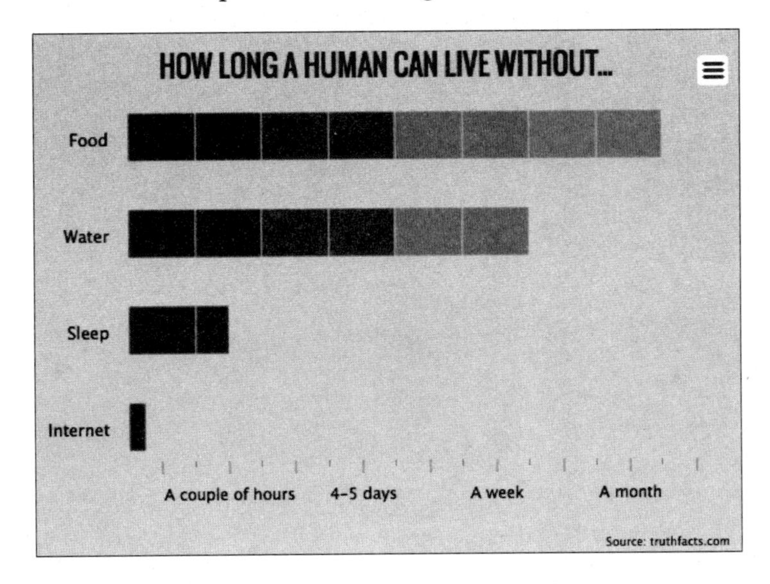

Summary

In this chapter, we have learned how to plot waterfall and funnel charts with fabricated sales data. We familiarized ourselves with and tested the flexibilities of Highcharts by reconstructing a pyramid chart from a financial report. We examined the construction of a heatmap chart and studied the color axis property with different examples.

In the next chapter, we will investigate that long-awaited Highcharts feature, 3D charts. We will explore how to apply 3D orientation on charts and plot a gallery of various series charts in 3D.

9

3D Charts

3D charts have been the most long awaited and the most desired feature in Highcharts. A feature that has been in the subject of heated debate within the users' community. There are users who agree, and those that disagree with the use of 3D charts. The pro camp argue for the use of 3D charts, for example in sales and marketing reports or infographics posters. The con party argue that 3D charts are misleading in terms of accuracy in data visualization and comparison, so the development resources should be focused elsewhere. As for Highcharts, it takes the pro-choice direction.

This chapter focuses on the basics of 3D configuration and shows a gallery of charts presented in 3D, as well as several specific options in each supported series. In this chapter, we will learn:

- What a 3D chart in Highcharts is and isn't
- Experimenting with 3D chart orientation: `alpha`, `beta`, `depth`, and `viewDistance`
- Configuring the 3D chart background
- Plotting the column, pie, donut, and scatter series in 3D charts
- Navigating with 3D charts

What a Highcharts 3D chart is and isn't

At the time of writing, the 3D chart feature is to display the 2D charts in a real 3D presentation, that is, we can rotate and tilt the charts in different dimensions and adjust the view distance. The 3D feature is currently only available for the column, pie, and scatter series.

What 3D charts cannot do is to construct a real three dimensional, *x*, *y*, and *z* axes chart, except for the scatter plot; we will even see some shortcomings in the scatter plot later. As this is the first major release of 3D charts, this limitation may be short lived.

Experimenting with 3D chart orientation

To enable the 3D feature, we first need to include the extension as follows:

```
<script
  src="http://code.highcharts.com/highcharts-3d.js"></script>
```

In Highcharts, there are two levels of 3D options: chart and series. The chart level options are located in `chart.options3d`, which mainly deals with the orientation and the frame around the plotting area, whereas the 3D options for the series remain in the usual `plotOptions` area, such as `plotOptions.column.depth`.

Currently, Highcharts supports 2 axles of rotation: horizontal and vertical, which are the `alpha` and `beta` options in `chart.options3d` respectively. The values for these options are in degrees. The following diagram illustrates with arrows the direction of chart rotation as the degree value increases and decreases:

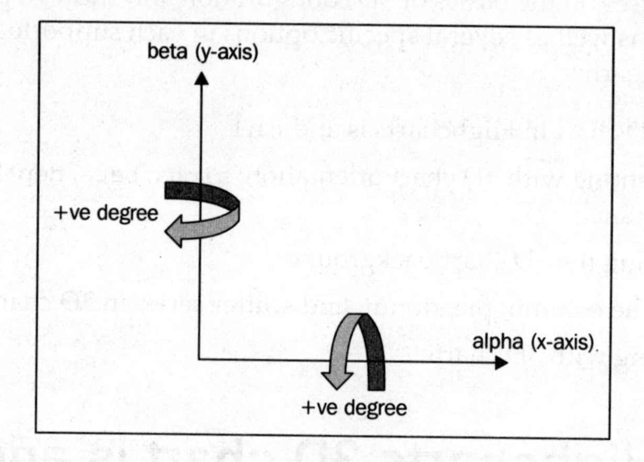

Alpha and beta orientations

In the previous diagram, we showed the direction of orientation. Let's experiment with a column chart. We are going to use a chart from *Chapter 4, Bar and Column Charts*, which looks as follows:

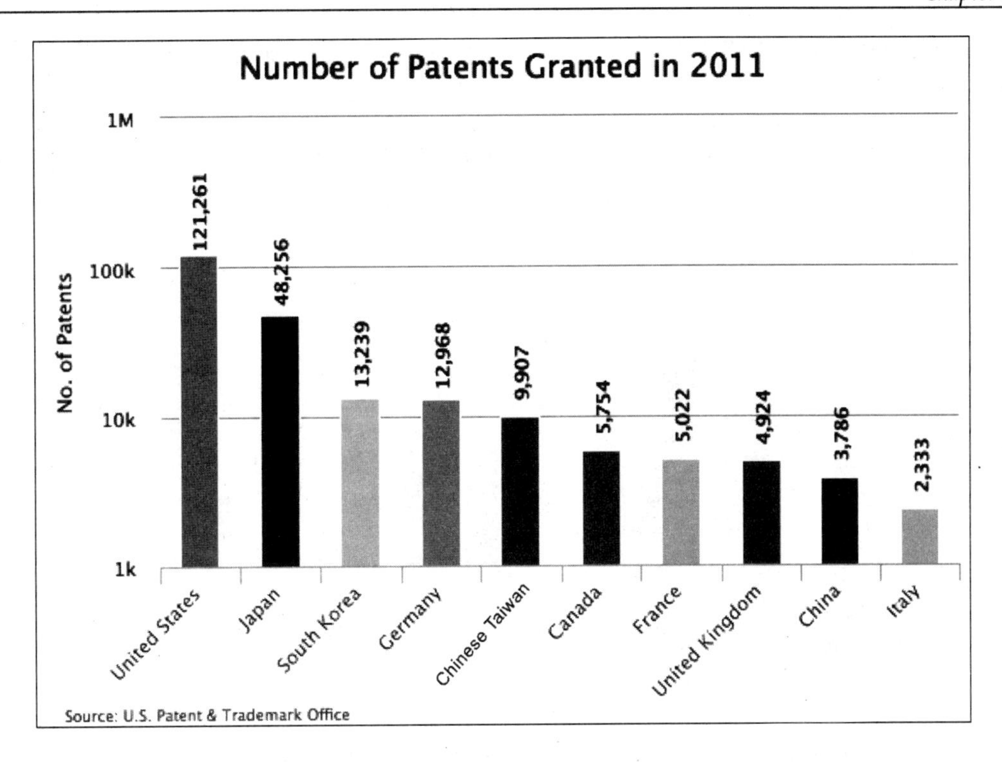

We then add the 3D options to the chart as follows:

```
chart: {
    renderTo: 'container',
    type: 'column',
    borderWidth: 1,
    options3d: {
        alpha: 0,
        beta: 0,
        enabled: true
    }
},
```

We switch the chart to 3D display mode with the enabled option and no rotation in both axles, which produces the following chart:

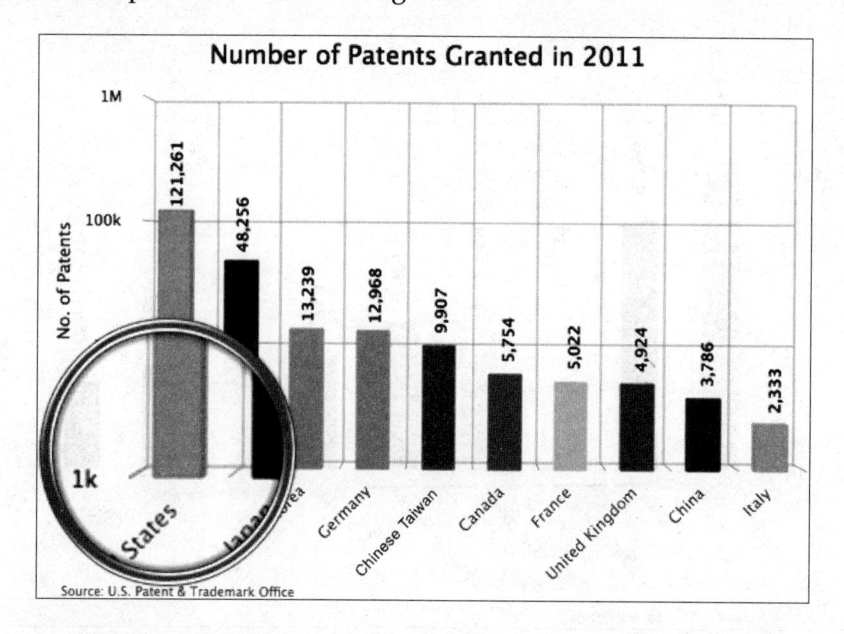

Notice that without any rotation, the chart looks almost the same as the original, as expected. When we look closer, the columns are actually constructed in 3D, with depth. Let's set the alpha axle to 30 degrees and the beta axle to remain at zero:

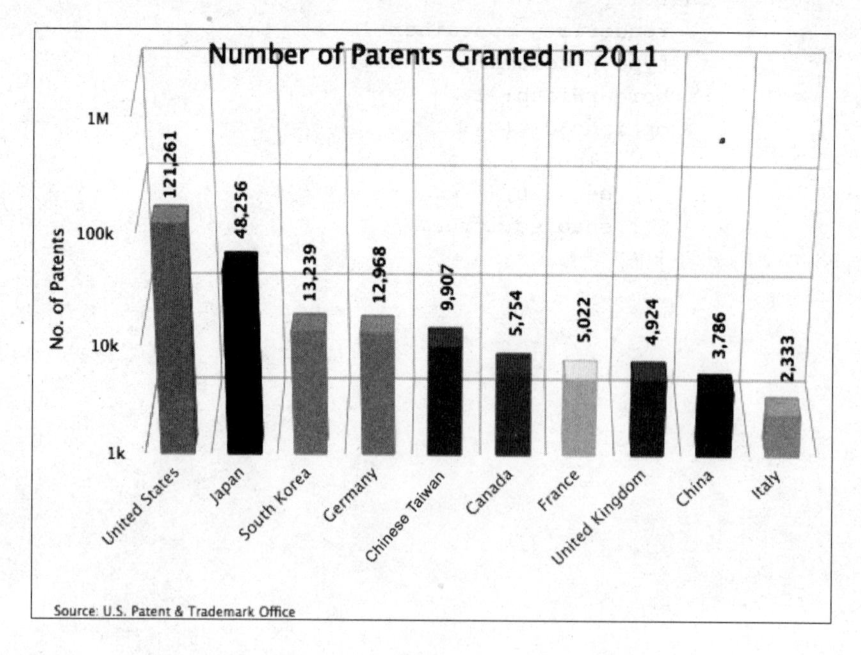

We can see the 3D structure more clearly as the alpha axle increases, as we are leaning towards a top-down view of the chart. Let's reset alpha back to zero and set the beta axle to 30 degrees:

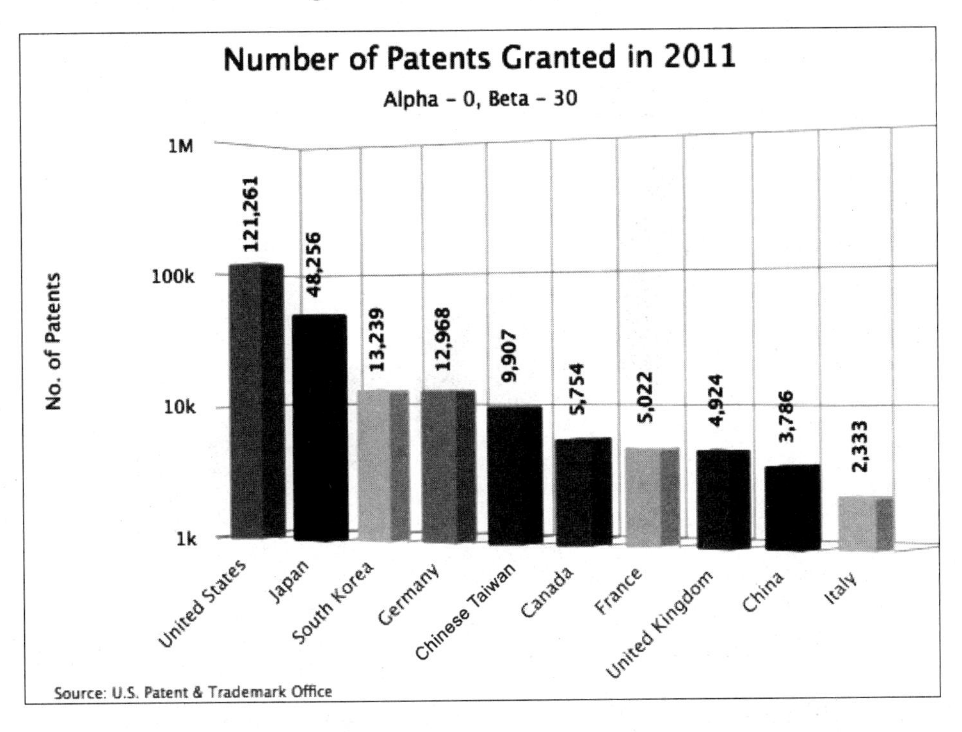

The chart is rotated horizontally on a vertical principal axis. Note that once the chart is rotated on the beta axle, a large gap appears between the title and the *y* axis. We can remove the gap by positioning the title with `x` and `y` options:

```
yAxis: {
    title: {
        text: 'No. of Patents',
        x: 35,
        y: -80
    }
},
```

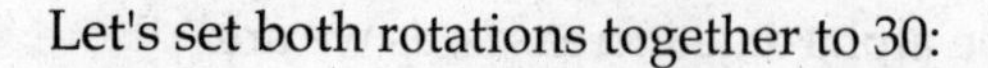

Let's set both rotations together to 30:

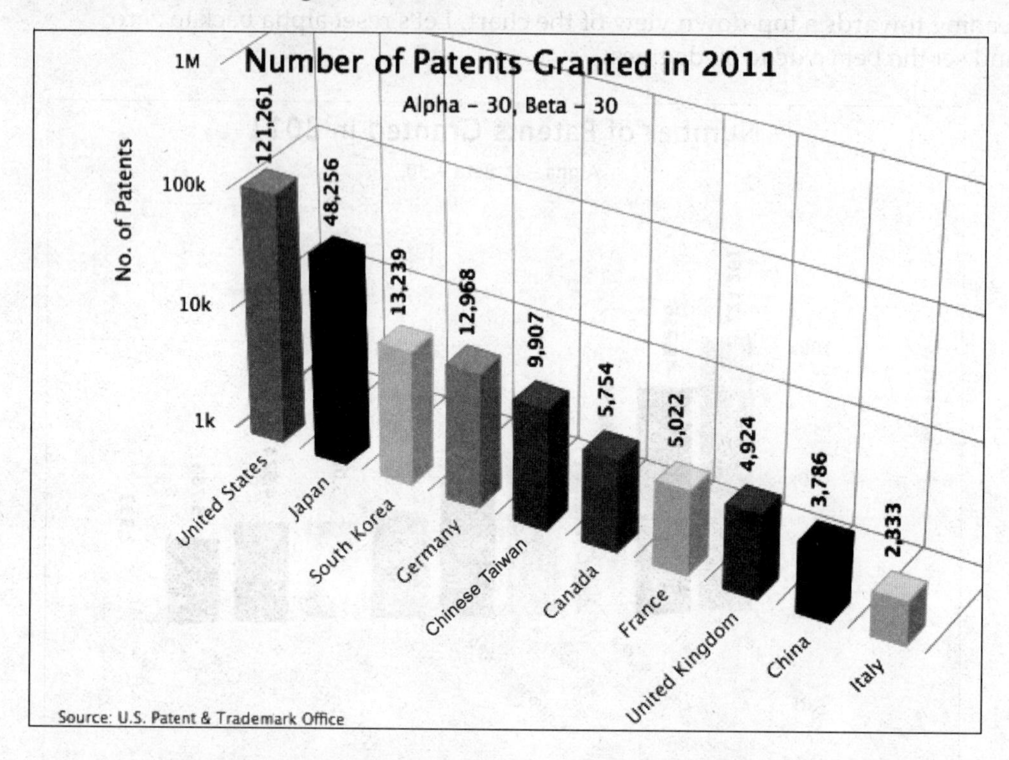

Notice that when we rotate a 3D chart, the plot area can interfere with other components, such as chart title and spacing on each side. In this case, we need to use the margin settings to manually accommodate the chart properly, depending on the size and view angle. Here is an example:

```
var chart = new Highcharts.Chart({
    chart: {
        ....,
        marginTop: 105,
        marginLeft: 35,
        spacingBottom: 25
    }
    ....
```

The depth and view distance

Apart from alpha and beta options, we can use the depth option to control the distance between the data display and the 3D background. Let's set the depth option to be exactly the same as the default depth of the 3D columns, which is 25 pixels deep. We also specify the viewDistance option to the default value, 100 (at the time of writing, changing the depth value also changes the viewDistance value), as follows:

```
options3d: {
    alpha: 30,
    beta: 30,
    enabled: true,
    viewDistance: 100,
    depth: 25
}
```

As we can see, the columns are backed up against the background frame:

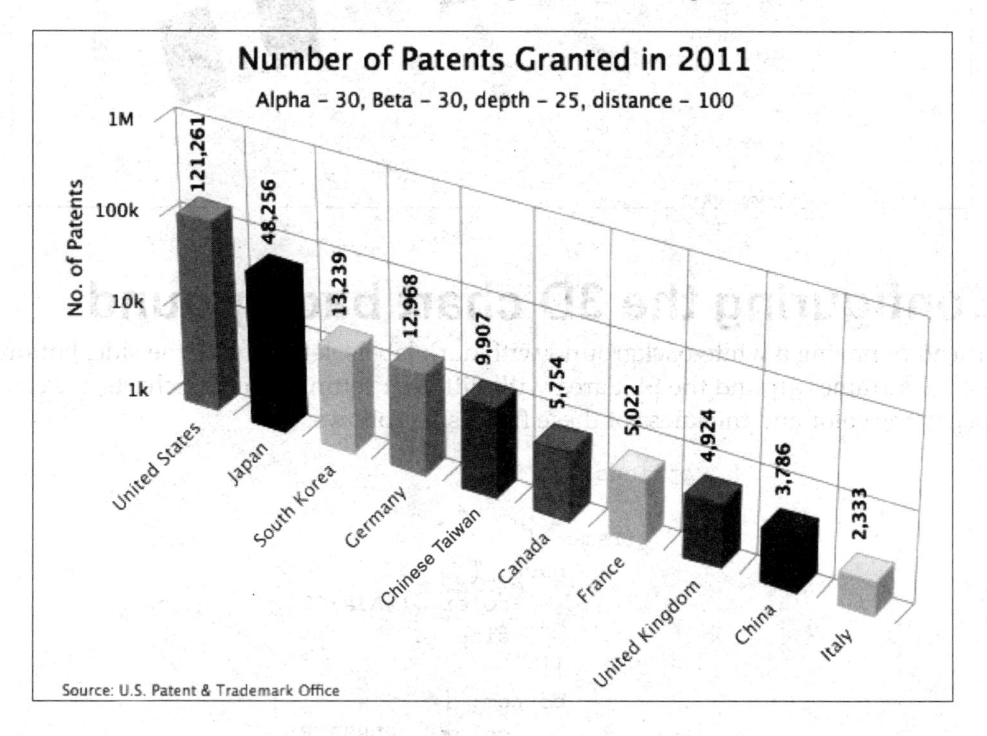

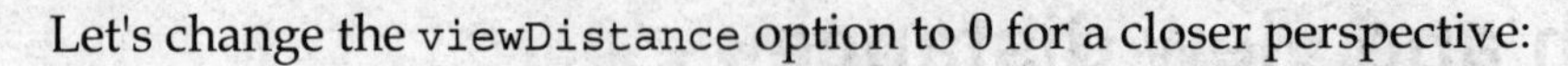

Let's change the `viewDistance` option to 0 for a closer perspective:

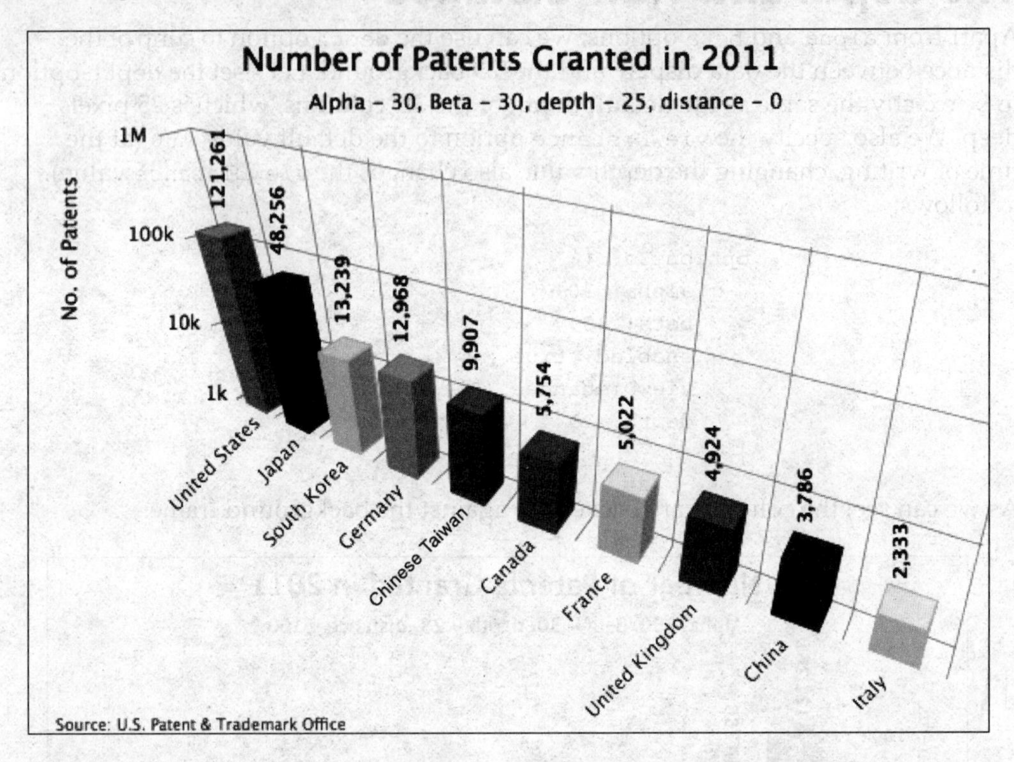

Configuring the 3D chart background

Instead of having a white background with dark labels, let's replace the side, bottom, and back frames around the plot area with different settings. In Highcharts, we can specify the color and thickness of these frames, as follows:

```
options3d: {
    ....,
    frame: {
        back: {
            color: '#A3A3C2',
            size: 4
        },
        bottom: {
            color: '#DBB8FF',
            size: 10
        },
```

```
                side: {
                    color: '#8099E6',
                    size: 2
                }
            }
        }
```

Meanwhile, we also change the data label color to white which contrasts nicely with the new background color:

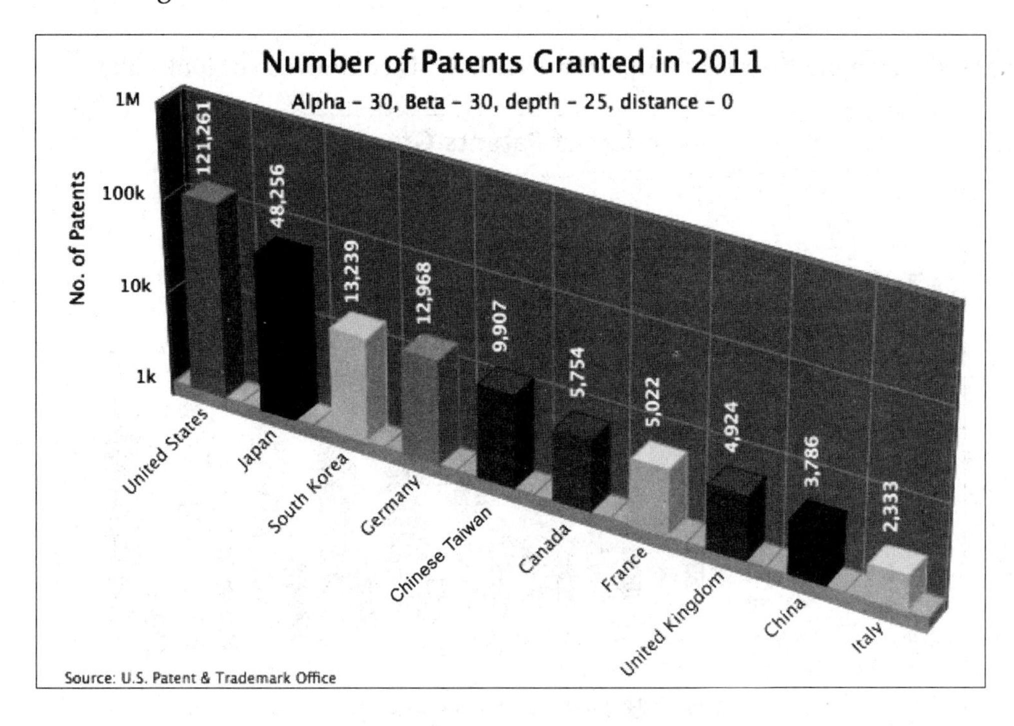

Note that if we apply a background image to the 3D chart, Highcharts doesn't automatically angle the image along the background.

Plotting the column, pie, donut, and scatter series in 3D charts

In this section, we will plot a gallery of column, pie, donut, and scatter series which are currently supported for 3D presentation. We will also examine the 3D options specific to each series. Some of the examples used here are taken from previous chapters.

3D columns in stacked and multiple series

Let's start with a multi-series stacked column chart embedded with the `options3d` setting:

```
options3d: {
    alpha: 10,
    beta: 30,
    enabled: true
}
```

Here is what a multi-series grouped and stacked 3D columns chart looks like:

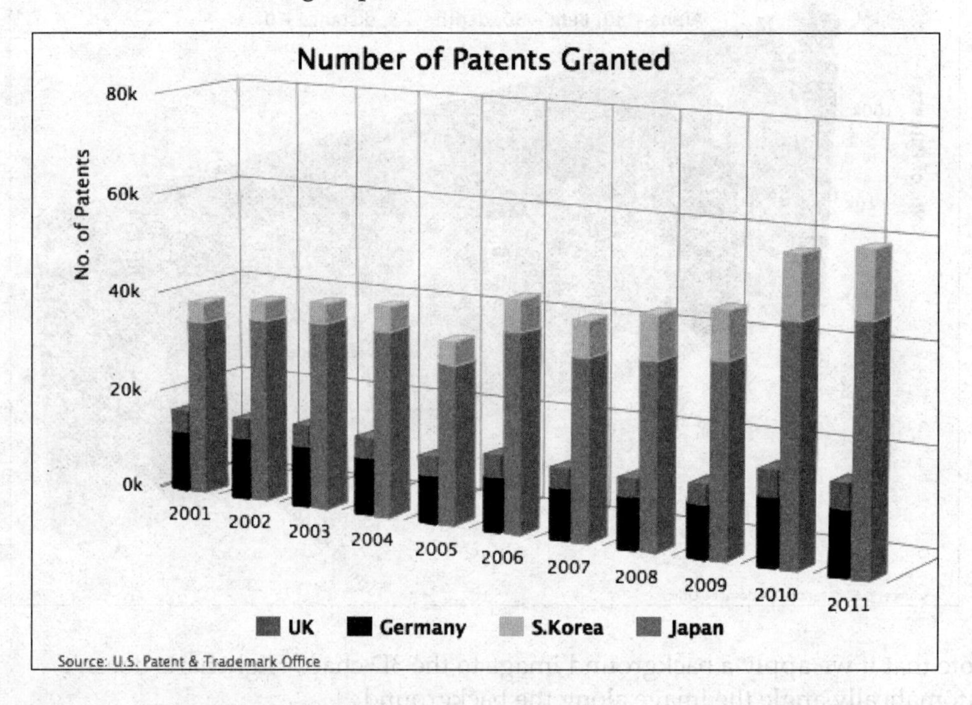

Column depth and Z-padding

Notice from the preceding chart, the sides of the UK/Germany stacked columns are covered by the S. Korea/Japan columns. Suppose that we want to show part of the sides of the UK/Germany columns. In order to do that, we can reduce the thickness of the S. Korea/Japan columns with the `plotOptions.column.depth` option, such as:

```
series: [{
    name: 'UK',
    data:  ....,
```

```
            stack: 'Europe'
        }, {
            name: 'Germany',
            data:  ....,
           stack: 'Europe'
        }, {
            name: 'S.Korea',
            data:  ....,
            stack: 'Asia',
            depth: 12
        }, {
            name: 'Japan',
            data: ....,
            stack: 'Asia',
            depth: 12
        }]
```

Here, we reduce the thickness of the S. Korea/Japan columns to 12 pixels which is around half the depth of the default value. The following chart shows one group of columns thinner than the other:

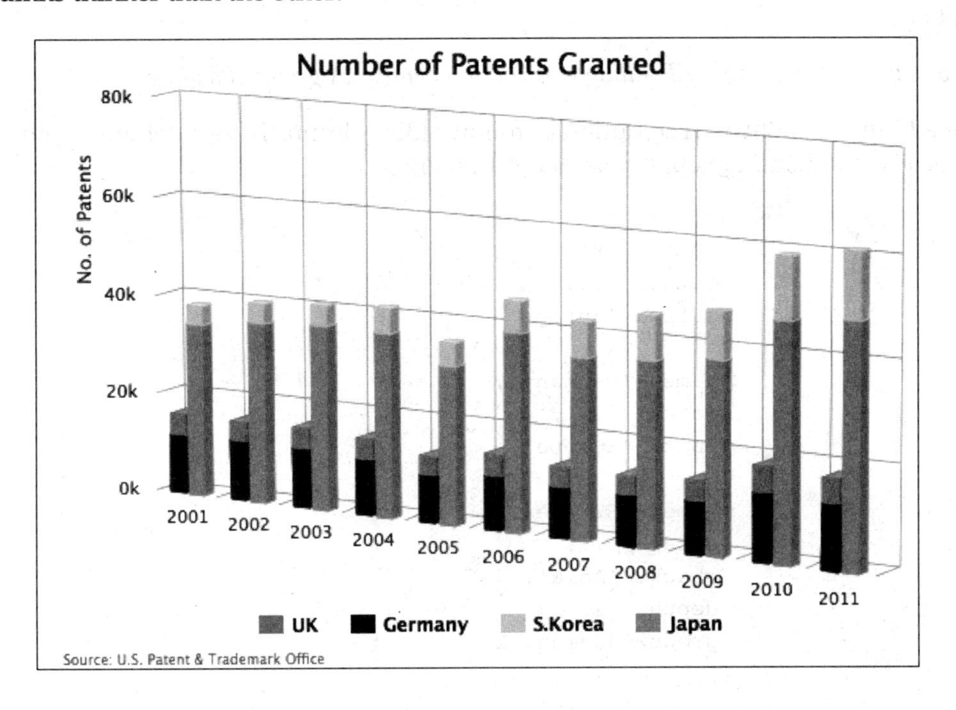

Notice that even when we have reduced the depth of one group of columns, the sides of the UK/Germany columns are still blocked. This is because the surface of the 3D columns are by default aligned along each other in Highcharts. To change that behavior, we use the `groupZPadding` option. In order to understand the concept of the `groupZPadding` option, it is best to illustrate it with a top-down view of the 3D columns in multiple series:

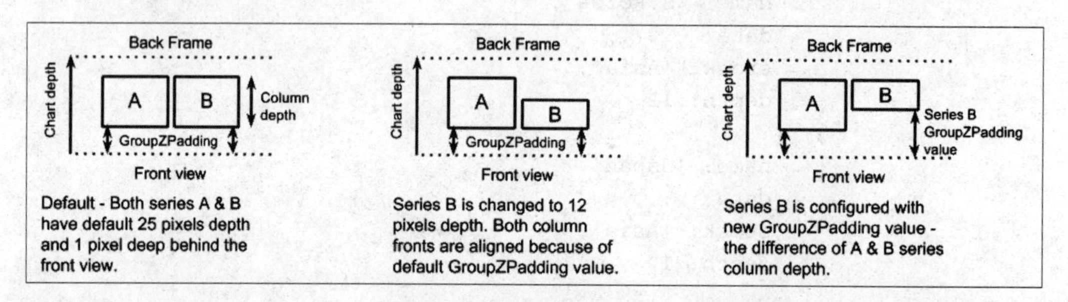

The `groupZPadding` option is the distance between the front view and the face of the column. In order to have the back of both series columns in line with each other, we need to increase the `groupZPadding` value for series B which has reduced depth, that is:

B column depth + B groupZPadding = A column depth + A groupZPadding

We add the `groupZPadding` option as follows (3D columns have a default depth of 25 pixels and default `groupZPadding` of 1 pixel):

```
series: [{
     name: 'UK',
     data:  ....,
     stack: 'Europe',
  }, {
     name: 'Germany',
     data:  ....,
     stack: 'Europe',
  }, {
     name: 'S.Korea',
     data:  ....,
     stack: 'Asia',
     depth: 12,
     groupZPadding: 14
  }, {
     name: 'Japan',
     data: ....,
```

```
            stack: 'Asia',
            depth: 12,
            groupZPadding: 14
        }]
```

As expected, the S. Korea/Japan columns are pushed back and show the column sides of the other series:

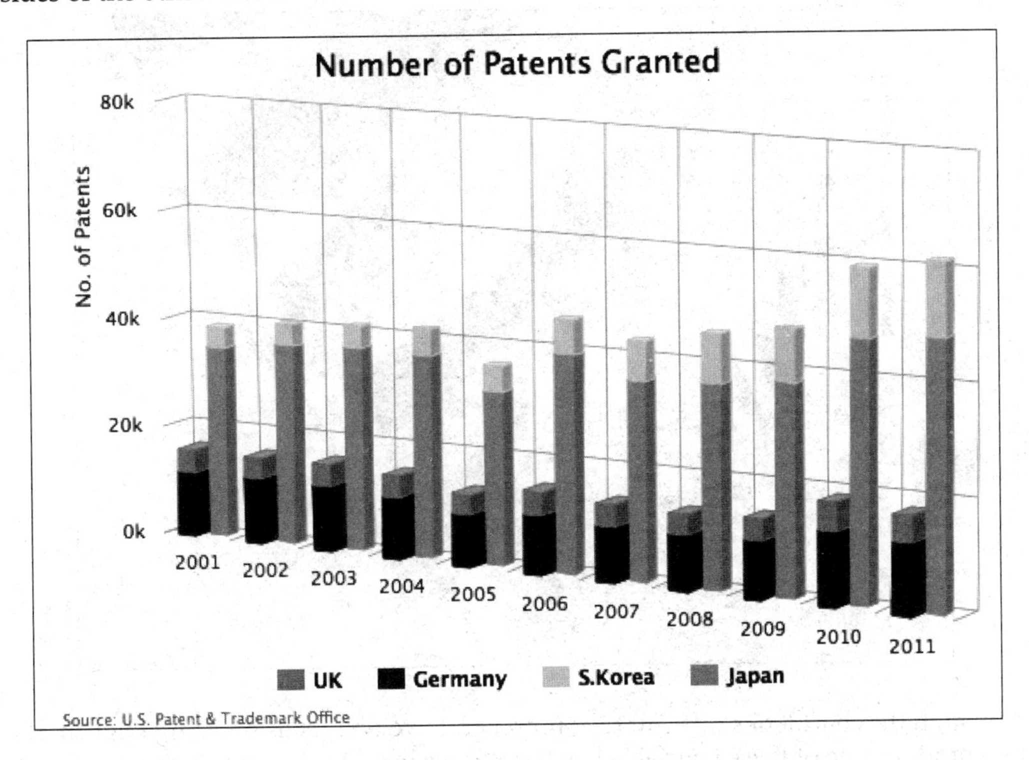

Plotting the infographic 3D columns chart

Let's see whether we can use Highcharts to plot an infographic style 3D chart. In this section, we will use the infographics designed by Arno Ghelfi and published in the Wired magazine, Geekiness at Any Price (See `http://starno.com/client/wired/#geekiness-at-any-price-wired`). Here is part of the infographics poster:

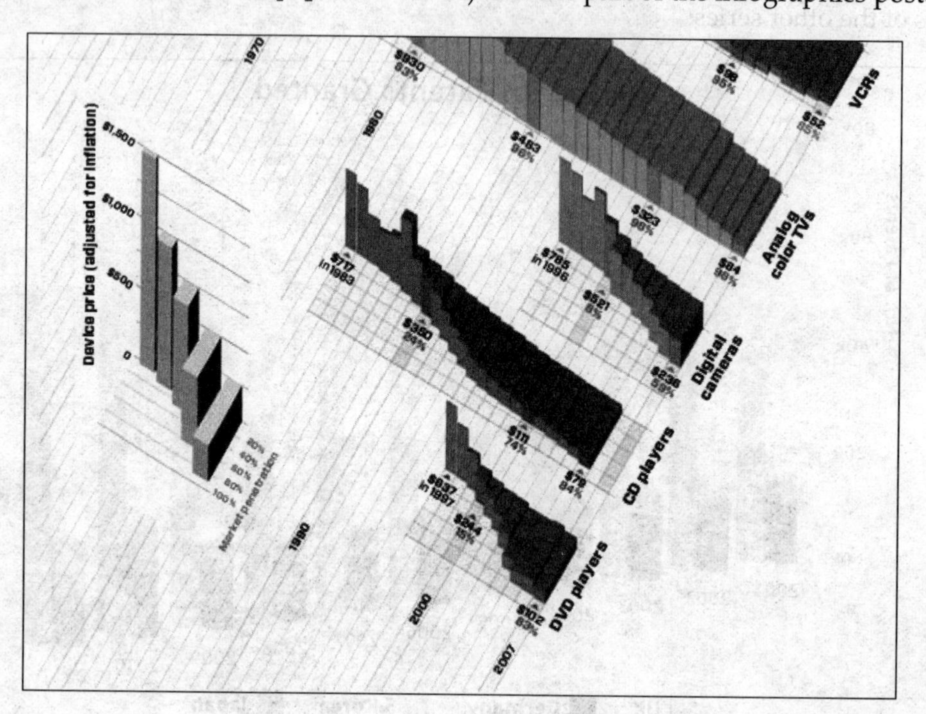

Although the chart looks difficult to comprehend, we will focus on using Highcharts to reproduce one of these special 3D style columns here. From the previous exercise, we know that we can set up the columns with increasing depth, or with decreasing `groupZPadding` values to achieve the same effect. This is because the depth and `groupZPadding` options are from the `plotOptions.column` configuration which is designed on a per-series basis. Therefore, we need to put each item of data as a separate series, in order to have columns with various depths and `groupZPadding` values, like the following:

```
series: [{
    data: [ 1500 ],
    depth: 5,
    groupZPadding: 95
}, {
    data: [ 1300 ],
    depth: 10,
```

```
            groupZPadding: 90
        }, {
            data: [ 1100 ],
            depth: 15,
            groupZPadding: 85
        }, {
```

Second, the columns in the example chart have no space between them. We can accomplish the same result by setting the padding spaces to zero with the `groupPadding` and `pointPadding` options. Then, we set all the columns to have the same color with the color option. Without space and with the same color for the columns will make them look indistinguishable. Fortunately, there is another specific option for 3D columns, `edgeColor`, which is for the color along the column edges. Here is the outcome of the configuration:

```
        plotOptions: {
            column: {
                pointPadding: 0,
                groupPadding: 0,
                color: '#C5542D',
                edgeColor: '#953A20'
            }
        },
```

Here is our attempt to create an infographic style chart with Highcharts:

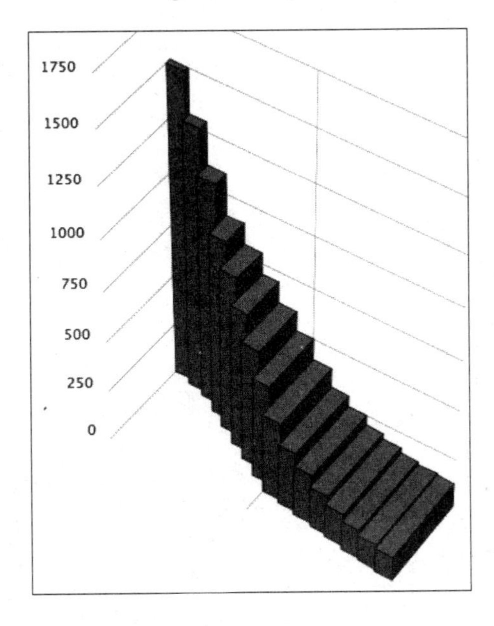

Plotting 3D pie and donut charts

Plotting 3D pie and donut charts follows the same principle as the column chart. Simply include the `options3d` configuration for the orientation. To control the thickness of the pie chart, we use `plotOptions.pie.depth` in a similar fashion.

Let's use some live data to plot a 3D pie and donut charts. First, we borrow some code from a Highcharts online demo (`http://www.highcharts.com/demo/pie-gradient`) which turns the Highcharts standard series colors into some gradient style colors. This automatically makes the chart more appealing with light shading:

```
Highcharts.getOptions().colors =
    Highcharts.map(Highcharts.getOptions().colors, function(color)
      {
            return {
              radialGradient: { cx: 0.5, cy: 0.3, r: 0.7 },
              stops: [
                  [0, color],
                  [1, Highcharts.Color(color).
                   brighten(-0.3).get('rgb')] // darken
              ]
          };
    });
```

In this pie chart, we set the pie `depth` and `alpha` rotation options to 50 and 55 respectively:

```
chart: {
    ....,
    options3d: {
        alpha: 55,
        beta: 0,
        enabled: true
    }
  },
....,
plotOptions: {
    pie: {
        center: [ '50%', '45%' ],
        depth: 50,
        slicedOffset: 40,
        startAngle: 30,
        dataLabels: {
            ....
        },
```

```
            size: "120%"
        }
    },
    series: [{
        type: 'pie',
        data: [ {
            name: 'Sleep',
            y: 8.74,
            sliced: true,
        }, {
            name: 'Working and work-related activities',
            y: 3.46
        }, {
            ....
        }]
    }]
```

As we can see, the `depth` option produces a rather thick 3D pie chart with nice color shading:

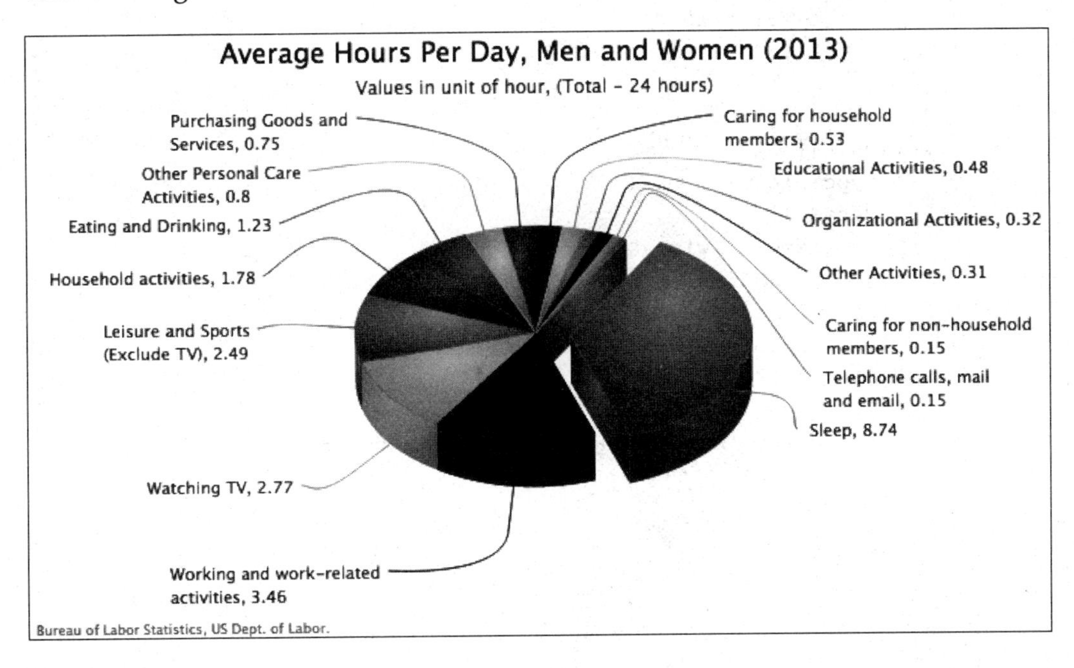

In the next donut chart, we set up all the sections to be separated with the `sliced` option set to `true`, and apply orientations:

```
chart: {
    ....
    options3d: {
        alpha: 50,
        beta: 40,
        enabled: true
    }
},
....,
plotOptions: {
    pie: {
        depth: 40,
        center: [ '50%', '44%' ],
        slicedOffset: 15,
        innerSize: '50%',
        startAngle: 270,
        size: "110%",
        ....
    }
},
series: [{
    type: 'pie',
    data: [{
        name: 'Swiss & UK',
        y: 790,
        sliced: true,
        // color, dataLabel decorations
        ....
    }, {
        name: 'Australia, U.S & South Africa',
        y: 401,
        ....
    }, {
        ....
    }]
}]
```

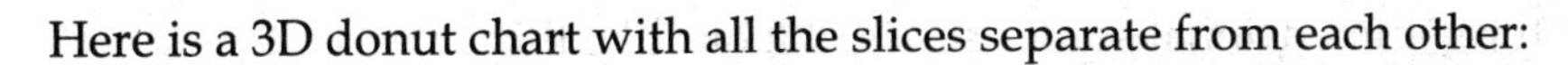

Here is a 3D donut chart with all the slices separate from each other:

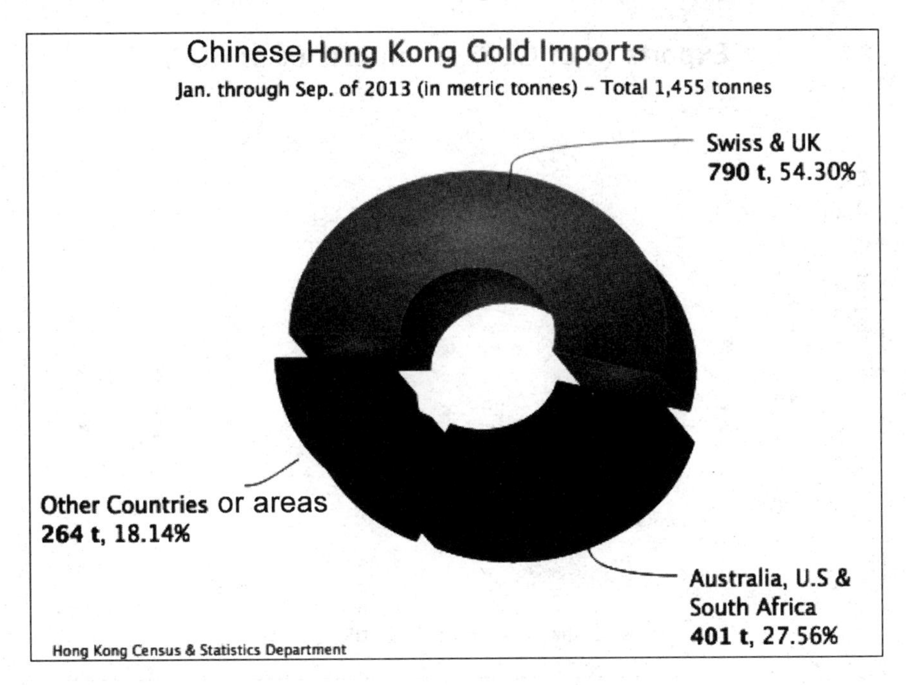

The 3D scatter plot

Highcharts supports real three-dimensional scatter series, although they are rather visually obscure. Despite this, we can use the `zAxis` option to define the third axis range. The chart doesn't display any details along the *z* axis like others. Let's construct the chart with some 3D data. In order to plot a three-dimensional scatter series, apart from the `options3d` configuration, we need to define the data series in an array of triplets:

```
series: [{
    name: 'China',
    data: [ [ 2000, 23.32, 20.91 ],
            [ 2001, 22.6, 20.48 ],
            [ 2002, 25.13, 22.56 ],
            ....,
```

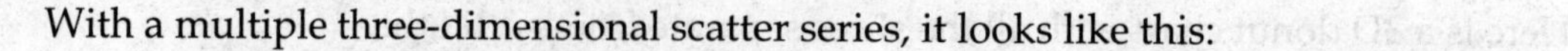

With a multiple three-dimensional scatter series, it looks like this:

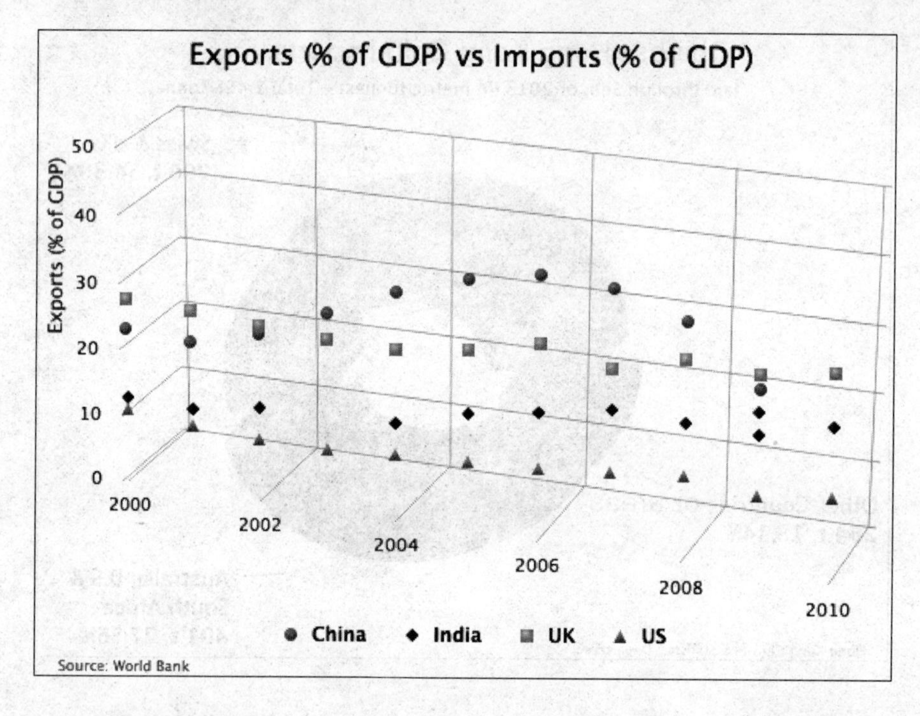

As we can see, it is difficult to make sense of the data points hanging in 3D space. Note that Highcharts currently doesn't show any intervals along the *z* axis. Let's add some navigation to assist the data visualization:

```
var orgColors = Highcharts.getOptions().colors;
// Apply color shading to Highcharts.getOptions().colors
....
chart: ....,
....,
tooltip: {
 crosshairs: [ { width: 2, color: '#B84DDB' },
               { width: 2, color: '#B84DDB' },
               { width: 2, color: '#B84DDB' } ],
 formatter: function() {
     var color = orgColors[this.series.index];
     return '<span style="color:' + color + '">\u25CF</span> ' +
            this.series.name + ' - In <b>' + this.point.x +
            '</b>: Exports: <b>' + this.point.y +
            '%</b>, Imports: <b>' + this.point.z + '%</b><br/>';
 },
 shape: 'square',
```

```
    positioner: function(width, height, point) {
        var x = chart.plotLeft + chart.plotWidth - width + 20;
        var y = 40;
        return { x: x, y: y };
    }
}
```

We change the tooltip into crosshairs for all dimensions. In order to avoid the tooltip blocking the crosshair lines, we fix the tooltip position to the top-right of the chart with the `positioner` option. Next, we use `formatter` to style the tooltip content. We borrow the tooltip formatting code from a Highcharts online demo (`http://www.highcharts.com/demo/3d-scatter-draggable`); the `\u25CF` is a dot Unicode symbol which is used as a bullet point with the series color. Since the tooltip is away from the hovered point, it is less confusing to have the tooltip in a rectangular shape rather than the default speech balloon style (`callout` in Highcharts). The following screenshot shows the chart with a hovered over crosshair:

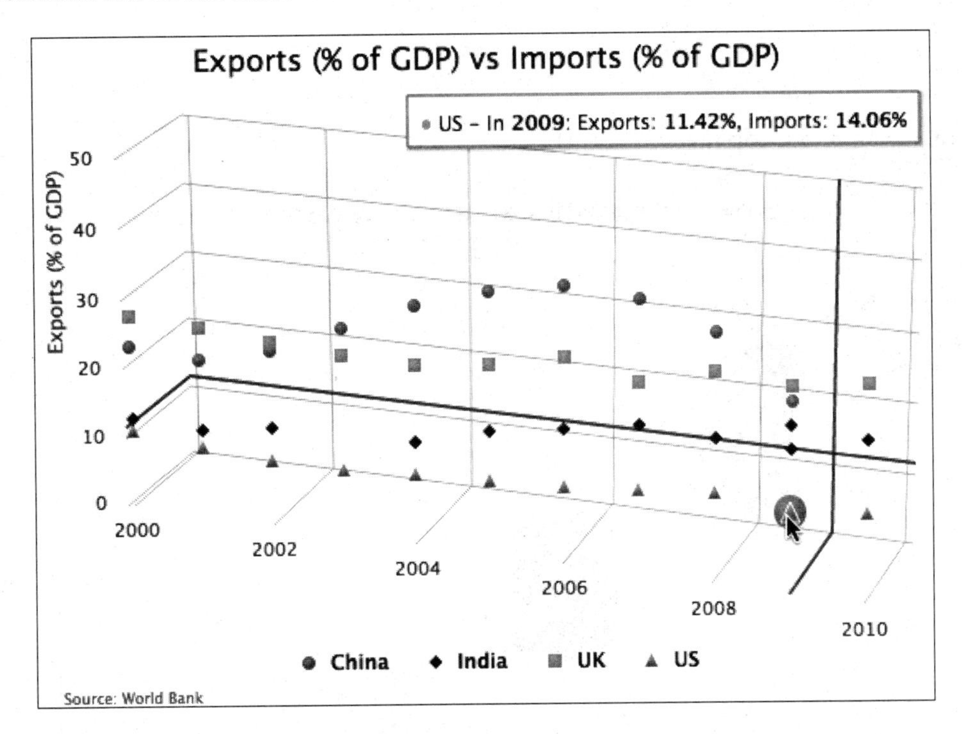

As we can see, even if we specify the crosshair in the z dimension, there is no crosshair line projected along the axis.

Navigating with 3D charts

In this section, we will investigate how you can interact with 3D charts. In *Chapter 2, Highcharts Configurations*, we have already explored the drill down feature and we will revisit it briefly in 3D charts. Another interaction specific to 3D charts is the impressive click-and-drag feature.

Drilldown 3D charts

Let's convert our previous drill down example into a 3D chart. First, we add the `options3d` option (and other positioning options):

```
options3d: {
    enabled: true,
    alpha: 25,
    beta: 30,
    depth: 30,
    viewDistance: 100
},
```

Here, we have our top level 3D chart:

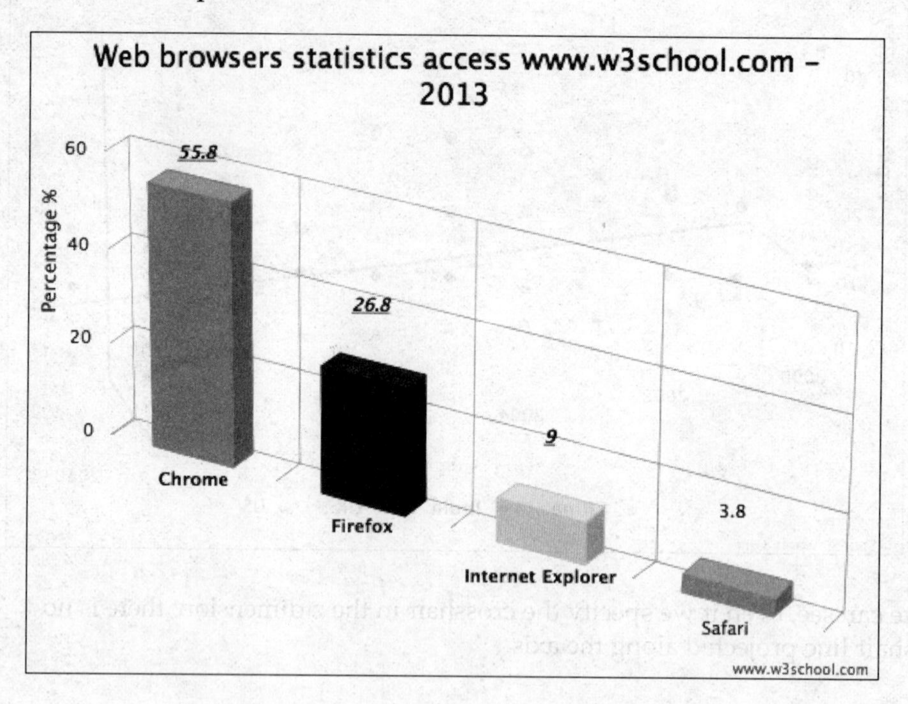

The next step is to improve our drill down pie chart in 3D with different orientation. Since the top level column is already configured in 3D, the drill down pie chart (defined in the `drilldown.series` option with matching `id` value) will also follow suit. However, we won't notice the pie in 3D because the default depth is 0. So, we need to set the thickness of the pie chart with the `depth` option:

```
plotOptions: {
    pie: {
       depth: 30,
       ....,
```

In order to set the drill down chart to a different rotation, we change the chart options in the `drilldown` event callback:

```
chart: {
    renderTo: 'container',
    ....,
    events: {
        drilldown: function(e) {
           ....
           if (e.point.name == 'Internet Explorer') {
               // Create the '9%' string in the center of
               // the donut chart
               pTxt = chart.renderer.text('9%',
                  (chart.plotWidth / 2) + chart.plotLeft - 25,
                     (chart.plotHeight / 2) + chart.plotTop + 25).
               css({
                  // font size, color and family
                  ....,
                  '-webkit-transform':
                        'perspective(600) rotateY(50deg)'
               }).add();

               chart.options.chart.options3d.alpha = 0;
               chart.options.chart.options3d.beta = 40;
           }
        },
        drillup: function() {
            // Revert to original orientation
            chart.options.chart.options3d.alpha = 25;
            chart.options.chart.options3d.beta = 30;
            ....
        }
    }
```

When the `drilldown` event is triggered from the `Internet Explorer` column, which zooms down to a donut chart, we set the chart with new `alpha` and `beta` orientations to 0 and 40 respectively. We use the CSS3 `-webkit-transform` setting on the `9%` sign to make it appear to have the same rotation as the donut chart. Finally, we reset the chart to its original orientation in the `drillup` callback, which is triggered when the user clicks on the **Back to ....** button. Here is the display of the zoom down donut chart:

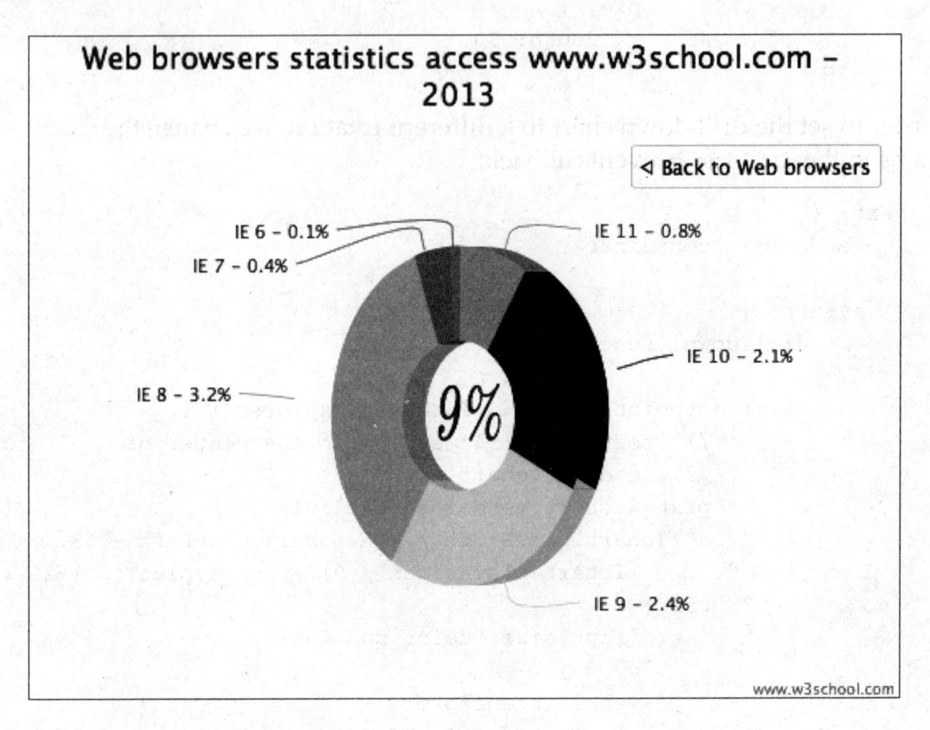

Click and drag 3D charts

Another impressive interaction with 3D charts is the click-and-drag function which we can use to drag the chart into any orientation. This interaction is actually from the Highcharts 3D scatter online demo, (`http://www.highcharts.com/demo/3d-scatter-draggable`). Here, we are going to explore how this is achieved. Before we do that, let's reuse our infographics example and copy the click-and-drag jQuery demo code into this exercise. The following is the click-and-drag code with minor modifications:

```
$(function () {
    $(document).ready(function() {
        document.title = "Highcharts " + Highcharts.version;
```

```
var chart = new Highcharts.Chart({
    ....
});
// Add mouse events for rotation
$(chart.container).bind('mousedown.hc touchstart.hc',
  function (e) {
    e = chart.pointer.normalize(e);

    var posX = e.pageX,
        posY = e.pageY,
        alpha = chart.options.chart.options3d.alpha,
        beta = chart.options.chart.options3d.beta,
        newAlpha,
        newBeta,
        sensitivity = 5; // lower is more sensitive

    $(document).bind({
        'mousemove.hc touchdrag.hc':
           function (e) {
              // Run beta
              newBeta =
                beta + (posX - e.pageX) / sensitivity;
              newBeta =
                Math.min(100, Math.max(-100,
                        newBeta));
              chart.options.chart.options3d.beta =
                newBeta;

            // Run alpha
            newAlpha =
               alpha + (e.pageY - posY) / sensitivity;
            newAlpha = Math.min(100, Math.max(-100,
                       newAlpha));
            chart.options.chart.options3d.alpha =
                newAlpha;

            // Update the alpha, beta and viewDistance
            // value in subtitle continuously
            var subtitle = "alpha: " +
                Highcharts.numberFormat(newAlpha, 1) +
                ", beta: " +
                Highcharts.numberFormat(newBeta, 1) +
                ", viewDistance: " +
```

```
                    Highcharts.numberFormat(
            chart.options.chart.options3d.viewDistance, 1);
                    chart.setTitle(null,
                        { text: subtitle }, false);

                    chart.redraw(false);
                },

                'mouseup touchend': function () {
                    $(document).unbind('.hc');
                }
            });
        });
```

The chart container's element is bound with the mousedown and touchstart events. The event names suffixed by '.hc' mean that the event handlers are grouped into the same name space, '.hc'. This is later used to unbind the event handlers declared under the name space.

So when the user performs a mousedown or touchstart event in the chart container, it executes the handler. The function first normalizes the event object for cross-browser event compatibility. Then it records the current pointer position values (pageX, pageY) as well as alpha and beta values under the mousedown event. Further under the mousedown or touchstart events, we bind an additional handler with the mousemove and touchdrag events in the same '.hc' name space. In other words, this means that under the mousedown or touchstart action, moving the mouse and dragging via touch will bind the second handler to the container.

The second handler implementation is to calculate the movement in x and y directions by comparing the current movement coordinates to the initial position recorded in the mousedown handler. Then, it transforms the scroll distance into orientation and updates the new alpha and beta values. The Math.max and Math.min expressions are to limit the alpha and beta into a range between -100 and 100. Note that the code doesn't restrict us to only using pageX/Y from the event object. We can use other similar properties such as screenX/Y or clientX/Y, as long as both handlers are referring to the same.

Finally, we call chart.draw(false) to redraw the chart with the new orientation but without animation. The reason for a lack of animation is that the movement handler is being called frequently with scroll action and animation, which will require extra overhead, degrading the responsiveness of the display.

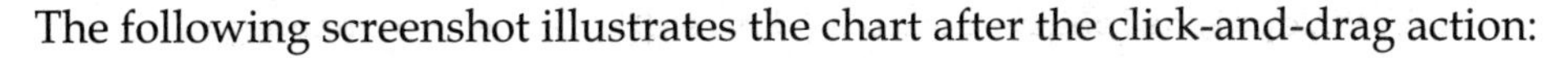

The following screenshot illustrates the chart after the click-and-drag action:

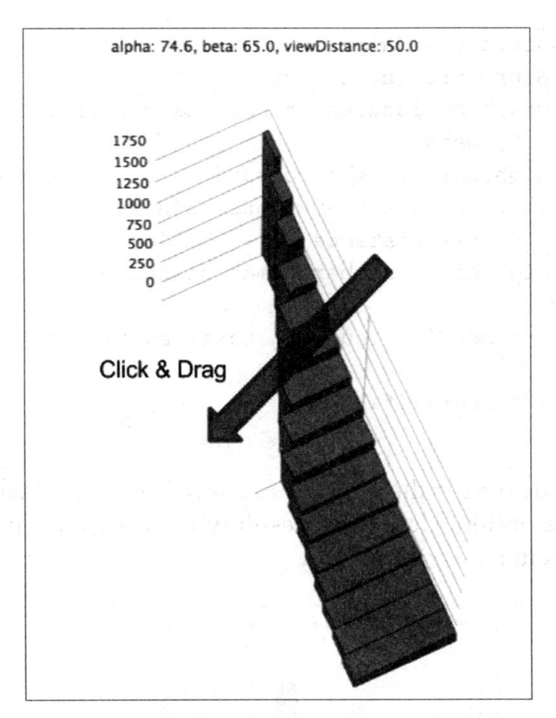

Mousewheel scroll and view distance

We can improve the experience by introducing another interaction, `mousewheel`, so that the view distance of the chart responds to the `mousewheel` actions. For the sake of a cross-browser compatibility solution, we use a jQuery mousewheel plugin by Brandon Aaron (`http://github.com/brandonaaron/jquery-mousewheel/`).

The following is the handler code:

```
// Add mouse events for zooming in and out view distance
$(chart.container).bind('mousewheel',
     function (e) {
         e = chart.pointer.normalize(e);
         var sensitivity = 10; // lower is more sensitive
         var distance =
             chart.options.chart.options3d.viewDistance;

         distance += e.deltaY / sensitivity;
         distance = Math.min(100, Math.max(1, distance));
```

```
            chart.options.chart.options3d.viewDistance = distance;

            var subtitle = "alpha: " +
                 Highcharts.numberFormat(
                 chart.options.chart.options3d.alpha, 1) +
                   ", beta: " +
                 Highcharts.numberFormat(
                chart.options.chart.options3d.beta, 1) +
                   ", viewDistance: " +
                 Highcharts.numberFormat(distance, 1);

              chart.setTitle(null, { text: subtitle }, false);

              chart.redraw(false);
        });
```

In a similar fashion, `deltaY` is the `mousewheel` scroll value and we apply the change to the `viewDistance` option. Here is the result when we apply click-and-drag and mousewheel scroll actions:

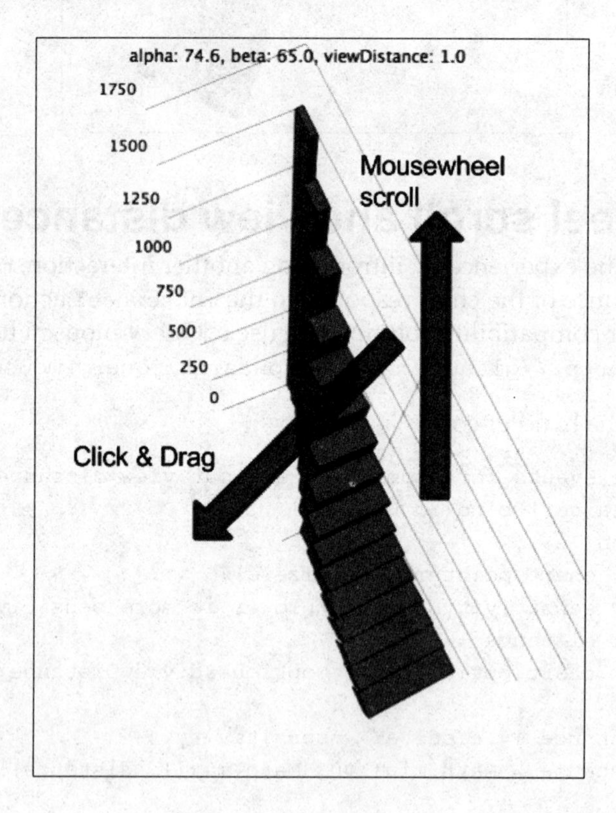

Summary

In this chapter, we learned how to create 3D column, scatter, and pie charts, and how to manipulate the orientation and configurations specific to each series. We tested these options by producing a gallery of 3D charts. On top of that, we explored a number of user interactions for 3D charts in Highcharts.

In the next chapter, we will explore the Highcharts APIs, which are responsible for making a dynamic chart, such as using Ajax queries to update the chart content, accessing components in Highcharts objects, and exporting charts to SVG.

10
Highcharts APIs

Highcharts offers a small set of APIs designed for plotting charts with dynamic interactions. In order to understand how the APIs work, we must first familiarize ourselves with the chart's internal objects and how they are organized inside a chart. In this chapter, we will learn about the chart class model and how to call the APIs by referencing the objects. Then we build a simple stock price application with PHP, jQuery, and jQuery UI to demonstrate the use of Highcharts APIs. After that, we turn our attention to four different ways of updating a series. We experiment with all the series update methods with the purpose of building an application to illustrate variations in visual effects, and the difference in CPU performance between them. Finally, we investigate performance when updating a series with popular web browsers, in terms of different sizes of datasets. In this chapter, we will cover the following topics:

- Understanding the Highcharts class model
- Getting data in Ajax and displaying new series with `Chart.addSeries`
- Displaying multiple series with simultaneous Ajax calls
- Using `Chart.getSVG` to format SVG data into an image file
- Using the `Chart.renderer` methods
- Exploring different methods to update series and their performance
- Experimenting with Highcharts performance with large datasets

Understanding the Highcharts class model

The relationship between Highcharts classes is very simple and obvious. A chart is composed of five different classes—`Chart`, `Axis`, `Series`, `Point`, and `Renderer`. Some of these classes contain an array of lower-level components and an object property to back-reference to a higher level-owner component. For example, the `Point` class has the `series` property pointing back to the owner `Series` class. Each class also has a set of methods for managing and displaying its own layer. In this chapter, we will focus on the set of APIs for dynamically modifying charts. The following class diagram describes the association between these classes:

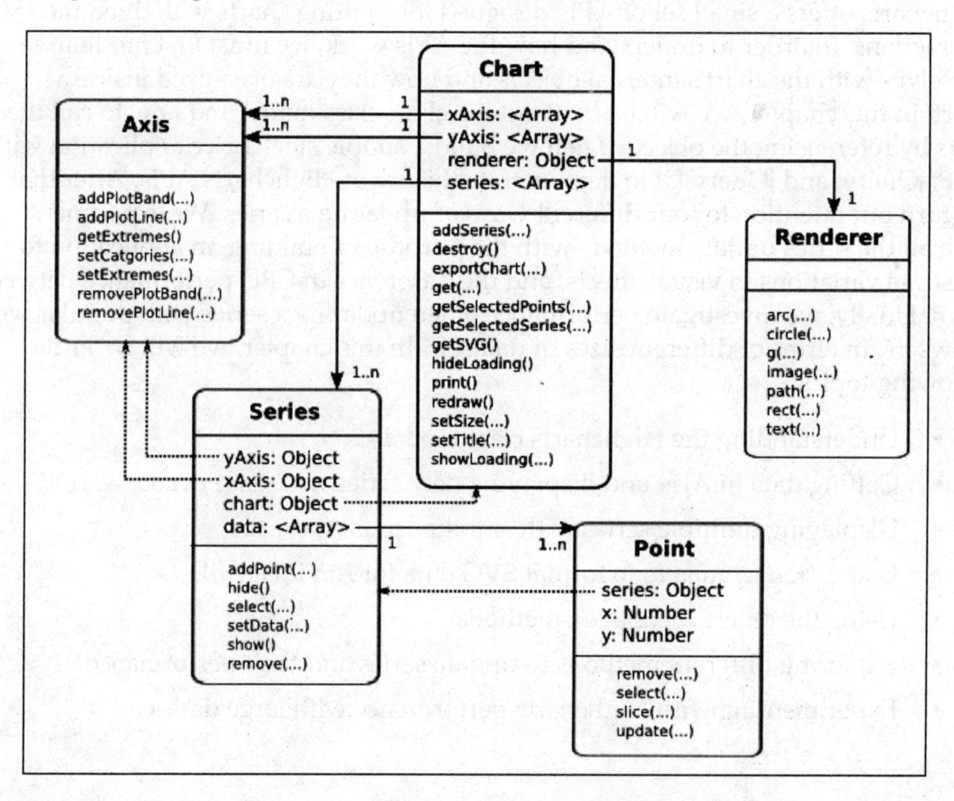

The `Chart` class is the top level class representing the whole chart object. It contains method calls to operate the chart as a whole—for example, exporting the chart into SVG or other image formats and setting the dimension of the chart. The `Chart` class has multiple arrays of `Axis` and `Series` objects; that is, a chart can have one or more *x* axes, *y* axes, and series. The `Renderer` class is a utility class that has a one-to-one relationship per chart and provides a common interface for drawing in SVG and VML-based browsers.

The `Series` class has an array of `Point` objects. The class has back-reference properties towards the `Chart` and `Axis` objects (see the dotted lines in the previous diagram) and provides functions for managing its list of `Point` objects. The `yAxis` and `xAxis` properties in the `Series` class are necessary, as a chart can have multiple axes.

The `Point` class is just a simple object containing x and y values and a back-reference to its series object (see the dotted line). The APIs are for managing the data points in the chart.

Highcharts constructor – Highcharts.Chart

Needless to say, the most important method in the APIs is the `Highcharts.Chart` method, with which we have seen plenty of action so far. However, there is more to this constructor call. `Highcarts.Chart` creates and returns a chart object but it also has a second optional parameter known as `callback`:

```
Chart(Object options, [ Function callback ])
```

The callback function is called when a chart is created and rendered. Inside the function, we can either call the component methods or access the properties inside the chart object. The newly created chart object is passed via the only callback function parameter. We can also use the '`this`' keyword inside the callback function, which also refers to the chart object. Instead of using the `Highcharts.Chart` callback parameter, we can achieve the same result by declaring our code inside the `chart.events.load` handler, which will be explored in the next chapter.

Navigating through Highcharts components

In order to use the Highcharts API, we must navigate to the right object inside the class hierarchy. There are several ways to traverse within the chart object: through the chart hierarchy model, retrieving the component directly with the `Chart.get` method, or a mixture of both.

Using the object hierarchy

Suppose that the chart object is created, as follows:

```
<script type="text/javascript">
  $(document).ready(function() {
    var chart = new Highcharts.Chart({
      chart: {
          renderTo: "container"
      },
      yAxis: [{
```

```
        min: 10,
        max: 30
      }, {
        min: 40,
        max: 60
      }],
      series: [{
        data: [ 10, 20 ]
      }, {
        data: [ 50, 70 ],
        yAxis: 1
      }],
      subtitle: {
          text: "Experiment Highcharts APIs"
      }

    });
  }, function() {
    ...
  });
</script>
```

We can then access the first position of the series' object with the index 0 from the chart. Inside the callback handler, we use the `this` keyword to refer to the chart object as follows:

```
var series = this.series[0];
```

Suppose there are two *y* axes in the configuration. To retrieve the second *y* axis, we can do as follows:

```
var yAxis = this.yAxis[1];
```

To retrieve the third data point object from the second series of the chart, type the following:

```
var point = this.series[1].data[2];
```

Supposing multiple charts are created on the page, a chart object can be accessed via the Highcharts namespace:

```
var chart = Highcharts.charts[0];
```

We can also retrieve the chart's container element through the container option:

```
var container = chart.container;
```

To examine the options structure of a created chart, use the options property:

```
// Get the chart subtitle

var subtitle = chart.options.subtitle.text;
```

Using the Chart.get method

Instead of cascading down the object hierarchy, we can directly retrieve the component using the `Chart.get` method (the `get` method is only available at the chart level, not in each component class). Assigning components with IDs will allow you to access them directly using the `get` method instead of traversing nodes in the object's hierarchical structure. The configuration uses the option key `id`, so this value must be unique.

Suppose we have created a chart with the following configuration code:

```
xAxis: {
  id: 'xAxis',
  categories: [ ... ]
},
series: [{
  name: 'UK',
  id: 'uk',
  data: [ 4351, 4190,
      { y: 4028, id: 'thirdPoint' },
      ... ]
}]
```

We can retrieve the components as follows:

```
var series = this.get('uk');
var point = this.get('thirdPoint');
var xAxis = this.get('xAxis');
```

If the `id` option is not previously configured, we can use the JavaScript `filter` method to search for the item based on the property:

```
this.series.filter(function(elt) {return elt.name == 'uk';})[0];
```

Using both the object hierarchy and the Chart.get method

It is cumbersome to define the `id` option for every component inside the chart. Alternatively, we can navigate through the components using both approaches, as follows:

```
var point = this.get('uk').data[2];
```

Using the Highcharts APIs

In this section, we will build an example using jQuery, jQuery UI, and Highcharts to explore each component's APIs. All the example code from here on will be using the object hierarchy to access chart components, that is, `chart.series[0].data[0]`. The user interface used here has a very minimal look and is far from perfect, as the main purpose of this exercise is to examine the Highcharts APIs.

First, let's see the usage of this user interface; then we will dissect the code to understand how the operations are performed. The following is the final UI screenshot that will be created in this section:

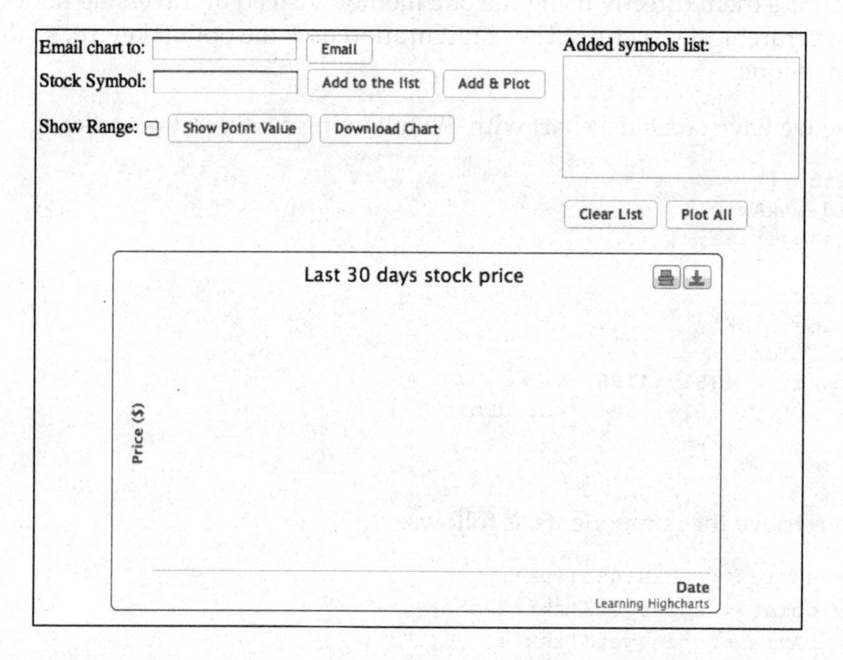

This is a simple web frontend for plotting the stock data chart for the past 30 days. The top part is a group of buttons for setting the stock symbols, getting the stock price, and retrieving the chart image by downloading it or via e-mail. The **Add to the list** button is for adding a stock symbol straight to the list without getting the stock prices and plotting the data. The **Plot All** button is for launching multiple stock price queries from the symbol list simultaneously, and to plot the data when all the results arrive. Alternatively, **Add & Plot** is a quick option for plotting a single stock symbol.

The bottom half contains a chart that we have already created. The chart is displayed with empty data and axes with titles (setting the `showAxes` option to `true`). The whole idea is to reuse the existing chart rather than recreating a new chart object every time new results arrive. Therefore, there is no flickering effect when the chart is destroyed and created and it appears as a smooth update animation. This also provides better performance without running extra code to regenerate the chart object.

This example is also available online at `http://www.joekuan.org/Learning_Highcharts/Chapter_10/Example_1.html`.

Chart configurations

The following is the chart configuration used for the example. Once a stock query is made, the server-side script will request the stock prices for the last 30 days and format the results into days in units of milliseconds. Hence, the *x* axis is configured as a `datetime` type with intervals on a daily basis:

```
var chart = new Highcharts.Chart({
  chart: {
    renderTo: 'container',
    showAxes: true,
    borderWidth: 1
  },
  title: { text: 'Last 30 days stock price' },
  credits: { text: 'Learning Highcharts' },
  xAxis: {
    type: 'datetime',
    tickInterval: 24 * 3600 * 1000,
    dateTimeLabelFormats: { day: '%Y-%m-%d' },
    title: {
      text: 'Date',
      align: 'high'
    },
    labels: {
      rotation: -45,
      align : 'center',
      step: 2,
      y: 40,
      x: -20
    }
  },
  yAxis: {
    title: { text: 'Price ($)' }
  },
```

```
    plotOptions: {
      line: { allowPointSelect: true }
    }
  });
```

Getting data in Ajax and displaying a new series with Chart.addSeries

Let's examine the action behind the **Add & Plot** button, defined as the following HTML syntax:

```
Stock Symbol: <input type=text id='symbol' />
<input type='button' value='Add to the list' id='addStockList' />
<input type='button' value='Add &amp; Plot' id='plotStock'>
...
Added symbols list:
<ol id='stocklist'>
</ol>
```

The jQuery code for the button action is listed, as follows:

```
$('#plotStock').button().click(
  function(evt) {

   // Get the input stock symbol, empty the
   // list andinsert the new symbol into the list
   $('#stocklist').empty();
   var symbol = $('#symbol').val();
        $('#stocklist').append($("<li/>").append(symbol));

   // Kick off the loading screen
   chart.showLoading("Getting stock data ....");

   // Launch the stock query
   $.getJSON('./stockQuery.php?symbol=' +
                symbol.toLowerCase(),
         function(stockData) {
            // parse JSON response here
            .....
         }
   );
  }
);
```

The previous code defines the event handler for the **Add & Plot** button's click event. First, it empties all the entries in the stock symbol list box that have IDs as `stocklist`. Then, it retrieves the stock symbol value from the input field `symbol` and appends the symbol to the list. The next step is to initiate a loading message screen on the chart by calling the `chart.showLoading` method. The following screenshot shows the loading message screen:

Last 30 days stock price

Getting stock data ....

The next call is to launch a jQuery Ajax call, `$.getJSON`, to query the stock price. The `stockQuery.php` server script (of course, any other server-side language can be used) does two tasks: it resolves the symbol into the full name of the organization, launches the symbol query from another website (`http://ichart.finance.yahoo.com/table.csv?s=BATS.L`) for the past stock price data, then packs the data into rows and encodes them into JSON format. The following is the code in the `stockQuery.php` file:

```
<?php
  $ch = curl_init();
  curl_setopt($ch, CURLOPT_RETURNTRANSFER, true);

  // Get the stock symbol name
  curl_setopt($ch, CURLOPT_URL, "http://download.finance.yahoo.com/d/
quotes.csv?s={$_GET['symbol']}&f=n");
  $result = curl_exec($ch);
  $name = trim(trim($result), '"');

  // Get from now to 30 days ago
  $now = time();
  $toDate = localtime($now, true);
  $toDate['tm_year'] += 1900;
  $fromDate = localtime($now - (86400 * 30), true);
  $fromDate['tm_year'] += 1900;
  $dateParams = "a={$fromDate['tm_mon']}&b={$fromDate['tm_
  mday']}&c={$fromDate['tm_year']}" ."&d={$toDate['tm_
  mday']}&e={$toDate['tm_mday']}&f={$toDate['tm_year']}";

  curl_setopt($ch, CURLOPT_URL, "http://ichart.finance.yahoo.com/
  table.csv?s={$_GET['symbol']}&{$dateParams}&g=d");
  $result = curl_exec($ch);
```

```
    curl_close($ch);

    // Remove the header row
    $lines = explode("\n", $result);
    array_shift($lines);

    $stockResult['rows'] = array();
    // Parse the result into dates and close value
    foreach((array) $lines as $ln) {
      if (!strlen(trim($ln))) {
        continue;
      }
      list($date, $o, $h, $l, $c, $v, $ac) =
        explode(",", $ln, 7);
      list($year, $month, $day) = explode('-', $date, 3);
      $tm = mktime(12, 0, 0, $month, $day, $year);
      $stockResult['rows'][] =
        array('date' => $tm * 1000,
            'price' => floatval($c));
    }

    $stockResult['name'] = $name;
    echo json_encode($stockResult);
  ?>
```

The following is the result returned from the server side in the JSON format:

```
  {"rows":[ {"date":1348138800000,"price":698.7},
        {"date":1348225200000,"price":700.09},
        ... ],
    "name": "Apple Inc."
  }
```

Once the JSON result arrives, the data is passed to the definition of the handler of `getJSON` and parsed into an array of rows. The following are the details of the handler code:

```
  $.getJSON('./stockQuery.php?symbol=' +
        symbol.toLowerCase(),
        function(stockData) {

            // Remove all the chart existing series
            while (chart.series.length) {
                chart.series[0].remove();
            }
```

```
            // Construct series data and add the series
            var seriesData = [];
            $.each(stockData.rows,
                function(idx, data) {
                    seriesData.push([ data.date, data.price ]);
                }
            );

            var seriesOpts = {
                name: stockData.name + ' - (' + symbol +')',
                data: seriesData,

                // This is to stop Highcharts rotating
                // the color and data point symbol for
                // the series
                color: chart.options.colors[0],
                marker: {
                    symbol: chart.options.symbols[0]
                }
            };

            chart.hideLoading();
            chart.addSeries(seriesOpts);
        }
    );
```

First of all, we remove all the existing series displayed in the chart by calling `Series.remove`. We then construct a series option with a data array of date (in UTC time) and price. We then remove the loading screen with `Chart.hideLoading` and display a new series with the `Chart.addSeries` methods. The only minor issue is that the default color and point marker for the series change when the series is reinserted; the internal indices in `chart.options.colors` and `chart.options.symbols` are incremented when a series is removed and added back to the chart. We can explicitly set the series color and point symbol to resolve this issue.

Alternatively, we can call `Series.setData` to achieve the same result but, once the name (subject) of a series is assigned and the series is created, it is not allowed to change. Therefore, we stick to `Chart.addSeries` and `Series.remove` in this example.

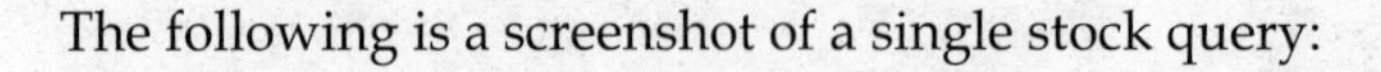

The following is a screenshot of a single stock query:

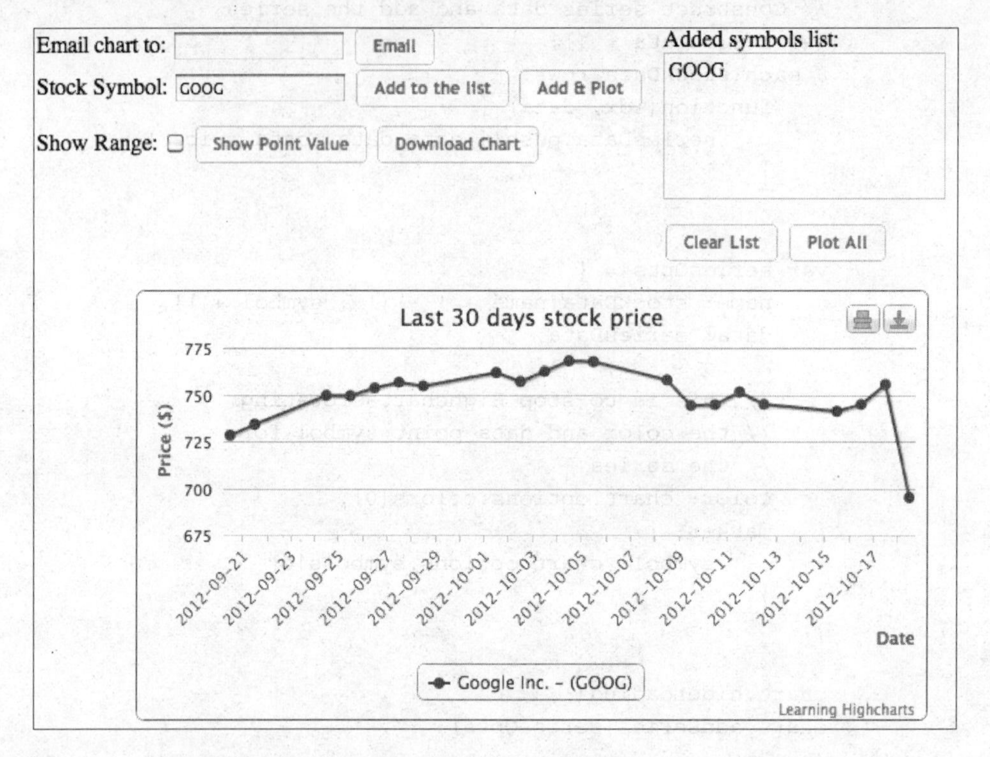

Displaying multiple series with simultaneous Ajax calls

The next part is to explore how to launch multiple Ajax queries simultaneously and plot series together when all the results have been returned. The implementation is pretty much the same as plotting a single stock query, except that we build up the series array option as we gather the results and plot them only when the last result arrives:

```
// Query all the stocks simultaneously and
// plot multipleseries in one go
$('#plotAll').button().click(

  function(evt) {

    // Kick off the loading screen
```

```
chart.showLoading("Getting multiple stock data ....");

// Get the list of stock symbols and launch
// the query foreach symbol
var total = $('#stocklist').children().length;

// start Ajax request for each of the items separately
  $.each($('#stocklist').children(),
  function(idx, item) {
   var symbol = $(item).text();
   $.getJSON('./stockQuery.php?symbol=' +
     symbol.toLowerCase(),
          function(stockData) {

      // data arrives, buildup the series array
      $.each(stockData.rows,
        function(idx, data) {
         $.histStock.push([ data.date,
                  data.price ]);
        }
      );

      seriesOpts.push({
       name: stockData.name + ' - (' +
          symbol +')',
       data: $.histStock,
       // This is to stop Highcharts
       // rotating the colorfor the series
       color: chart.options.colors[idx],
       marker: {
         symbol: chart.options.symbols[idx]
       }
      });

      // Plot the series if this result
      // is the last one
      if (seriesOpts.length == total) {

        // Remove all the chart existing series
        while (chart.series.length) {
          chart.series[0].remove()
        }

        chart.hideLoading();
```

```
            $.each(seriesOpts,
              function(idx, hcOption) {
               chart.addSeries(hcOption,
                        false);
              }
            );

            chart.redraw();
                } // else - do nothing,
          // not all results came yet
         } // function(stockData)
       ); // getJSON
    }); // $.each($('#stocklist')
  }); // on('click'
```

The second Boolean parameter of `Chart.addSeries`, `redraw`, is passed as `false`. Instead, we finalize all the updates in one single call, `Chart.redraw`, to save CPU time. The following is the screenshot for the multiple stock queries:

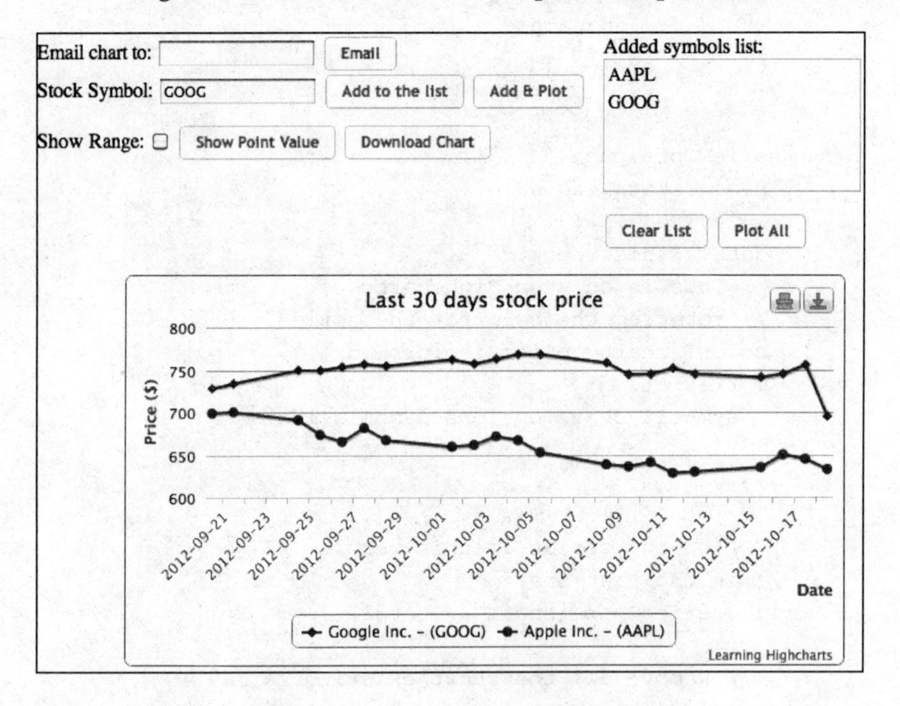

Extracting SVG data with Chart.getSVG

In this section, we will learn how to extract chart output and deliver it via e-mail or file download. Although we can rely on the exporting module and call the `exportChart` method to export the chart into the desired image format, it would be beneficial to see the whole process from formatting the original SVG content to creating an image file. After that, it is just a matter of calling different utilities to deliver the image file on the server side.

To extract the data underneath SVG from the displaying chart, the `getSVG` method is called, available when the exporting module is loaded. This method is similar to `exportChart`: it accepts the `chartOptions` parameter, which is used for applying configurations to the export chart output.

Here is the client-side jQuery code for handling both the **Download** and **Email** buttons.

Here, we use the `protocol` variable to specify the action for the chart and both buttons call the defined common function, `deliverChart`:

```
// Export chart into SVG and deliver it to the server
function deliverChart(chart, protocol, target) {

  // First extracts the SVG markup content from the
  // displayed chart
  var svg = chart.getSVG();

  // Send the whole SVG to the server and url
  $.post('./deliverChart.php', {
      svg: svg,
      protocol: protocol,
      target: target
    },
    function(result) {
      var message = null;
      var title = null;

      switch (protocol) {

        // Create a dialog box to show the
        // sent status
        case 'mailto':
          message = result.success ?
            'The mail has been sent successfully' :
            result.message;
```

```
            title = 'Email Chart';
            break;

          // Uses hidden frame to download the
          // image file created on the server side
          case 'file':
            // Only popup a message if error occurs
            if (result.success) {
              $('#hidden_iframe').attr("src",
                  "dlChart.php");
            } else {
              message = result.message;
              title = 'Download Chart';
            }
            break;
        }

        if (message) {
          var msgDialog = $('#dialog');
          msgDialog.dialog({ autoOpen: false,
            modal: true, title: title});
          msgDialog.text(message);
          msgDialog.dialog('open');
        }
    }, 'json');
}
```

The `deliverChart` method first calls the Highcharts API `getSVG` to extract the SVG content, then launches a `POST` call with both SVG data and action parameters. When `$.post` returns with a task status value, it shows a message dialog. As for the download chart, we create a hidden `<iframe>` to download the chart image file upon the successful return of the task status value.

The following is a simple server-side script for converting the SVG content and delivering the exported file:

```
<?php
$svg = $_POST['svg'];
$protocol = $_POST['protocol'];
$target = $_POST['target'];

function returnError($output) {
  $result['success'] = false;
  $result['error'] = implode("<BR/>", $output);
  echo json_encode($result);
```

```
  exit(1);
}

// Format the svg into an image file
file_put_contents("/tmp/chart.svg", $svg);
$cmd = "convert /tmp/chart.svg /tmp/chart.png";
exec($cmd, $output, $rc);
if ($rc) {
  returnError($output);
}

// Deliver the chart image file according to the url
if ($protocol == 'mailto') {

  $cmd = "EMAIL='{$target}' mutt -s 'Here is the chart' -a /tmp/chart.
png -- {$protocol}:{$target} <<.
Hope you like the chart
.";

  exec($cmd, $output, $rc);
  if ($rc) {
   returnError($output);
  }
  $result['success'] = true;

} else if ($protocol == 'file') {
  $result['success'] = true;
}

echo json_encode($result);
?>
```

The web server is running on a Linux platform (Ubuntu 12.04). As for the e-mail action, we use two command-line utilities to help us. The first is a fast image conversion tool, **convert**, that is part of the **ImageMagick** package (see the package website for more details at `http://www.imagemagick.org/script/index.php`). Inside the script, we save the SVG data from the `POST` parameter into a file and then run the convert tool to format it into a PNG image. The convert tool supports many other image formats and comes with a myriad of advanced features. Alternatively, we can use Batik to do a straightforward conversion by issuing the following command:

`java -jar batik-rasterizer.jar /tmp/chart.svg`

The given command also converts an SVG file and outputs `/tmp/chart.png` automatically. For the sake of implementing the e-mail feature quickly, we will launch an e-mail tool, **mutt** (see the package website for more details at `http://www.mutt.org`), instead of using the PHP mail extension. Once the PNG image file is created, we use mutt to send it as an attachment and use a heredoc to specify the message body.

> A **heredoc** is a quick way of inputting strings in a Unix command line with new lines and white spaces. See `http://en.wikipedia.org/wiki/Here_document`.

The following is the screenshot of the e-mail that is sent:

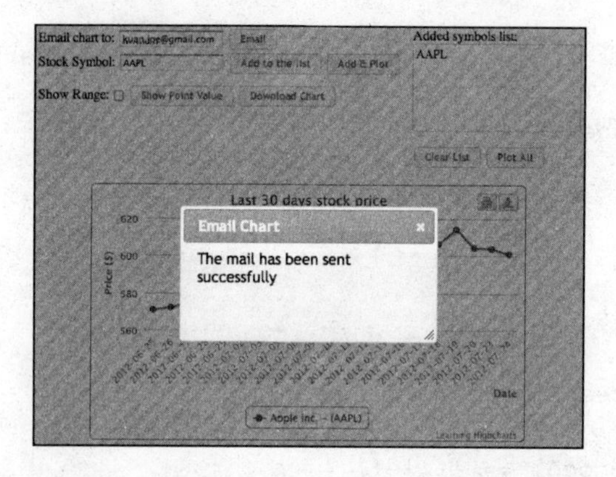

The following is the screenshot of the attachment e-mail that arrived in my e-mail account:

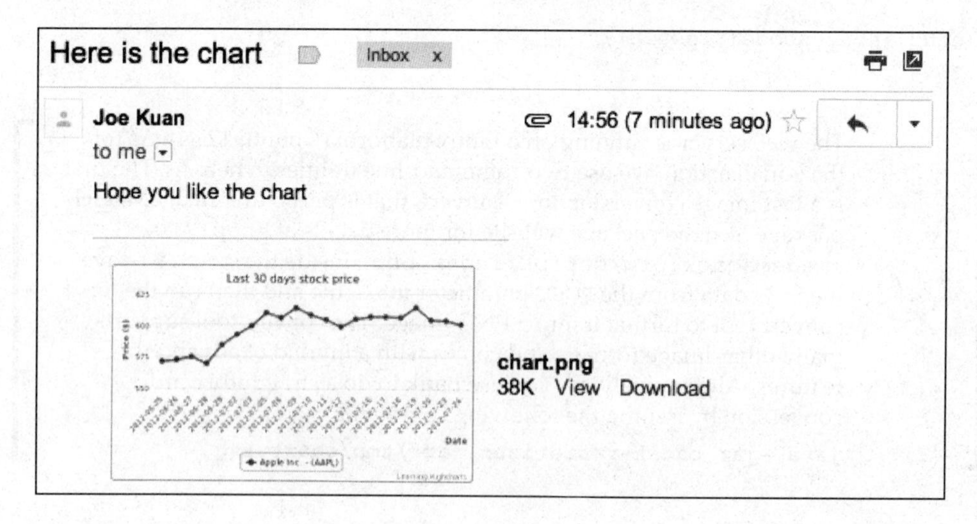

Selecting data points and adding plot lines

The next part is to implement the **Show Range** checkbox and the **Show Point Value** button. The **Show Range** option displays plot lines along the highest and lowest points in the chart, whereas **Show Point Value** displays a box with the value at the bottom left-hand side if a point is selected. The following screenshot demonstrates how both are enabled in the chart:

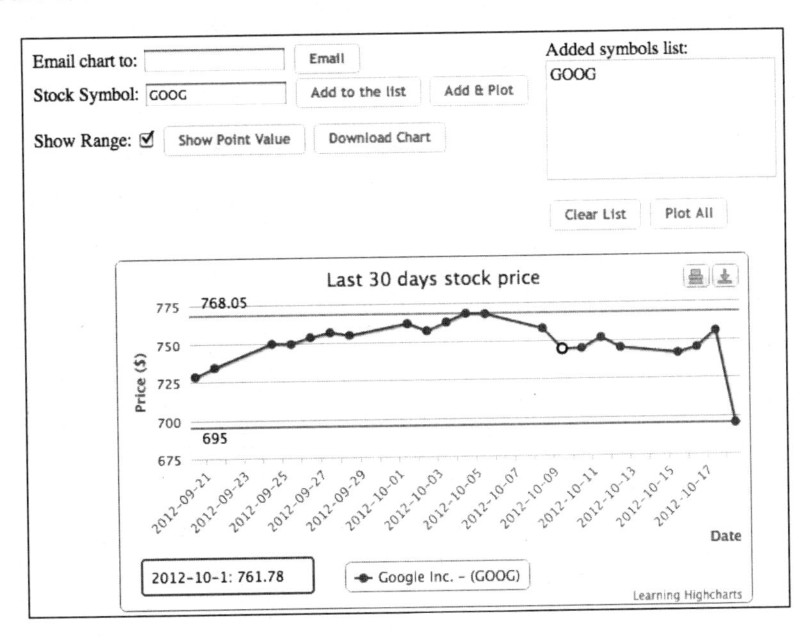

Although it is more natural for the **Show Point Value** checkbox to show the selected point display, this will become a callback implementation to every point select event. Instead, we use a button here, so that we can directly call the `Chart.getSelectedPoints` method.

Using Axis.getExtremes and Axis.addPlotLine

The `Axis.getExtremes` method not only returns the axis current minimum and maximum range in display, but also the highest and the lowest values for the data points. Here, we use the method to combine with the `Axis.addPlotLine` function to add a pair of plot lines along the *y* axis. The `addPointLine` routine expects a plot line configuration.

In this example, we specify a data label as well as an `id` name, so that we can remove lines at both high and low ends when the **Show Range** option is unchecked or plot lines need to be redisplayed with a new value. The following is the code for the **Show Range** action:

```
// Show the highest and lowest range in the plotlines.
var showRange = function(chart, checked) {
  if (!chart.series || !chart.series.length) {
     return;
  }

  // Checked or not checked, we still need to remove
  // any existing plot lines first
  chart.yAxis[0].removePlotLine('highest');
  chart.yAxis[0].removePlotLine('lowest');

  if (!checked) {
    return;
  }

  // Checked - get the highest & lowest points
  var extremes = chart.yAxis[0].getExtremes();

  // Create plot lines for the highest & lowest points
  chart.yAxis[0].addPlotLine({
    width: 2,
    label: {
      text: extremes.dataMax,
      enabled: true,
      y: -7
    },
    value: extremes.dataMax,
    id: 'highest',
    zIndex: 2,
    dashStyle: 'dashed',
    color: '#33D685'
  });

  chart.yAxis[0].addPlotLine({
    width: 2,
    label: {
      text: extremes.dataMin,
      enabled: true,
      y: 13
```

```
      },
      value: extremes.dataMin,
      zIndex: 2,
      id: 'lowest',
      dashStyle: 'dashed',
      color: '#FF7373'
    });
  };
```

Using the Chart.getSelectedPoints and Chart.renderer methods

The **Show Point Value** button makes use of the `Chart.getSelectedPoints` method to retrieve the data point that is currently selected. Note that this method requires the series option `allowPointSelect` to be enabled in the first place. Once a data point is selected and the **Show Point Value** button is clicked, we use functions provided by the `Chart.renderer` method to draw a tooltip-like box showing the selected value. We can use the `Renderer.path` or `Renderer.rect` methods to draw the rounded box, then `Renderer.text` for the data value.

Highcharts also supports multiple data point selection, which can be done by clicking on the left mouse button while holding down the *Ctrl* key.

Additionally, we use the `Renderer.g` routine to group the SVG box and value string together and add the resulting group element into the chart. The reason for that is so that we can re-display the box with a new value by removing the old group object as a whole instead of each individual element:

```
$('#showPoint').button().click(function(evt) {
  // Remove the point info box if exists
  chart.infoBox && (chart.infoBox =
          chart.infoBox.destroy());

  // Display the point value box if a data point
  // is selected
  var selectedPoint = chart.getSelectedPoints();
  var r = chart.renderer;
  if (selectedPoint.length) {
    chart.infoBox = r.g();
    r.rect(20, 255, 150, 30, 3).attr({
      stroke: chart.options.colors[0],
      'stroke-width': 2,
```

```
            fill: 'white'
        }).add(chart.infoBox);

        // Convert selected point UTC value to date string
        var tm = new Date(selectedPoint[0].x);
        tm = tm.getFullYear() + '-' +
          (tm.getMonth() + 1) + '-' + tm.getDate();
        r.text(tm + ': ' + selectedPoint[0].y,
          28, 275).add(chart.infoBox);
        chart.infoBox.add();
      }
    });
```

Highcharts' `Renderer` class also comes with other methods to draw simple SVG shapes on the chart, such as `arc`, `circle`, `image`, `rect`, `text`, `g`, and `path`. For more advanced shapes, we can use the `path` method, which accepts the SVG path syntax and has limited support on VML paths. Moreover, the `Renderer` class can be used independently from a chart—that is, we can call methods of the `Renderer` class without creating a chart beforehand and add SVG contents to an HTML element:

```
var renderer = new Highcharts.Renderer($('#container')[0],
               200, 100);
```

This creates a `Renderer` object that allows us to create SVG elements inside the `container` element, with an area 200 pixels wide and 100 pixels high.

Exploring series update

The series update is one of the most frequent tasks performed in charts. In this section, we investigate it in high definition. In Highcharts, there are several approaches to updating a series. Generally, we can update a series from a series or data point level. Then, the update method itself can either be actually changing the value, or reinserting it. We will discuss each approach and create a comprehensive example to experiment with all the techniques.

In order to compare each approach, we will continue to use stock market data but we will change the user interface this time to enable replaying the historical stock price. The following is the screenshot of the example in action:

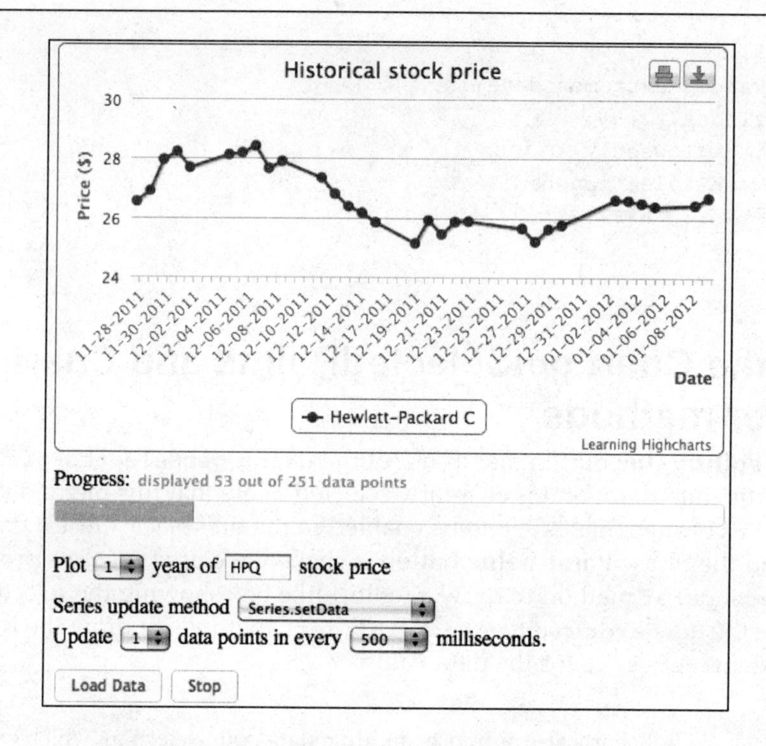

As we can see, there are multiple selection boxes to choose from: how many years of historical stock price to replay, how many data points to update in each iteration, and how long the wait is in between each update. Most importantly, we can choose which series update method should be used; it is interesting to observe the behavioral difference between them, especially during the whole replay. This demo is also available on my website at `http://www.joekuan.org/Learning_Highcharts/Chapter_10/Example_2.html`. I strongly recommend readers give it a go. Before we look into each update approach, let's find out how to construct this continuous series update process.

Continuous series update

Once we enter a stock symbol and select the number of years of stock prices to replay, we can click on the **Load Data** button to retrieve the price data. Once the data arrives, a confirmation dialog pops up with a **Start** button to kick-start the process. The following is the action code for the **Start** button:

```
// Create a named space to store the current user
// input field values and the timeout id
$.histStock = {};

$('#Start').button().click(function() {
```

```
    chart.showLoading("Loading stock price ... ");

    // Remove old timeout if exists
    $.histStock.timeoutID &&
      clearTimeout($.histStock.timeoutID);

    var symbol =
      encodeURIComponent($('#symbol').val().toLowerCase());
    var years = encodeURIComponent($('#years').val());

    // Remember current user settings and initialise values
    // for the run
    $.histStock = {
      // First loop start at the beginning
      offset: 0,
      // Number of data pts to display in each iteration
      numPoints: 30,
      // How long to wait in each iteration
      wait: parseInt($('#updateMs').val(), 10),
      // Which Highcharts method to update the series
      method: $('#update').val(),
      // How many data points to update in each iteration
      update: parseInt($('#updatePoints').val(), 10)
    };

    // Clean up old data points from the last run
    chart.series.length && chart.series[0].setData([]);

    // Start Ajax query to get the stock history
    $.getJSON('./histStock.php?symbol=' + symbol +
          '&years=' + years,
      function(stockData) {
        // Got the whole period of historical stock data
        $.histStock.name = stockData.name;
        $.histStock.data = stockData.rows;

        chart.hideLoading();
        // Start the chart refresh
        refreshSeries();
      }
    );
  })
```

We first create a variable, `histStock`, under the jQuery namespace, that is accessed by various parts within the demo. The `histStock` variable holds the current user's inputs and the reference to the refresh task. Any changes from the user interface update `$.histStock`, so the series update responds accordingly.

Basically, when the **Start** button is clicked, we initialize the `$.histStock` variable and start an Ajax query with the stock symbol and number-of-years parameters. Then, when the stock price data returns from the query, we store the result into the variable. We then call `refreshSeries`, which calls itself by the setting via a timer routine. The following code is the simplified version of the method:

```
var refreshSeries = function() {
  var i = 0, j;

  // Update the series data according to each approach
  switch ($.histStock.method) {
    case 'setData':
      ....
    break;
    case 'renewSeries':
      ....
    break;
    case 'update':
      ....
    break;
    case 'addPoint':
      ....
    break;
  }

  // Shift the offset for the next update
  $.histStock.offset += $.histStock.update;

  // Update the jQuery UI progress bar
  ....

  // Finished
  if (i == $.histStock.data.length) {
    return;
  }

  // Setup for the next loop
  $.histStock.timeoutID =
       setTimeout(refreshSeries, $.histStock.wait);
};
```

Inside `refreshSeries`, it inspects the settings inside the `$.histStock` variable and updates the series depending on the user's choice. Once the update is done, we increment the `offset` value, which is at the start position for copying the stock result data into the chart. If the counter variable `i` hits the end of the stock data, then it simply exits the method. Otherwise, it will call the JavaScript timer function to set up the next loop. The next goal is to review how each update method is performed.

Testing the performance of various Highcharts methods

There are four techniques for updating the series data: `Series.setData`, `Series.remove/Chart.addSeries`, `Point.update`, and `Series.addPoint`. We measure the performance for all four techniques in terms of CPU and memory usage with the Resource Monitor tool. Each method is timed when replaying the stock prices for the past year along with 0.5 seconds of waiting time between each update. We repeated the same run twice and recorded the average. The experiment is repeated on a selection of browsers: Firefox, Chrome, Internet Explorer 8 and 11, and Safari. Although IE 8 does not support SVG and only supports VML, it is important to use it in the experiment because Highcharts' implementation is compatible with IE 8. One thing that we instantly notice is the same chart on IE8 is not as appealing as in SVG.

> The whole experiment is running on a PC with Windows 7 Ultimate installed, and the hardware is 4GB RAM Core 2 Duo 3.06 GHz with an Intel G41 Graphics chipset.
>
> The browser versions are Firefox 31.0, Chrome 36.0.1985, IE11 11.0.9600, Safari 5.1.7, and IE8 8.0.6001. Safari may not be a true performance indicator as it is rather old for a PC platform.
>
> It is no longer possible to install/run IE8 on Windows 7, as Microsoft has discontinued support. Although we can set the user agent on IE11 to IE8 and conduct the experiment, it doesn't offer a true reflection of IE8 performance. Therefore, we set up another system running Windows XP with IE8 on identical hardware.

In the following sections, each series update approach is explained and a performance comparison is presented between the browsers. Readers must not use the result as a guide to the browser's general performance, which is derived from running a myriad of tests in a number of areas. What we are experimenting with here is simply how Highcharts performs on each browser in terms of SVG animations.

Note that the results are different compared to what was presented in the previous edition. This is mainly due to using a more up-to-date version of Highcharts: we use 4.0.3 in this experiment, whereas 2.2.24 is documented in the previous edition.

Applying a new set of data with Series.setData

We can apply a new set of data to an existing series using the `Series.setData` method:

```
setData (Array<Mixed> data, [Boolean redraw])
```

The data can be an array of one dimensional data, an array of x and y value pairs, or an array of data point objects. Note that this method is the simplest form of all the approaches, and doesn't provide any animation effects at all. Here is how we use the `setData` function in our example:

```
case 'setData':
  var data = [];

  // Building up the data array in the series option
  for (i = $.histStock.offset, j = 0;
     i < $.histStock.data.length &&
     j < $.histStock.numPoints; i++, j++) {
    data.push([
      $.histStock.data[i].date,
      $.histStock.data[i].price ]);
  }

  if (!chart.series.length) {

    // Insert the very first series
    chart.addSeries({
      name: $.histStock.name,
      data: data
    });
  } else {

    // Just update the series with
    // the new data array
    chart.series[0].setData(data, true);
  }
  break;
```

There are two sets of animations appearing in the chart: *x*-axis labels moving from the center of the chart and the data points in the series. Although the series scrolls smoothly, the movement of the *x*-axis labels appears too quickly and becomes choppy. The following graph shows the performance comparison when using the `setData` method across the browsers:

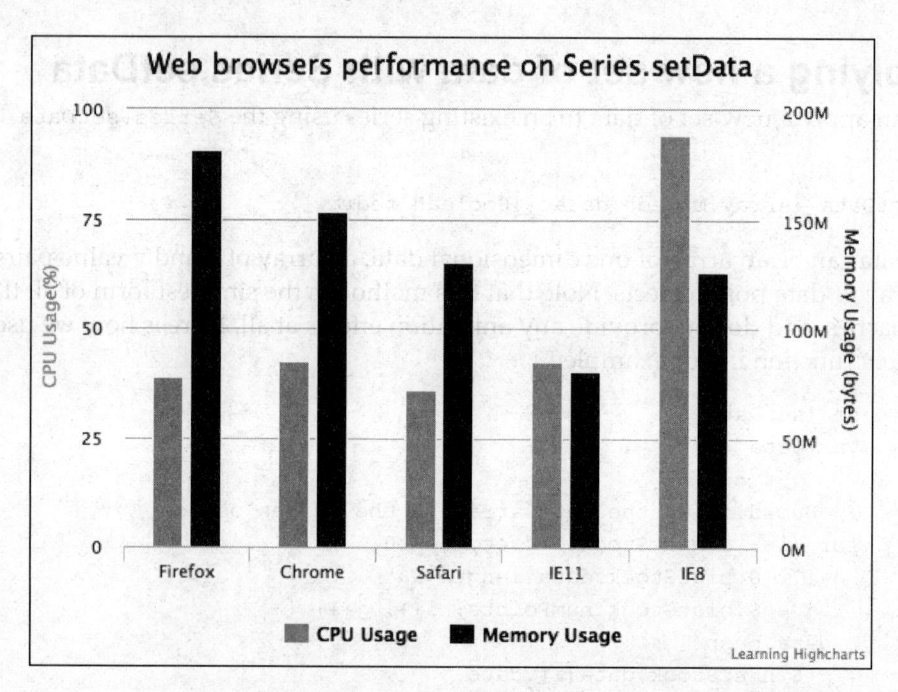

There are minor differences in terms of CPU usage, except for IE8 that runs on VML instead of SVG. IE8 consumed much higher CPU usage and took much longer to finish. The animation lagged throughout the experiment. Among the browsers, Safari is marginally the best. Out of all the browsers, Firefox has the highest memory footprint, whereas IE 11 had the smallest. Perhaps a slight surprise is that Safari has a better performance than Firefox and is also very close to Chrome.

Using Series.remove and Chart.addSeries to reinsert series with new data

Alternatively, we can remove the whole series with the `Series.remove` method, then rebuild the series options with the data and reinsert a new series using `Chart.addSeries`. The downside of this approach is that the internal index for the default colors and point symbols is incremented, as we found in the earlier example. We can compensate for that by specifying the color and the marker options. Here is the code for the `addSeries` method:

```
    case 'renewSeries':
      var data = [];
      for (i = $.histStock.offset, j = 0;
          i < $.histStock.data.length &&
          j < $.histStock.numPoints; i++, j++) {
        data.push([ $.histStock.data[i].date,
              $.histStock.data[i].price ]);
      }
      // Remove all the existing series
      if (chart.series.length) {
        chart.series[0].remove();
      }

      // Re-insert a new series with new data
      chart.addSeries({
        name: $.histStock.name,
        data: data,
        color: chart.options.colors[0],
        marker: {
          symbol: chart.options.symbols[0]
        }
      });
      break;
```

In this experiment, we use the refresh rate for every half-second, which is shorter than the time span of the default animation. Therefore the series update appears erratic without much animation, as in `setData`. However, if we change the refresh rate to 3 seconds or more, then we can see the series being redrawn from the left-hand to the right-hand side in each update. Unlike other methods, the x-axis labels are updated without any animations:

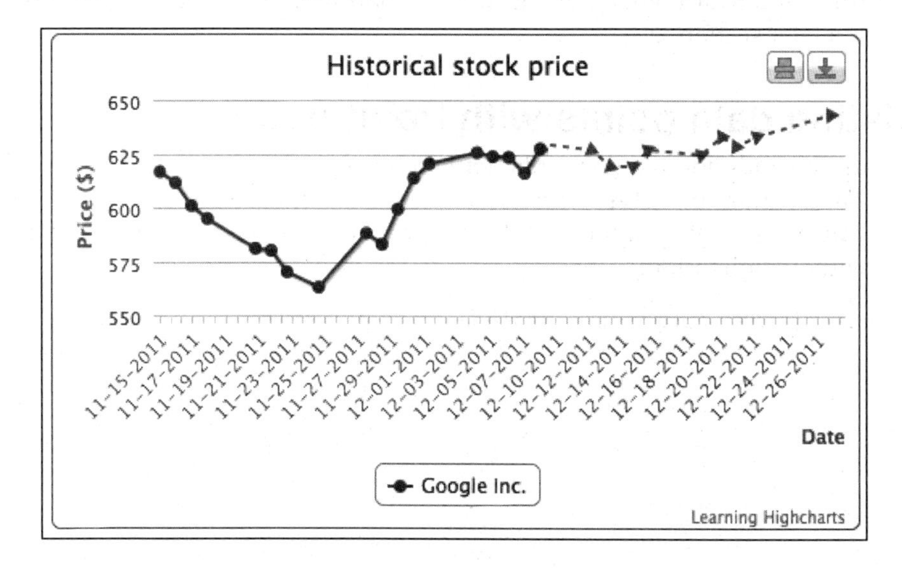

The following graph shows the performance comparison when using the `addSeries` method across the browsers:

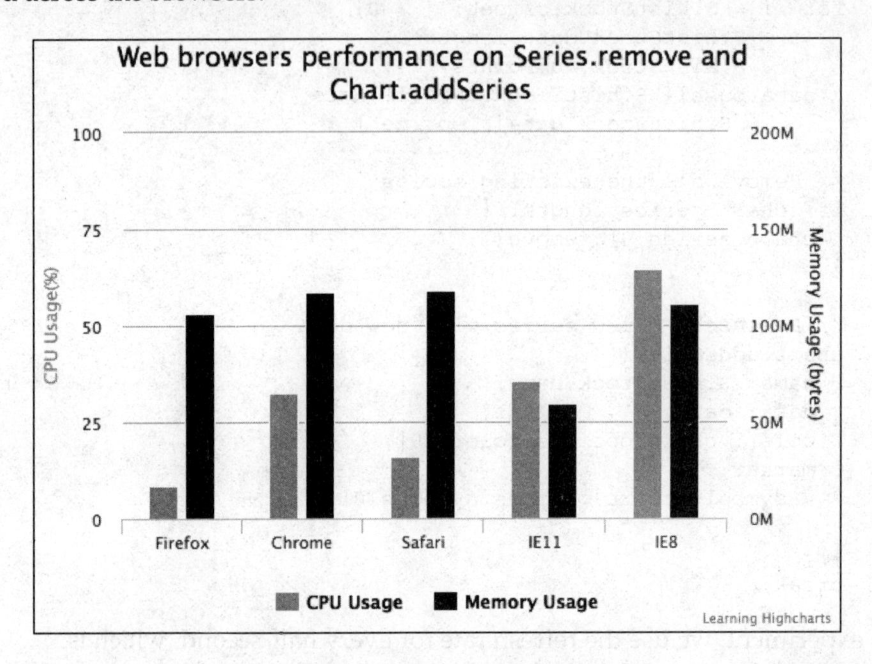

Since this approach seems to have the least animation, the CPU usage levels are relatively low across all the browsers, as is the memory usage. As expected, IE8 consumes the most resources. Next is IE11, which is roughly on a par with Chrome. The most unusual result is that Firefox requires significantly less CPU usage than both Chrome and Safari when there is little animation. We will investigate this further in a later section.

Updating data points with Point.update

We can update individual data points with the `Point.update` method. The update method has a similar prototype to `setData`, which accepts a single value, an array of x and y values, or a data point object. Each update call can be redrawn into the chart with or without animation:

```
update ([Mixed options], [Boolean redraw], [Mixed animation])
```

Here is how we use the `Point.update` method: we traverse through each point object and call its member function. In order to save CPU time, we set the `redraw` parameter to `false` and call `Chart.redraw` after the last data point is updated:

```
case 'update':
  // Note: Series can be already existed
  // at start if we click 'Stop' and 'Start'
  // again
  if (!chart.series.length ||
      !chart.series[0].points.length) {
    // Build up the first series
    var data = [];
    for (i = $.histStock.offset, j = 0;
       i < $.histStock.data.length &&
       j < $.histStock.numPoints; i++, j++) {
      data.push([
        $.histStock.data[i].date,
        $.histStock.data[i].price ]);
    }

    if (!chart.series.length) {
      chart.addSeries({
        name: $.histStock.name,
        data: data
      });
    } else {
      chart.series[0].setData(data);
    }

  } else {
    // Updating each point
    for (i = $.histStock.offset, j = 0;
       i < $.histStock.data.length &&
       j < $.histStock.numPoints; i++, j++) {
      chart.series[0].points[j].update([
        $.histStock.data[i].date,
        $.histStock.data[i].price ],
      false);
    }
    chart.redraw();
  }
  break;
```

`Point.update` animates each data point vertically. It gives a wavy effect overall as the graph is progressively updated. In the same way as the `setData` method, the labels approach the x-axis line diagonally. The following graph shows the performance comparison of the `Point.update` method across the browsers:

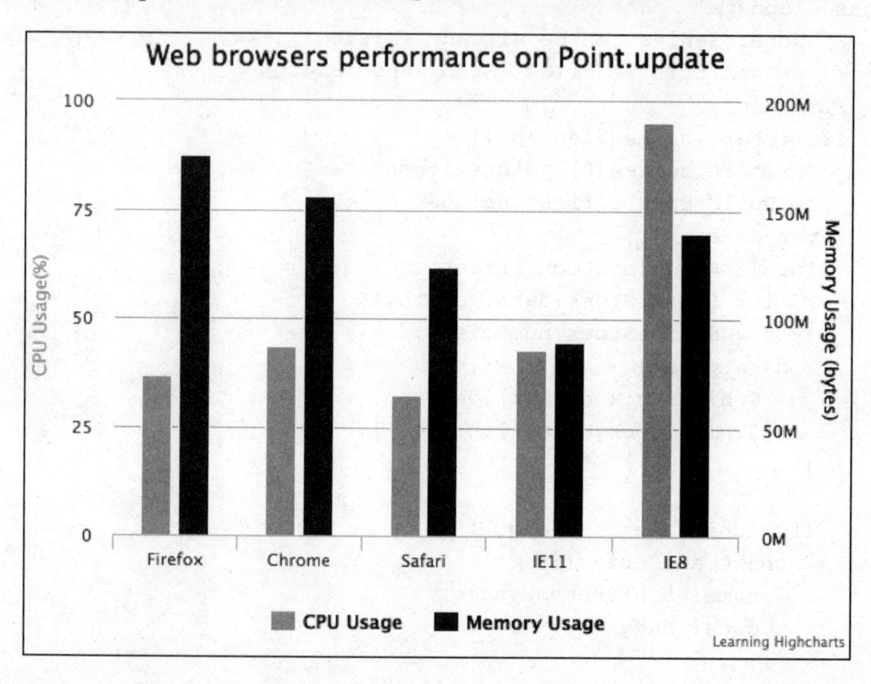

As the animations are pretty much the same as the `setData` approach, the performance shown in the preceding chart is very close to the results of the `setData` performance experiment.

Removing and adding data points with Point.remove and Series.addPoint

Instead of updating each individual data point, we can use `Point.remove` to remove data points within the `series.data` array and use `Series.addPoint` to add new data points back into the series:

```
remove ([Boolean redraw], [Mixed animation])
addPoint (Object options, [Boolean redraw], [Boolean shift],

         [Mixed animation])
```

As for the time series data, we can use `addPoint` along with the `shift` parameter set to `true`, which will automatically shift the series point array:

```
case 'addPoint':
  // Note: Series can be already existed at
  // start if we click 'Stop' and 'Start' again
  if (!chart.series.length ||
      !chart.series[0].points.length) {

    // Build up the first series
    var data = [];
    for (i = $.histStock.offset, j = 0;
        i < $.histStock.data.length &&
      j < $.histStock.numPoints; i++, j++) {
      data.push([
        $.histStock.data[i].date,
        $.histStock.data[i].price ]);
    }

    if (!chart.series.length) {
      chart.addSeries({
        name: $.histStock.name,
        data: data
      });
    } else {
      chart.series[0].setData(data);
    }

    // This is different, we don't redraw
    // any old points
    $.histStock.offset = i;

  } else {

    // Only updating the new data point
    for (i = $.histStock.offset, j = 0;
      i < $.histStock.data.length &&
      j < $.histStock.update; i++, j++) {
      chart.series[0].addPoint([
        $.histStock.data[i].date,
        $.histStock.data[i].price ],
        false, true );
    }
    chart.redraw();
  }
  break;
```

The following graph shows the performance comparison of the `addPoint` method across the browsers:

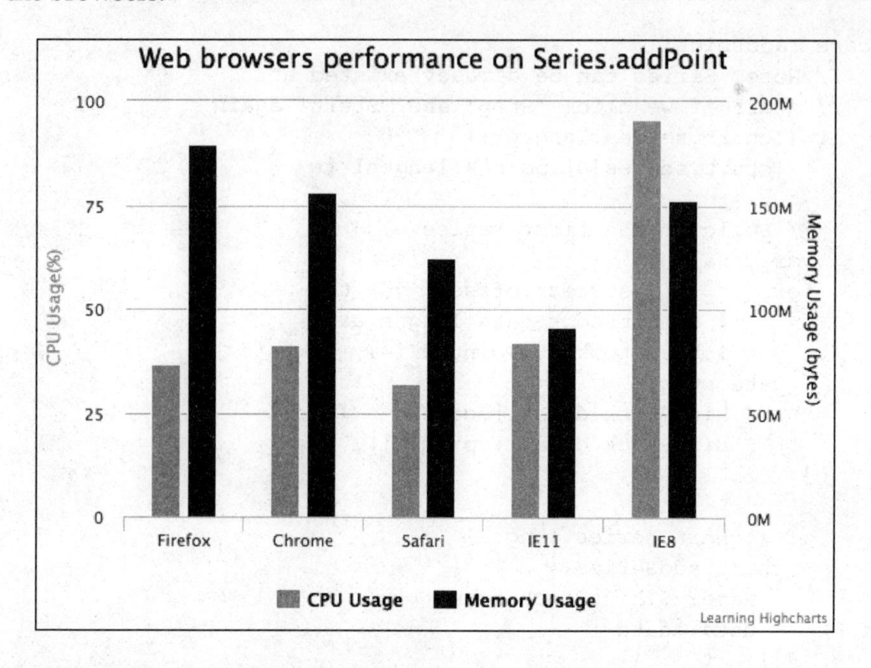

There is hardly any difference between the `addPoint` method, `setData`, and `update` in terms of both CPU and memory usage.

Exploring SVG animation performance on browsers

So far, we have seen that the level of CPU usage increased with animations. However, the question left unanswered is why Safari had lower CPU consumption than Chrome and Firefox. A number of browser benchmark suites have been run on the test machine to confirm the general consensus that the Firefox and Chrome browsers have overall better performance than Safari.

> All browsers were benchmarked with SunSpider `http://www.webkit.org/perf/sunspider/sunspider.html`, Google's V8 Benchmark suite `http://octane-benchmark.googlecode.com/svn/latest/index.html`, and Peacekeeper `http://peacekeeper.futuremark.com/`.

Nonetheless, there is one particular area where Safari has better performance than the other browsers: SVG animations; this is reflected in our previous experiments. Here, we use a benchmark test, written by Cameron Adams, that is especially designed to measure SVG animations with bouncing particles in frames per second. The test (HTML5 versus Flash: Animation Benchmarking `http://www.themaninblue.com/writing/perspective/2010/03/22/`) was originally written to compare various HTML5 animation technologies against Flash. Here, we run the SVG test with the Chrome and Safari browsers. The following is a Safari screenshot running with a 500-particle test:

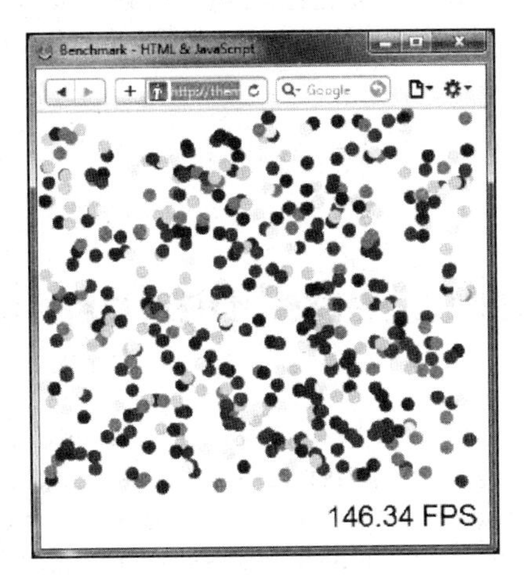

As for Chrome, the test is running at around 165 FPS. We repeat the assessment with various numbers of particles on both browsers. The following graph summarizes the performance difference with regard to SVG animations:

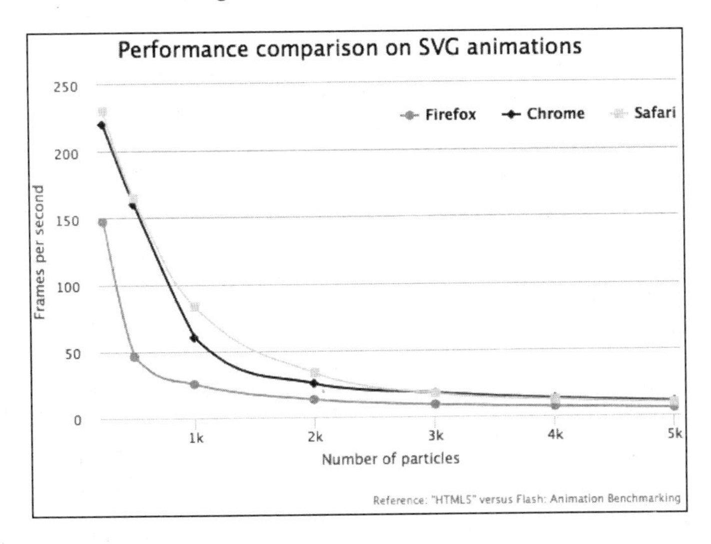

As we can see, Safari manages a higher frame rate with particles fewer than 3,000. After that, the Safari performance starts to degrade in parallel with Chrome. Firefox has a consistently lower frame rate and the frame rate drops considerably compared to the others.

This leads to another inevitable question: why is there such a difference, given that both browsers run with the same code base of webkit? It is difficult to pinpoint where the discrepancy lies. However, one of the few differences between both products is the JavaScript engines, which may affect that area, or possibly it is caused by the minor difference in the webkit version. In addition, other specific SVG performance tests in `http://jsperf.com` have also been run, in which Safari again had a higher score than Chrome.

In the next section, we will see how Highcharts' performance corresponds to the data size.

Comparing Highcharts' performance on large datasets

Our final test is to observe how Highcharts performs with large data sets. In this experiment, we are going to plot scatter series across various data sizes and observe the time taken to display the data. We chose to use the scatter series because, when there is a very large data set with tens of thousands of samples, the user is likely to plot only data points on the chart. Here is the simplified code illustrating how we do it:

```
var data = [];
// Adjust for each experiment
var num = 0;
if (location.match(/num=/)) {
   var numParam = location.match(/num=([^&]+)/)[1];
   num = parseInt(numParam, 10);
}
for (var i = 0; i < num; i ++) {
    data.push([ Math.random() * 1000000000, Math.random() * 50000 ]);
}

var start = new Date().getTime();
var chart = new Highcharts.Chart({
    chart: {
         renderTo: 'container',
         showAxes: true,
         animation: false,
         ....
    },
    series: [{
```

```
            type: 'scatter',
            data: data
        }],
        tooltips: {
            enabled: false
        },
        plotOptions: {
            series: {
                turboThreshold: 1,
                animation: false,
                enableMouseTracking: false
            }
        }
        }, function() {
            var stop = new Date().getTime();
            // Update the time taken label
            $('#time').append(((stop - start) / 1000) + " sec");
        });
```

The page is loaded with URL parameters to specify the dataset size. We start timing before the chart object is created and stop at the callback handler in the constructor, `Chart`, method. We repeat the experiment with the same dataset size on each browser used in the previous benchmarking experiments. Once the page is loaded on a browser, the dataset is randomly generated. Then, timing begins just before the chart object is constructed and stops at the callback method in Highcharts. The chart function is executed when the chart is finally displayed onto the screen. The following screenshot shows the time taken to display 3,000 data points on the Safari browser:

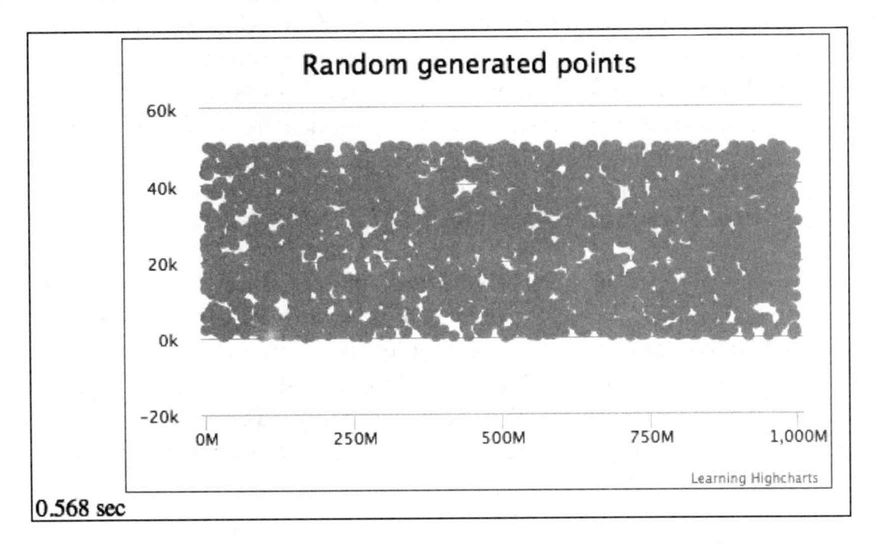

The following graph illustrates Highcharts' performance on different browsers with various dataset sizes. The lower the line, the less time it takes to display the number of data points, which indicates better performance:

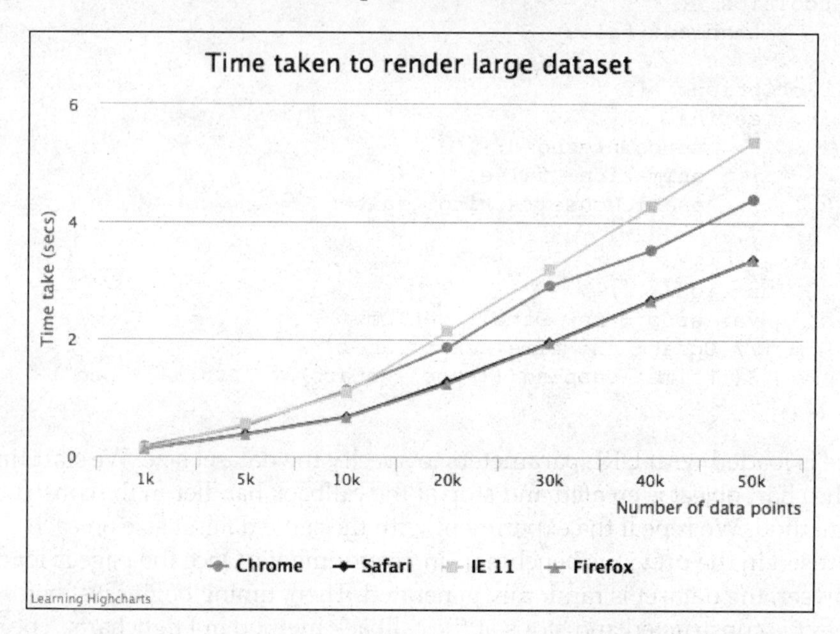

None of the browsers take particularly long to display large datasets. This shows the impressive scalable performance of Highcharts. Even IE 11, which is consistently slower in all tests, only takes 5.3 seconds to render 50,000 data points. As for smaller datasets, there is hardly any difference in timing. Both Firefox and Safari have very similar results, although Firefox is marginally better. This is due to the fact that there is no animation in scatter series and Firefox generally performs better without SVG animation. This also supports our findings in the least animated `addSeries` experiment, where Firefox required less CPU usage. As for Chrome, its performance is roughly equidistant between Safari/Firefox and IE 11.

From all these experiments, we can conclude that Safari performs the best with Highcharts (on a Windows PC), particularly when there are a lot of SVG animations in the chart. If a chart is static, then Firefox gives the best performance. IE 11 has the best memory utilization among the browsers but runs slower. Chrome achieves a consistent mid-range performance. We must stress that SVG is only one of the many areas of performance testing and Chrome has outperformed other browsers in the benchmark score.

Summary

In this chapter, we studied the Highcharts APIs, from the class model to applying them in applications. Then we performed a comprehensive study of the different techniques for updating a chart series in Highcharts, and carried out an experiment to analyze the difference in their performance. Finally, the chapter finished by analyzing the speed of different web browsers when rendering data points with regard to different sizes of large datasets.

In the next chapter, we will look into Highcharts events handling, which is closely related to Highcharts APIs.

11
Highcharts Events

In the previous chapter, we learned about the Highcharts API. In this chapter, we will go through Highcharts events handling. We will start the chapter by introducing the set of events supported by Highcharts. Then, we will build two web applications to cover most of the events; each one explores a different set of events. Although the applications are far from perfect and there is plenty of room for improvement, the sole purpose is to demonstrate how Highcharts events work. In this chapter, we will cover the following topics:

- Launching an Ajax query with a chart `load` event
- Activating the user interface with a chart `redraw` event
- Selecting and unselecting a data point with the point `select` and `unselect` events
- Zooming the selected area with the chart selection event
- Hovering over a data point with the point `mouseover` and `mouseout` events
- Using the chart `click` event to create plot lines
- Launching a dialog with the series `click` event
- Launching a pie chart with the series `checkboxClick` event
- Editing the pie chart with the point `click`, `update`, and `remove` events

Introducing Highcharts events

By now, we have gone through most of the Highcharts configurations, but there is one area not yet covered: event handling. Highcharts offers a set of event options in several areas such as chart events, series events, and axis base events; they are triggered by API calls and user interactions with the chart.

Highcharts events can be specified through object configuration while creating a chart or through APIs that accept object configurations, such as Chart.addSeries, Axis.addPlotLine, and Axis.addPlotBand.

An event object is passed by an event handler that contains mouse information and specific action data related to the event action. For example, event.xAxis[0] and event.yAxis[0] are stored in the event parameter for the chart.events.click handler. Inside each event function, the 'this' keyword can be used and refers to a Highcharts component where the event function is based. For example, the 'this' keyword in chart.events.click refers to the chart object, and the 'this' keyword in plotOptions.series.events.click refers to the series object being clicked.

The following is a list of Highcharts events:

- chart.events: addSeries, click, load, redraw, selection, drilldown, and drillup
- plotOptions.<series-type>.events: click, checkboxClick, hide, mouseover, mouseout, show, afterAnimate, and legendItemClick

Alternatively, we can specify event options specifically to a series in the series array, such as: series[{ events: click: function { ... }, ... }].

- plotOptions.<series-type>.point.events: click, mouseover, mouseout, remove, select, unselect, update, and legendItemClick

We can define point events for a specific series, as follows: series[{ point : { events: { click: function() { ... } }, ... }].

As for defining events for a particular data point in a series, we can specify them as follows: series[{ data: [{ events: { click: function() { ... } }], ... }].

- x/yAxis.events: setExtremes, and afterSetExtremes
- x/yAxis.plotBands[x].events and x/yAxis.plotLines[x].events: click, mouseover, mousemove, and mouseout

The Highcharts online documentation provides a comprehensive reference and plenty of mini examples; you are strongly advised to refer to that. There is not much point in repeating the same exercise. Instead, we will build two slightly sizable examples to utilize most of the Highcharts events and demonstrate how these events can work together in an application. Since the complete example code is too long to list in this chapter, only the relevant parts are edited and shown.

The full demo and source code can be found at http://www.joekuan.org/Learning_Highcharts/Chapter_11/chart1.html.

Portfolio history example

This application extends the historical stock chart in the previous chapter with an additional investment portfolio feature. The frontend is implemented with jQuery and jQuery UI, and the following events are covered in this example:

- chart.events: click, load, redraw, and selection
- plotOptions.series.points.events: mouseover, mouseout, select, and unselect
- xAxis/yAxis.plotLines.events: mouseover and click

The following is the startup screen of the demo, with the components labeled:

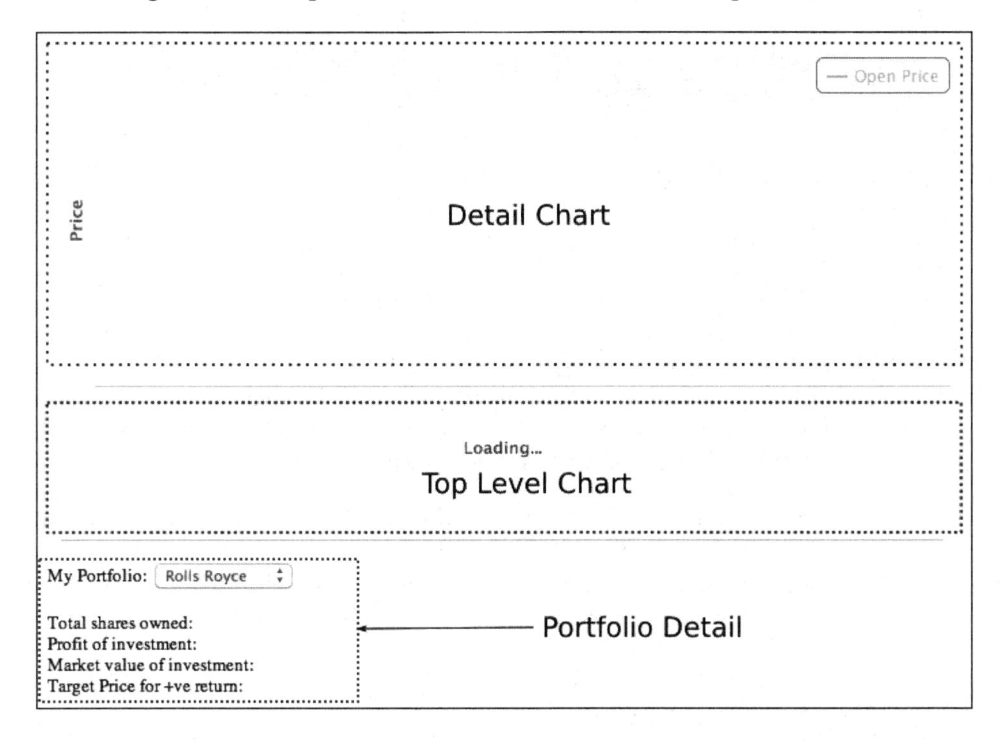

The application contains a pair of time series charts. The bottom chart is the top-level graph that shows the entire historic price movement and points to when company shares are bought and sold. The top chart is the detail chart that zooms in to the finer details when a selected area is made in the bottom graph.

As soon as the web application is loaded in a browser, both charts are created. The top-level chart is configured with a load event that automatically requests a stock historic price and portfolio history from the web server.

The following screenshot shows a graph after the top-level chart is auto-loaded:

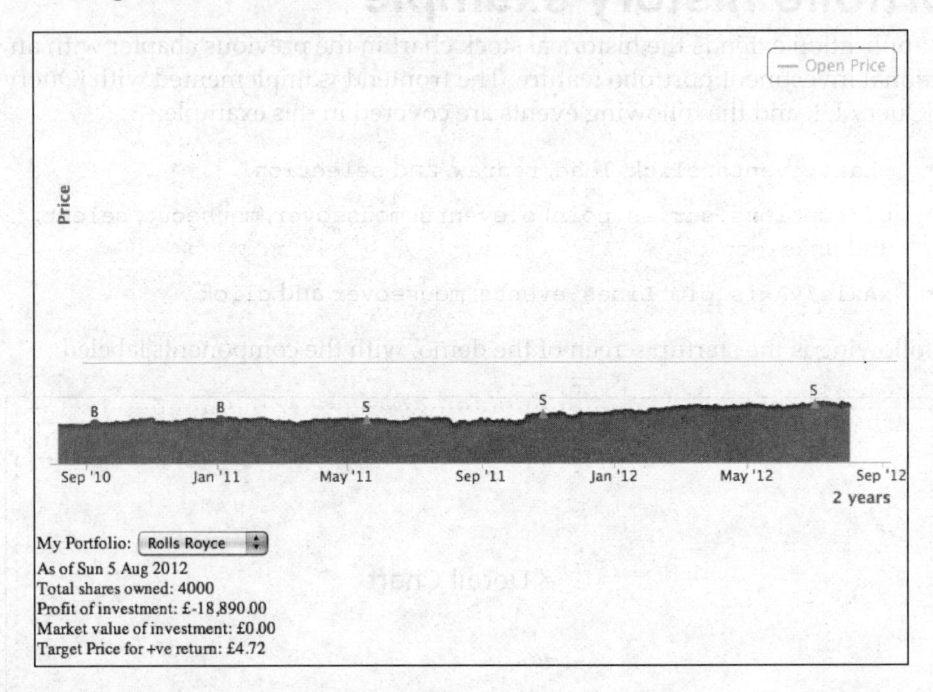

There are circular and triangular data points on top of the top-level chart. These denote the trade history. The **B** symbol indicates when the shares have been bought, whereas **S** signifies when they are sold. The information below the top-level chart is the portfolio detail for the stock as of the current date.

If we click on one of these trade history points, the portfolio detail section is updated to reflect the investment history as of the selected date. Moreover, when we select an area, it zooms in and displays the stock price movement in the detail chart. There are other features involved in event handling and we will discuss them in later sections.

The top-level chart

The following is the configuration code for the top-level chart (the bottom chart shows the entire historic price movement) and we store the `chart` object in the `myApp` namespace, as follows:

```
$.myApp.topChart = new Highcharts.Chart({
    chart: {
        zoomType: 'x',
        spacingRight: 15,
        renderTo: 'top-container',
        events: {
            // Load the default stock symbol of
            // the portfolio
            load: function() {  ....  },

            // The top level time series have
            // been redrawn, enable the portfolio
            // select box
            redraw: function() { .... },

            // Selection - get all the data points from
            // the selection and populate into the
            // detail chart
            selection: function(evt) {  ....  },
        }
    },
    title: { text: null },
    yAxis: {
        title: { text: null },
        gridLineWidth: 0,
        labels: { enabled: false }
    },
    tooltip: { enabled: false },
    xAxis: {
        title: { text: null },
        type: 'datetime'
    },
    series: [ ... ],
    legend: { enabled: false },
    credits: { enabled: false }
});
```

There is a lot going on in this configuration. The chart is defined with most of the features disabled, such as legend, title, tooltip, and *y*-axis label. More importantly, the chart is configured with a `zoomType` option, which enables the chart to be zoomable along the *x*-axis direction; hence, we can use the `select` event. The series array is composed of multiple series that also contain event configurations.

Constructing the series configuration for a top-level chart

In the series array, multiple series are defined with close and open price, bought and sold trade dates, and a hidden series for tracking mouse movement in the detail chart:

```
series: [{
    // Past closed price series
    type: 'areaspline',
    marker: { enabled: false },
    enableMouseTracking: false
}, {
    // This is the open price series and never shown
    // in the bottom chart. We use it to copy this
    // to the detail chart
    visible: false
}, {
    // Series for date and price when shares
    // are bought
    type: 'scatter',
    allowPointSelect: true,
    color: $.myApp.boughtColor,
    dataLabels: {
        enabled: true,
        formatter: function() { return 'B'; }
    },
    point: {
        events: { .... }
    }
}, {
    // Series for date and price when shares are sold
    type: 'scatter',
    allowPointSelect: true,
    color: $.myApp.soldColor,
    dataLabels: {
        enabled: true,
        formatter: function() { return 'S'; }
    },
    point: {
        events: { .... }
    }
}, {
    // This is the tracker series to show a single
    // data point of where the mouse is hovered on
    // the detail chart
```

```
            type: 'scatter',
            color: '#AA4643'
        }]
```

The first series is the historic stock price series and is configured without data point markers. The second series is hidden and acts as a placeholder for historic open price data in the detail chart. The third (bought) and fourth (sold) series are the scatter series revealing the dates when shares have been traded. Both series are set with the `allowPointSelect` option, so that we can define the `select` and `unselect` events in the `point.events` option. The final series is also a scatter series to reflect the mouse movement in the detail chart using the `mouseover` and `mouseout` events; we will see how all these are implemented later on.

Launching an Ajax query with the chart load event

As mentioned earlier, once the top-level chart is created and loaded on to the browser, it is ready to fetch the data from the server. The following is the chart's `load` event handler definition:

```
chart: {
    events: {
        load: function() {
            // Load the default stock symbol of
            // the portfolio
            var symbol = $('#symbol').val();
            $('#symbol').attr('disabled', true);
            loadPortfolio(symbol);
        },
```

We first retrieve the value from the **My Portfolio** selection box and disable the selection box during the query time. Then, we call a predefined function, `loadPortfolio`. The method performs several tasks, as follows:

1. Launching an Ajax call, `$.getJSON`, to load the past stock price and portfolio data.
2. Setting up a handler for the returned Ajax result that further executes the following steps:
    1. Hiding the chart loading mask.
    2. Unpacking the returned data and populating series data with it using the `Series.setData` method.
    3. Updating the data in the **Portfolio Detail** section to show how much the investment is worth as of the current date.

Activating the user interface with the chart redraw event

Once the top-level chart is populated with data, we can then enable the **My Portfolio** selection box on the page. To do that, we can rely on the `redraw` event, which is triggered by the `Series.setData` call in sub-step 2 inside step 2:

```
redraw: function() {
    $('#symbol').attr('disabled', false);
},
```

Selecting and unselecting a data point with the point select and unselect events

The bought and sold series share the same events handling; the only differences between them are the color and the point marker shape. The idea is that, when the user clicks on a data point in these series, the **Portfolio Detail** section is updated to show the investment detail for the stock as of the trade date. The following screenshot shows the effect after the first bought trade point is selected:

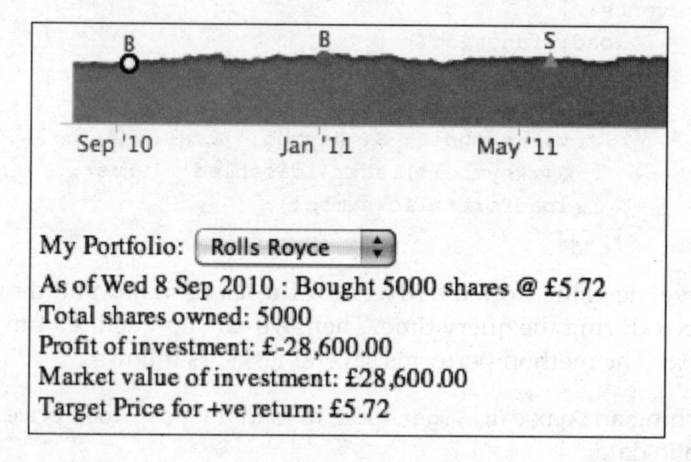

In order to keep the data point selected, we will use the `allowPointSelect` option, which allows us to define the `select` and `unselect` events. The following is the events configuration for the bought and sold series:

```
point: {
    events: {
        select: function() {
            updatePortfolio(this.x);
        },
        unselect: function() {
```

```
                    // Only default back to current time
                    // portfolio detail when unselecting
                    // itself
                    var selectPt =
                        $.myApp.topChart.getSelectedPoints();
                    if (selectPt[0].x == this.x) {
                        updatePortfolio(new Date().getTime());
                    }
                }
            }
        }
```

Basically, the `select` event handler calls a predefined function, `updatePortfolio`, that updates the **Portfolio Detail** section based on the selected data point time: `this.x`. The `'this'` keyword in the handler refers to the selected point object, where `x` is the time value.

Unselecting the data point will call the `unselect` event handler. The preceding implementation means that, if the unselected data point (`this.x`) is the same as the previously selected point, then it indicates that the user has unselected the same point, so we want to show the portfolio detail as of the current date. Otherwise it will do nothing because it means the user has selected another trade data point; thus, another `select` event call is made with a different date.

Zooming the selected area with the chart selection event

The `selection` event forms the bridge between the top-level chart and the detail chart. When we select an area in the top-level chart, the selected area is highlighted and the data is zoomed in the detail chart. This action triggers the `selection` event and the following is the cut-down code of the event handler:

```
selection: function(evt) {
    // Get the xAxis selection
    var selectStart = Math.round(evt.xAxis[0].min);
    var selectEnd   = Math.round(evt.xAxis[0].max);

    // We use plotBand to paint the selected area
    // to simulate a selected area
    this.xAxis[0].removePlotBand('selected');
    this.xAxis[0].addPlotBand({
        color: 'rgba(69, 114, 167, 0.25)',
        id: 'selected',
        from: selectStart,
```

```
            to: selectEnd
        });
        for (var i = 0;
             i < this.series[0].data.length; i++) {
           var pt = this.series[0].data[i];
           if (pt.x >= selectStart &&
               pt.x <= selectEnd) {
               selectedData.push([pt.x, pt.y]);
           }

           if (pt.x > selectEnd) {
               break;
           }
        }

        // Update the detail serie
        var dSeries = $.myApp.detailChart.series[0];
        dSeries.setData(selectedData, false);
        ....

        // Update the detail chart title & subtitle
        $.myApp.detailChart.setTitle({
            text: $.myApp.stockName + " (" +
                  $.myApp.stockSymbol + ")",
            style: { fontFamily: 'palatino, serif',
                     fontWeight: 'bold' }
            }, {
            text: Highcharts.dateFormat('%e %b %y',
                  selectStart) + ' -- ' +
                  Highcharts.dateFormat('%e %b %y',
                  selectEnd),
            style: { fontFamily: 'palatino, serif' }
        });

        $.myApp.detailChart.redraw();
        return false;
    }
```

There are several steps taken in the handler code. First, we extract the selected range values from the handler parameters—`evt.xAxis[0].min` and `evt.xAxis[0].max`. The next step is to make the selected area stay highlighted in the top-level chart. To do that, we create a plot band using `this.xAxis[0].addPlotBand` over the same area to simulate the selection.

The 'this' keyword refers to the top-level chart object. The next task is to give a fixed id, so that we can remove the old selection and highlight a new selection. Additionally, the plot band should have the same color as the selection being dragged on the chart. All we need to do is to assign the plot band color to be the same as the default value of the chart.selectionMarkerFill option.

After that, we copy the data within the selected range into an array and pass it to the detail chart using Series.setData. Since we called the setData method a couple of times, it is worth setting the redraw option to false to save resources and then calling the redraw method.

Finally, the most important step is to return false at the end of the function. Returning the false Boolean value tells Highcharts not to take the default action after the selection has been made. Otherwise the whole top-level chart is redrawn and stretched (alternatively, we can call event.preventDefault()).

The following screenshot zooms and displays the detail in another chart:

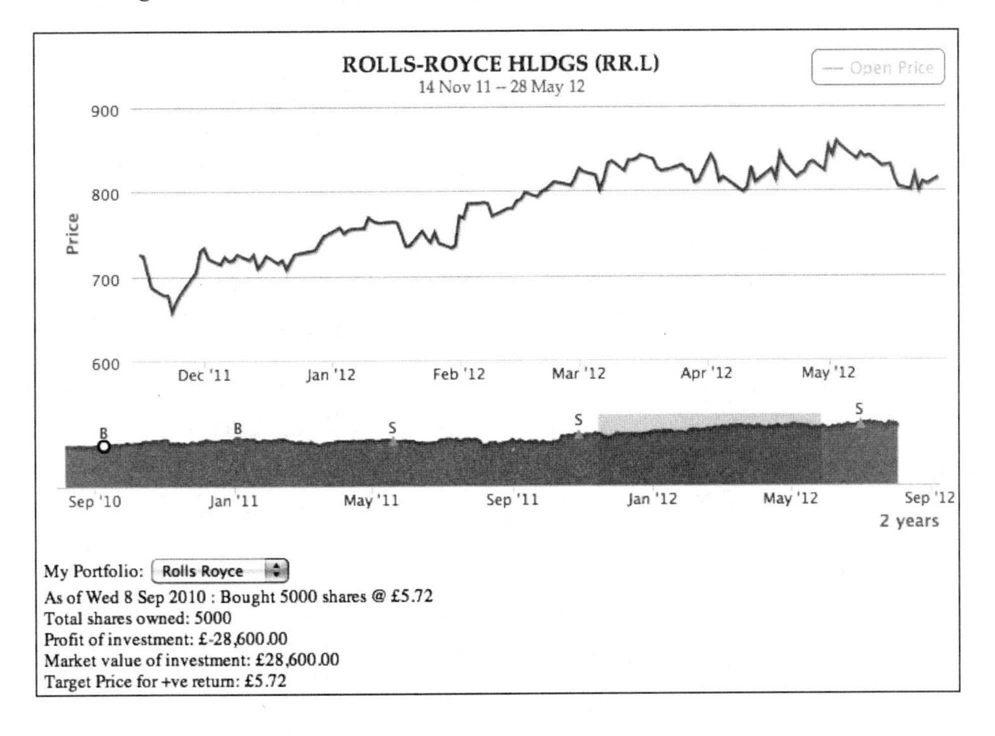

The detail chart

The detail chart is simply a line chart showing the selected region from the top-level chart. The chart is configured with a tool tip fixed in the upper-left corner and a number of events that we will discuss later:

```
$.myApp.detailChart = new Highcharts.Chart({
    chart: {
        showAxes: true,
        renderTo: 'detail-container',
        events: {
            click: function(evt) {
                // Only allow to prompt stop order
                // dialog if the chart contains future
                // time
                ....
            }
        },
    },
    title: {
        margin: 10,
        text: null
    },
    credits: { enabled: false },
    legend: {
        enabled: true,
        floating: true,
        verticalAlign: 'top',
        align: 'right'
    },
    series: [ ... ],
    // Fixed location tooltip in the top left
    tooltip: {
        shared: true,
        positioner: function() {
            return { x: 10, y: 10 }
        },
        // Include 52 week high and low
        formatter: function() {  .... }
    },
    yAxis: {
        title: { text: 'Price' }
    },
    xAxis: { type: 'datetime' }
});
```

The following is a screenshot showing a data point being hovered over and the tool tip shown in the upper-left corner:

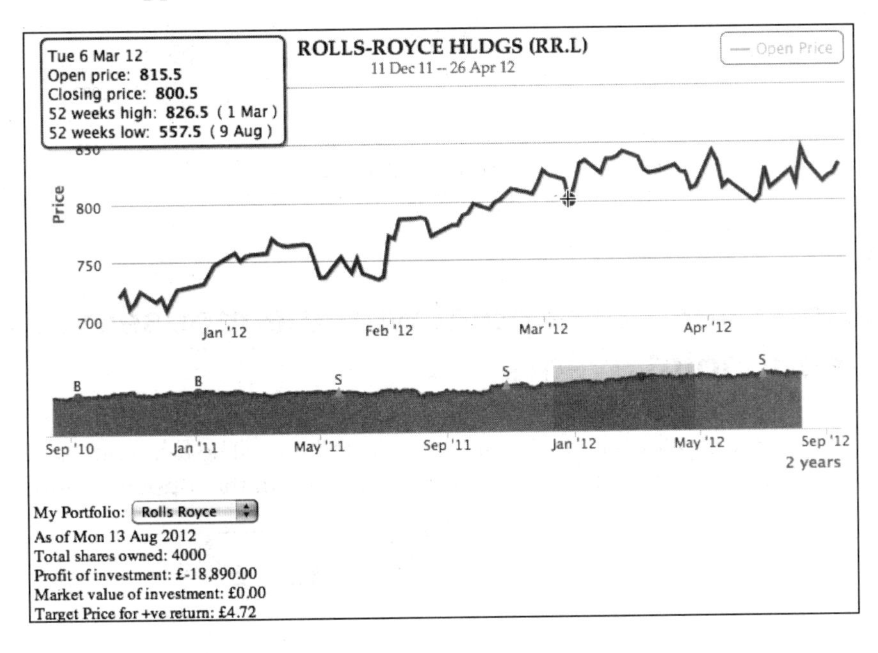

Constructing the series configuration for the detail chart

There are two series configured in the detail chart. The main focus is the first series, which is the stock closed price. The series is defined without data point markers and has `'crosshair'` as the `cursor` option, as we can see in the preceding screenshot. In addition, the `mouseout` and `mouseover` events are defined for the data points that create a marker to the tracker series in the top-level chart. We will go through these events in the next section. The series array is defined as follows:

```
series: [{
    marker: {
        enabled: false,
        states: {
            hover: { enabled: true }
        }
    },
    cursor: 'crosshair',
    point: {
        events: {
            mouseOver: function() { ... },
```

```
                mouseOut: function() { ... }
            }
        },
        stickyTracking: false,
        showInLegend: false
    }, {
        name: 'Open Price',
        marker: { enabled: false },
        visible: false
    }],
```

Hovering over a data point with the mouseover and mouseout point events

When we move the mouse pointer along the series in the detail chart, the movement is also reflected in the top-level chart within the selected area. The following screenshot shows the tracker point (the inverted triangle) displayed in the top-level chart:

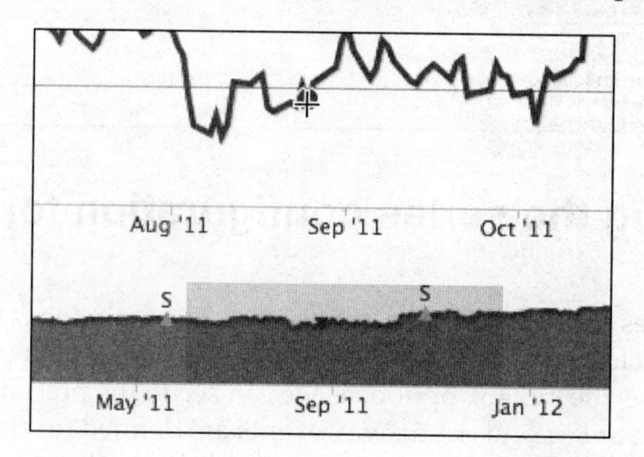

The inverted triangle indicates where we are browsing in the top-level chart. To do that, we will set up the `mouseOut` and `mouseOver` point event options in the detail chart series, as follows:

```
point: {
    events: {
        mouseOver: function() {
            var series = $.myApp.topChart.series[4];
            series.setData([]);
            series.addPoint([this.x, this.y]);
        },
        mouseOut: function() {
             var series = $.myApp.topChart.series[4];
```

```
                    series.setData([]);
                }
            }
        },
```

Inside the `mouseOver` handler, the `'this'` keyword refers to the hovered data point object and the `x` and `y` properties refer to the time and price values. Since both the top-level and detail charts share the same data type along both x and y axes, we can simply add a data point into the tracker series in the top-level chart. As for the `mouseOut` event, we reset the series by emptying the data array.

Applying the chart click event

In this section, we will apply the chart click event to create a *stop order* for investment portfolios. **Stop order** is an investment term for selling or buying a stock when it reaches the price threshold within a specified date/time range in the future. It is generally used to limit a loss or protect a profit.

Notice that there is an empty space at the right-hand side of the top-level chart. In fact, this is deliberately created for the next 30-day range from the current date. Let's highlight that area, so that the future date appears in the detail chart:

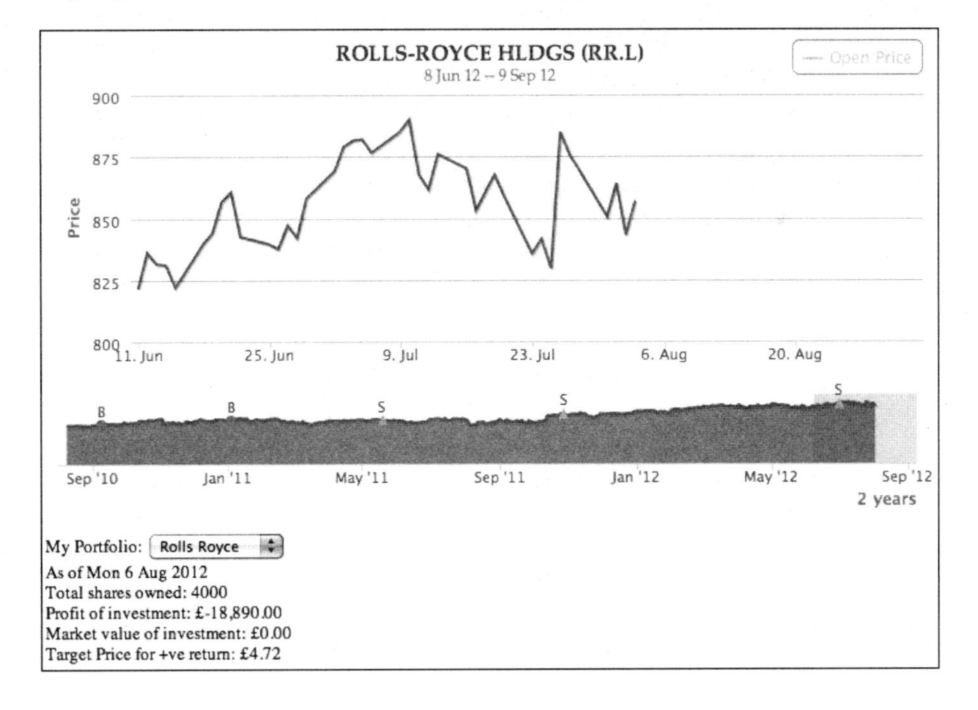

As we can see, the line series in the detail chart stops as soon as it hits the current date. If we click on the zone for future dates in the detail chart, a **Create Stop Order** dialog box appears. The x, y position of the click on the chart is then converted into date and price, which then populates the values into the dialog box. The following is the screenshot of the dialog box:

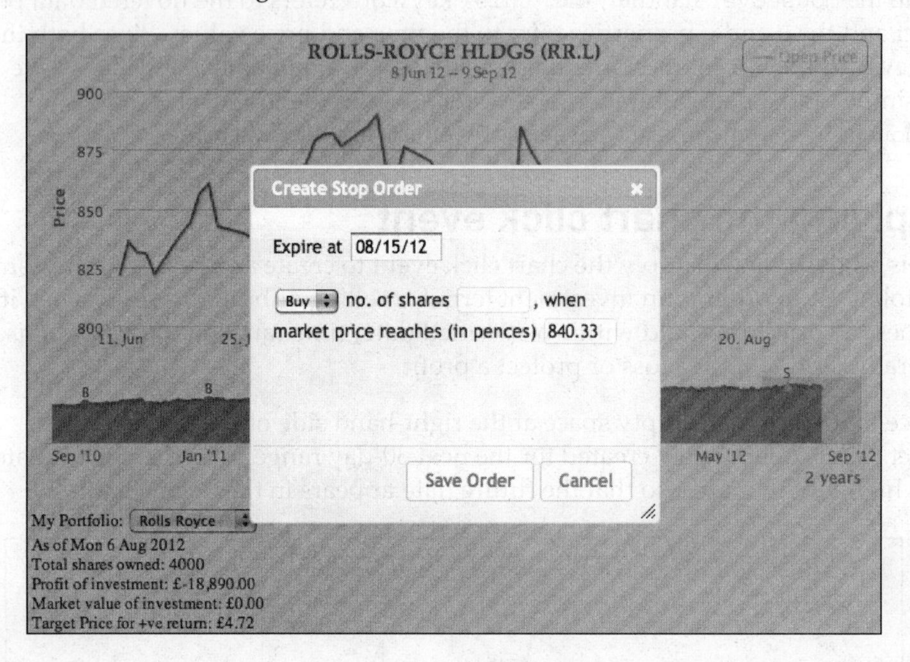

The expiry date and price fields can be further adjusted if necessary. Once the **Save Order** button is clicked, a stop order is created and a pair of x and y plot lines are generated to mark the chart. The following is a screenshot showing two stop orders on the chart:

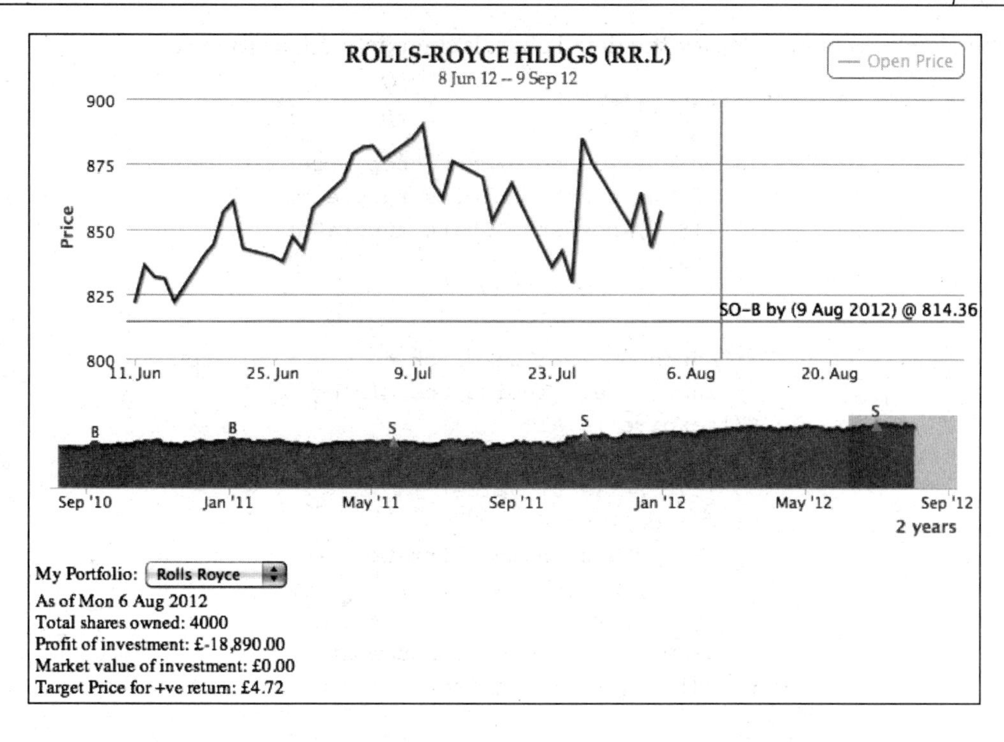

Let's see how all these actions can be derived from the code. First, the jQuery UI dialog is created based on an HTML form declared on the page:

```
<div id='dialog'>
   <form>
      <fieldset>
         <label for="expire">Expire at</label>
         <input type=text name="expire" id="expire" size=9
><br/><br/>
         <select name='stopOrder' id='stopOrder'>
            <option value='buy' selected>Buy</option>
            <option value='sell'>Sell</option>
         </select>
         <label for="shares">no. of shares</label>
         <input type="text" name="shares" id="shares" value=""
size=7 class="text ui-widget-content ui-corner-all" />,
         <label for="price">when market price reaches (in pences)</
label>
         <input type="text" name="price" id="price" value="" size=7
class="text ui-widget-content ui-corner-all" />
      </fieldset>
   </form>
</div>
```

The `click` event handler for the detail chart is then defined, as follows:

```
click: function(evt) {

    // Only allow to prompt stop order dialog
    // if the chart contains future time
    if (!$.myApp.detailChart.futureDate) {
        return;
    }

    // Based on what we click on the time, set
    // input field inside the dialog
    $('#expire').val(
        Highcharts.dateFormat("%m/%d/%y",
        evt.xAxis[0].value));
    $('#price').val(
        Highcharts.numberFormat(
        evt.yAxis[0].value, 2));

    // Display the form to setup stop order
    $('#dialog').dialog("open");
}
```

The first guard condition is to see whether the detail chart contains any future dates. If a future date exists, then it extracts the x and y values from the `click` event and assigns them into the form input fields. After that, it calls the jQuery UI dialog method to lay out the HTML form in a dialog box and displays it.

The following code snippet shows how we define the jQuery UI dialog box and its action buttons. The code is edited for readability:

```
// Initiate stop order dialog
$( "#dialog" ).dialog({
   // Dialog startup configuration -
   // dimension, modal, title, etc
   .... ,
   buttons: [{
      text: "Save Order",
      click: function() {
          // Check whether this dialog is called
          // with a stop order id. If not, then
          // assign a new stop order id
          // Assign the dialog fields into an
          // object - 'order'
          ....
```

```
                    // Store the stop order
                    $.myApp.stopOrders[id] = order;
                    // Remove plotlines if already exist.
                    // This can happen if we modify a stop
                    // order point
                    var xAxis = $.myApp.detailChart.xAxis[0];
                    xAxis.removePlotLine(id);
                    var yAxis = $.myApp.detailChart.yAxis[0];
                    yAxis.removePlotLine(id);

                    // Setup events handling for both
                    // x & y axis plotlines
                    var events = {
                        // Change the mouse cursor to pointer
                        // when the mouse is hovered above
                        // the plotlines
                        mouseover: function() { ... },

                        // Launch modify dialog when
                        // click on a plotline
                        click: function(evt) { ... }
                    };

                    // Create the plot lines for the stop
                    // order
                    xAxis.addPlotLine({
                        value: order.expire,
                        width: 2,
                        events: events,
                        color: (order.stopOrder == 'buy') ? $.myApp.
boughtColor : $.myApp.soldColor,
                        id: id,
                        // Over both line series and
                        // plot line
                       zIndex: 3
                    });

                    yAxis.addPlotLine({
                        value: order.price,
                        width: 2,
                        color: (order.stopOrder == 'buy') ? $.myApp.
boughtColor : $.myApp.soldColor,
                        id: id,
                        zIndex: 3,
```

```
                    events: events,
                    label: {
                        text: ((order.stopOrder == 'buy') ?
'SO-B by (' : 'SO-S by (')  + Highcharts.dateFormat("%e %b %Y",
parseInt(order.expire)) + ') @ ' + order.price,
                        align: 'right'
                    }
                });

                $('#dialog').dialog("close");
            }
        }, {
            text: "Cancel",
            click: function() {
                $('#dialog').dialog("close");
            }
        }]
    });
```

The dialog box setup code is slightly more complicated. In the **Save Order** button's handler, it performs several tasks, as follows:

1. It extracts the input values from the dialog box.
2. It checks whether the dialog box is opened with a specific stop order `id`. If not, then it assigns a new stop order `id` and stores the values with `id` into `$.myApp.stopOrders`.
3. It removes any existing plot lines that match with `id`, in case we modify an existing stop order.
4. It sets up the `click` and `mouseover` events handling for both x- and y-axis plot lines.
5. It creates x and y plot lines in the detail chart with the events definitions constructed in step 4.

One scenario with stop orders is that users may want to change or delete a stop order before the condition is fulfilled. Therefore, in step 4 the purpose of the `click` event on plot lines is to bring up a modify dialog box. Additionally, we want to change the mouse cursor to a pointer when hovering over the plot lines to show that it is clickable.

Changing the mouse cursor over plot lines with the mouseover event

To change the mouse cursor over the plot lines, we define the `mouseover` event handler, as follows:

```
mouseover: function() {
    $.each(this.axis.plotLinesAndBands,
        function(idx, plot) {
           if (plot.id == id) {
               plot.svgElem.element.style.cursor =
                  'pointer';
               return false;
           }
        }
    );
},
```

The `'this'` keyword contains an axis object that the hovered plot line belongs to. Since there can be multiple plot lines in each axis, we need to loop through the array of plot lines and plot bands that can be found in the `plotLinesAndBands` property inside the axis object. Once we have found the target plot line by matching `id`, we will dig into the internal element and set the cursor style to `'pointer'`. The following screenshot shows a mouse cursor hovered over the plot line:

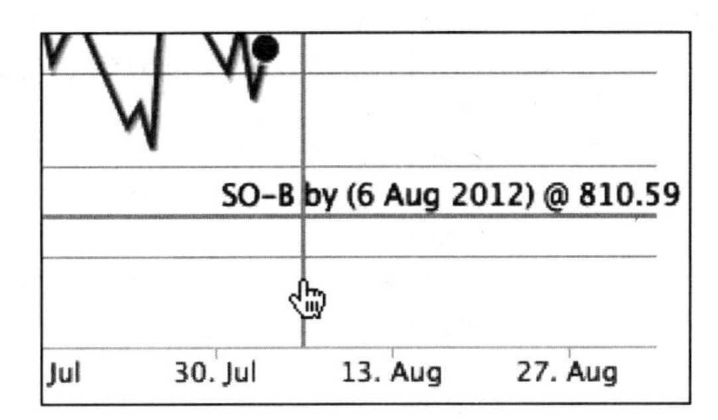

Setting up a plot line action with the click event

The `click` event for plot lines launches the **Modify Stop Order** dialog box for a stop order:

```
// Click on the prompt line
click: function(evt) {
    // Retrieves the stop order object stored in
    // $.myApp.stopOrders
```

```
        $('#dialog').dialog("option",
                         "stopOrderId", id);
        var stopOrder = $.myApp.stopOrders[id];

        // Put the settings into the stop order form
        $('#dialog').dialog("option", "title",
                         "Modify Stop Order");
        $('#price').val(
            Highcharts.numberFormat(
                     stopOrder.price, 2));

        $('#stopOrder').val(stopOrder.stopOrder);
        $('#shares').val(stopOrder.shares);
        $('#expire').val(
            Highcharts.dateFormat("%m/%d/%y",
                     stopOrder.expire));

        // Add a remove button inside the dialog
        var buttons =
            $('#dialog').dialog("option", "buttons");
        buttons.push({
            text: 'Remove Order',
            click: function() {
                // Remove plot line and stop order
                // settings
                delete $.myApp.stopOrders[id];
                var xAxis =
                    $.myApp.detailChart.xAxis[0];
                xAxis.removePlotLine(id);
                var yAxis =
                    $.myApp.detailChart.yAxis[0];
                yAxis.removePlotLine(id);

                // Set the dialog to original state
                resetDialog();
                $('#dialog').dialog("close");
            }
        });

        $('#dialog').dialog("option",
                         "buttons", buttons);

        $('#dialog').dialog("open");
    }
```

The `click` event handler simply retrieves the stop order settings and puts the values inside the **Modify Stop Order** dialog box. Before launching the dialog box, add a **Remove Order** button into the dialog box that the button handler calls `removePlotLine`, with the plot line `id`. The following is a screenshot of the **Create Stop Order** dialog box:

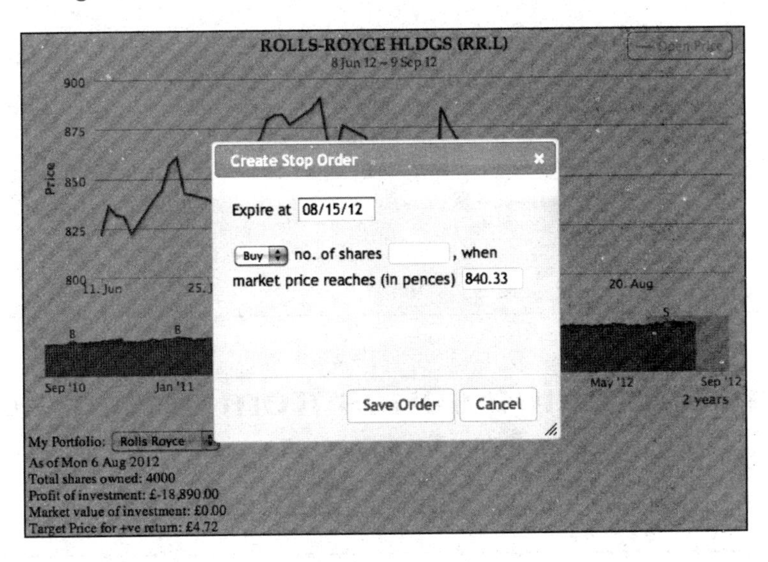

Stock growth chart example

Our next example (for the online demo, see `http://joekuan.org/Learning_Highcharts/Chapter_11/chart2.html`) is to demonstrate the following events:

- `chart.events`: `addSeries`
- `plotOptions.series.events`: `click`, `checkboxClick`, and `legendItemClick`
- `plotOptions.series.point.events`: `update` and `remove`

Suppose that we want to draft a long-term investment portfolio based on past stock growth performance as a reference. The demo contains a chart started with two series, Portfolio and Average growth, and a form to input stock symbols. Basically, we enter a stock symbol in this demo, and then a line series of stock growth is inserted into the chart.

So we can plot multiple stock yield trends and tweak their proportion in our portfolio to observe how the **Average** and **Portfolio** lines perform. The following screenshot shows the initial screen:

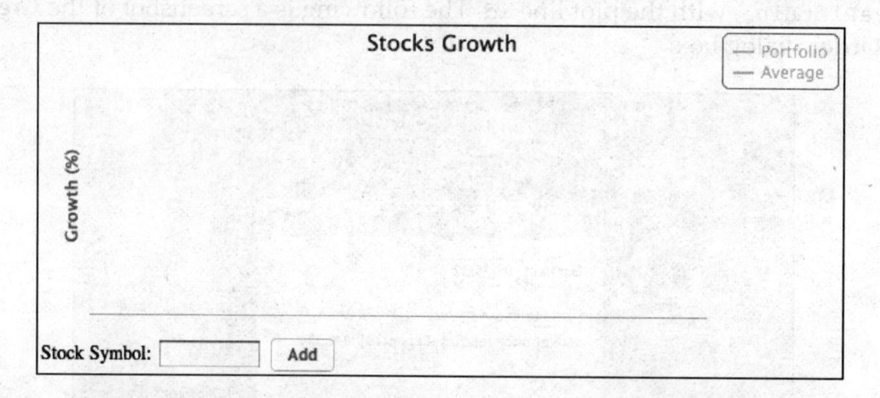

Plotting averaging series from displayed stock series

Let's query for two stocks and click on the **Average** legend to enable the series:

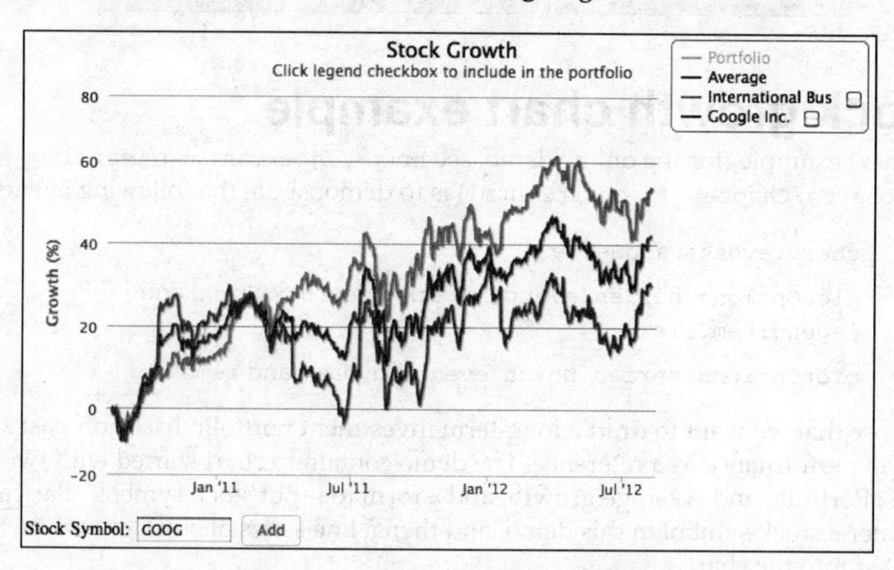

As expected, the **Average** line is plotted between the two stock lines. Assuming that future growth is similar to the past, this **Average** line projects future growth if we invest in both stocks equally for our portfolio. Let's add another stock symbol onto the chart:

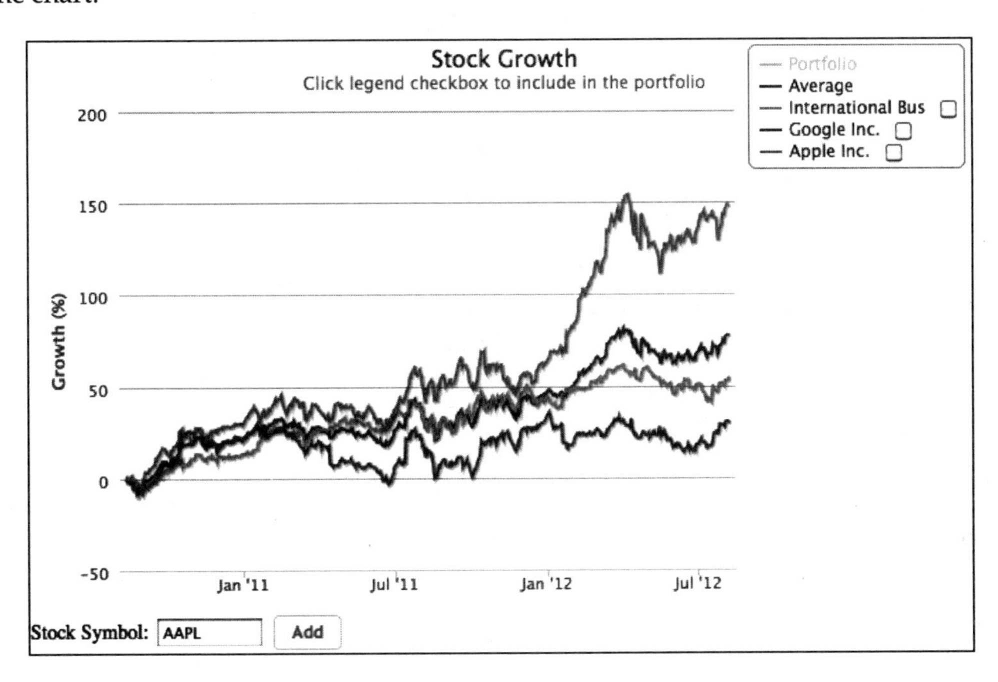

The new growth line generates a higher yield so that the **Average** line automatically re-adjusts itself and shifts to become the second line from the top. Let's see how it is implemented. The following is the chart configuration code:

```
$.myChart = new Highcharts.Chart({
    chart: {
        renderTo: 'container',
        showAxes: true,
        events: {
            addSeries: function() { ... }
        }
    },
    series: [{
        visible: false,
        name: 'Portfolio',
        color: $.colorRange.shift(),
        marker: { enabled: false },
        events: {
            legendItemClick: function(evt) { ... }
```

```
            }
        }, {
            name: 'Average',
            events: {
                legendItemClick: function(evt) { ... }
            },
            color: $.colorRange.shift(),
            visible: false,
            marker: { enabled: false }
        }, {
            visible: false,
            type: 'pie',
            point: {
                events: {
                    click: function(evt) { ... },
                        update: function(evt) { ... },
                        remove: function(evt) { ... }
                }
            },
            center: [ '13%', '5%' ],
            size: '30%',
            dataLabels: { enabled: false }
        }],
        title: { text: 'Stocks Growth' },
        credits: { enabled: false },
        legend: {
            enabled: true,
            align: 'right',
            layout: 'vertical',
            verticalAlign: 'top'
        },
        yAxis: {
            title: { text: 'Growth (%)' }
        },
        xAxis: { type: 'datetime' }
    });
```

The chart contains three series: Portfolio, Average, and a pie chart series to edit the portfolio distribution.

When we hit the **Add** button with a stock symbol, the `showLoading` method is called to put a loading mask in front of the chart, and then an Ajax connection is established with the server to query the stock yield data. We implement the Ajax handler by calling the `addSeries` function to insert a new series into the chart.

Once the `addSeries` event is triggered, it means that the data has been returned and is ready to plot. In this case, we can disable the chart loading mask, as follows:

```
chart: {
    .... ,
    events: {
        addSeries: function() {
            this.hideLoading();
        }
    },
    .... ,
```

The following is the implementation of the **Add** button action:

```
$('#add').button().on('click',
    function() {
        var symbol = $('#symbol').val().toLowerCase();
        $.myChart.showLoading();
        $.getJSON('./stockGrowth.php?symbol=' + symbol +
                  '&years=' + $.numOfYears,
            function(stockData) {
                // Build up the series data array
                var seriesData = [];

                if (!stockData.rows.length) {
                    return;
                }

                $.symbols.push({
                    symbol: symbol,
                    name: stockData.name
                });

                $.each(stockData.rows,
                    function(idx, data) {
                        seriesData.push([
                            data.date * 1000,
                            data.growth ]);
                });

                $.myChart.addSeries({
                    events: {
                        // Remove the stock series
                        click: { ... },
                        // Include the stock into portfolio
                        checkboxClick: { ... }
                    },
                    data: seriesData,
                    name: stockData.name,
```

```
                    marker: { enabled: false },
                    stickyTracking: false,
                    showCheckbox: true,

                    // Because we can add/remove series,
                    // we need to make sure the chosen
                    // color used in the visible series
                    color: $.colorRange.shift()
                }, false);

                updateAvg(false);
                $.myChart.redraw();
            }  // function (stockData)
        );  //getJSON
    });
```

We build a series configuration object from the Ajax returned data. Within this new series configuration, we set the `showCheckbox` option to `true` for a checkbox next to the legend item. A couple of events are also added into the configuration, `click` and `checkboxClick`, and are discussed later.

After the `addSeries` method call, we then call a predefined routine, `updateAvg`, that only recomputes and redraws the **Average** line if it is on display.

Recalling from the preceding Average series events definition, we use the `legendItemClick` event to capture when the Average series is clicked in the legend box:

```
series: [{
    ...
    }, {
    name: 'Average',
    events: {
        legendItemClick: function(evt) {
            if (!this.visible) {
                updateAvg();
            }
        }
    },
    .....
```

The preceding code means that, if the Average series is not currently in a visible state, then the series will be visible after this handler returns. Hence, it calculates the average values and shows the series.

Launching a dialog with the series click event

Instead of enabling or disabling a stock yield line by clicking on the legend item, we may want to completely remove the series line. In this scenario, we use the `click` event to do that, as follows:

```
$.myChart.addSeries({
    events: {
        // Launch a confirm dialog box to delete
        // the series
        click: function() {
            // Save the clicked series into the dialog
            $("#dialog-confirm").dialog("option",
               "seriesIdx", this.index);
            $("#dialog-confirm").dialog("option",
               "seriesName", this.name);
            $("#removeName").text(this.name);

            $("#dialog-confirm").dialog("open");
        },
        // Include the stock into portfolio
        checkboxClick: function(evt) { ... }
    },
    ....
});
```

The click action launches a confirmation dialog box for removing the series from the chart. We store the clicked series (the `'this'` keyword) information inside the dialog box. The **Remove** button's button handler uses that data to remove the series and recalculate the average series if it is shown. The following is the screenshot:

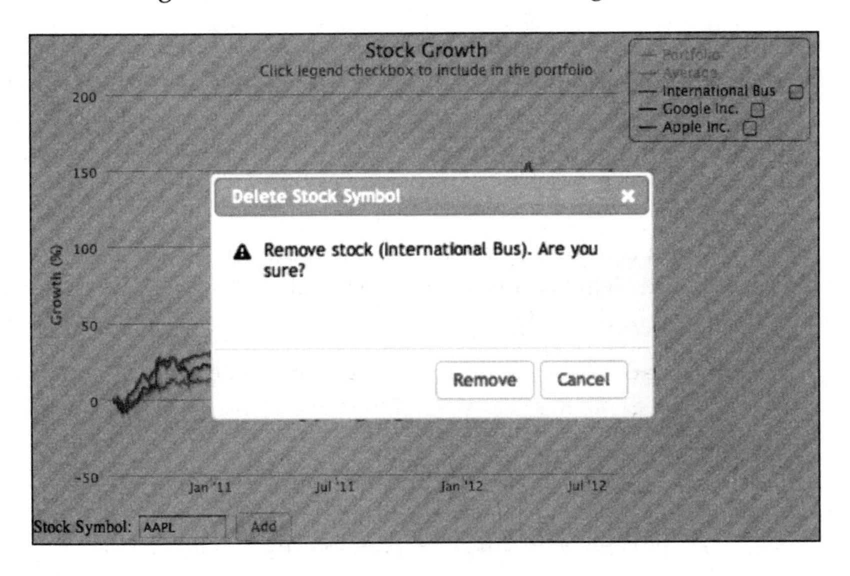

Launching a pie chart with the series checkboxClick event

Inside the legend box, each checkbox is used to include the stock in the portfolio. As soon as the checkbox is checked, a pie chart appears in the upper-left corner showing the distribution of the stock within the portfolio. Each slice in the pie chart shares the same color with the corresponding stock line. The following screenshot shows three growth lines and a portfolio pie chart equally distributed for each stock:

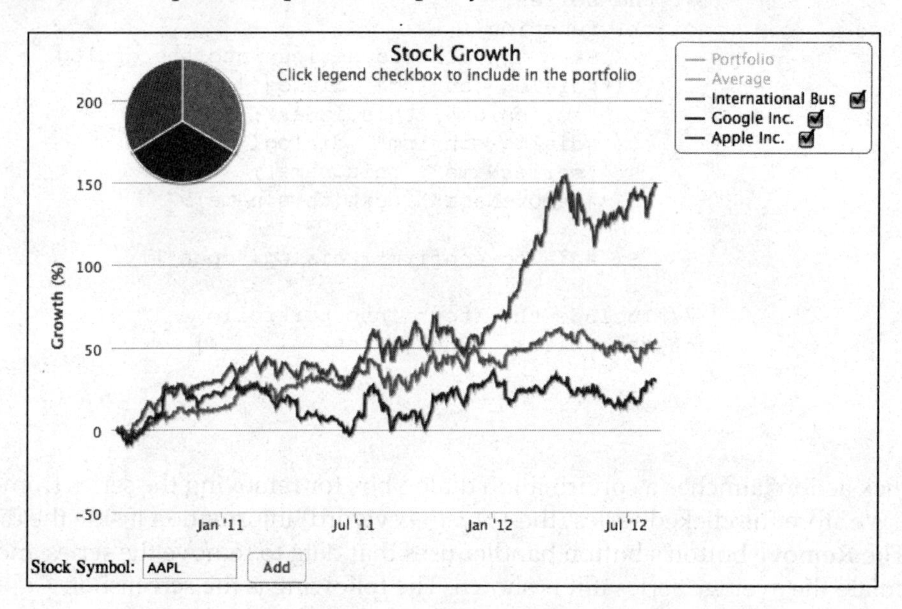

Since the growth line series is configured with the `showCheckbox` option, we can define the `checkboxClick` event to launch a pie chart when the checkbox is checked:

```
checkboxClick: function(evt) {
   updatePie(this, evt.checked);
}
```

The `updatePie` function is called in several places in this demo, for example, to remove a series, when the legend checkbox is checked, and so on. The following is the shortened version of the code:

```
var updatePie = function(seriesObj, checked) {

   var index = seriesObj.index;

   // Loop through the stock series. If checkbox
   // checked, then compute the equal distribution
   // percentage for the pie series data
```

```
            for (i = $.pfloIdx + 1;
                 i < $.myChart.series.length; i++) {
                var insert = (i == index) ? checked : $.myChart.
series[i].selected;
                if (insert) {
                    data.push({
                        name: $.myChart.series[i].name,
                        y: parseFloat((100 / count).toFixed(2)),
                        color: $.myChart.series[i].color
                    });
                }
            }

            // Update the pie chart series
            $.myChart.series[$.pfloIdx].setData(data, false);
            $.myChart.series[$.pfloIdx].show();
        };
```

The preceding code snippet basically loops through the stock series array and checks whether it is selected. If so, then it includes stock in the pie series in an equally distributed manner. Then the pie chart is displayed if there are one or more entries.

Editing the pie chart's slice with the data point's click, update, and remove events

It is unlikely that an investment portfolio will have an equal distribution of all stocks. Therefore, we can enhance the example by modifying portions within the pie chart. When a slice of the pie chart is clicked, a dialog box pops up. This allows us to adjust or remove the portion within the portfolio. The following screenshot shows this:

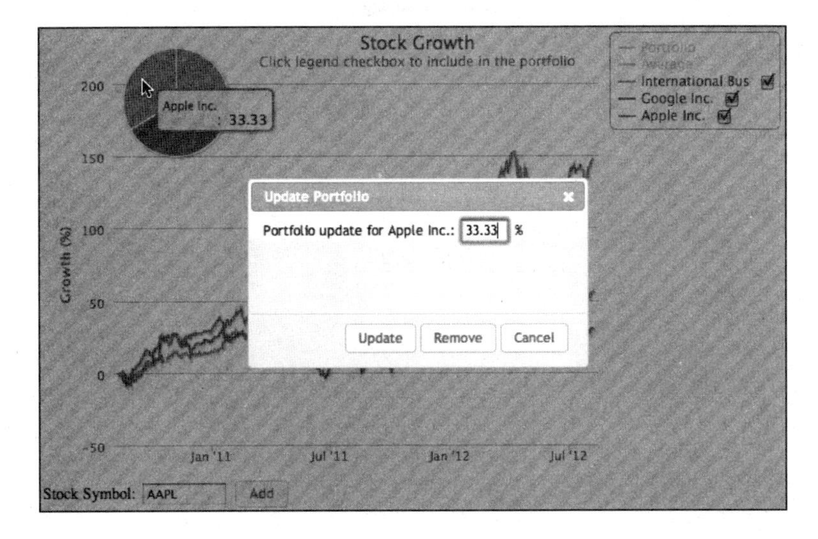

The **Update** button in the **Update Portfolio** dialog box updates the pie chart slice with the `Point.update` method, whereas the **Remove** button calls the `Point.remove` method. Both calls trigger the `update` and `remove` events respectively. Here, we define the data point's `click`, `update`, and `remove` events inside the pie chart:

```
series: [ {
    ....
    },
    visible: false,
    type: 'pie',
    point: {
        events: {
            // Bring up the modify dialog box
            click: function(evt) {
                // Store the clicked pie slice
                // detail into the dialog box
                $('#updateName').text(evt.point.name);
                $('#percentage').val(evt.point.y);
                $('#dialog-form').dialog("option",
                    "pieSlice", evt.point);

                $('#dialog-form').dialog("open");
            },
            // Once the Update button is clicked,
            // the pie slice portion is updated
            // Hence, this event is triggered and the
            // portfolio series is updated
            update: function(evt) {
                updatePortfolio();
            },
            // Pie slice is removed, unselect the series
            // in the legend checkbox and update the
            // portfolio series
            remove: function(evt) {
                var series = nameToSeries(this.name);
                series && series.select(false);
                updatePortfolio();
            }
        }
    }
```

The `click` event function stores the clicked slice (the point object) inside the Modify dialog box and launches it. Inside the dialog box, the **Update** and **Remove** buttons' button handlers then extract these stored point object, call the pie chart, and use the objects' update or remove method to reflect the change in the displayed pie chart. This subsequently triggers point `update` or `remove` event handlers and calls the predefined function, `updatePortfolio`, that recalculates the Portfolio series, with the new distribution among the included stocks. So let's update the distribution for the best past performance stock to an 80 percent ratio and the other two stocks to 10 percent each. The Portfolio series automatically readjusts itself from the `update` event, as shown in the following screenshot:

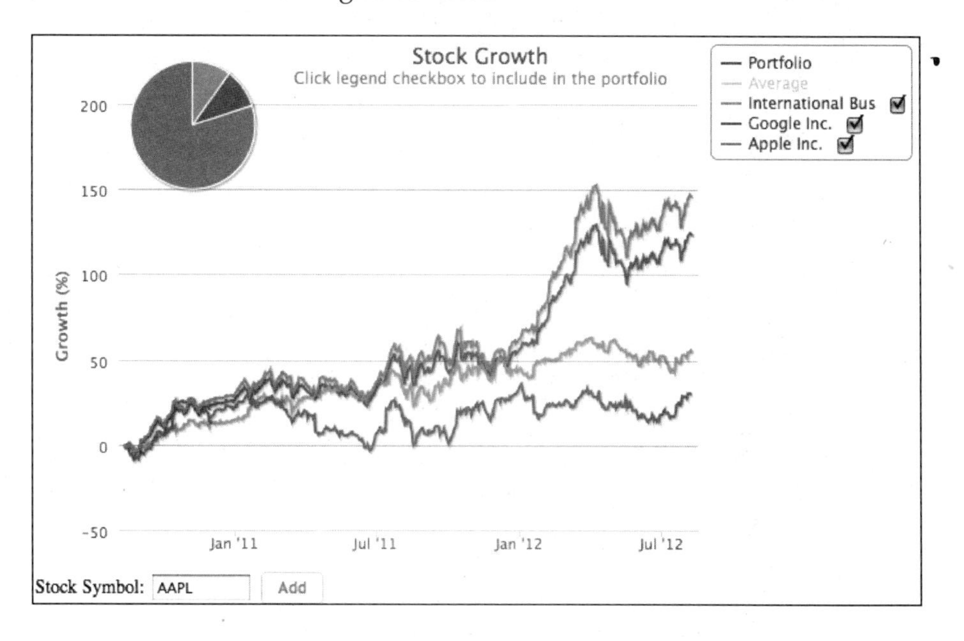

As we can see, the Portfolio series (the second line from the top) has been weighted towards the higher growth trend rather than being in the middle of all the stocks, like the Average series.

Summary

In this chapter, we covered the last part of Highcharts configuration: events handling. We built two share portfolio applications using jQuery and jQuery UI to demonstrate most of the Highcharts events.

In the next chapter, we will bring Highcharts to mobile devices with jQuery Mobile.

12

Highcharts and jQuery Mobile

Highcharts does not only work in desktop browsers; it also supports mobile platforms. In this chapter, we will explore how to deploy Highcharts to mobile platforms with a web mobile framework, jQuery Mobile, that is built on top of jQuery. A very brief introduction to jQuery Mobile is given. We will look into a couple of areas that are crucial to understanding the basics of the mobile framework. Then, we will integrate Highcharts and jQuery Mobile by building a mobile application using an Olympic 2012 medals table. We will demonstrate how to apply mobile events such as swipe, rotation, and pinch to navigate through the charts. In this chapter, we will cover the following topics:

- Introducing jQuery Mobile
- Understanding mobile page structure
- Understanding page initialization
- Linking between mobile pages
- Integrating Highcharts and jQuery Mobile
- Drilling down for data from one chart to another
- Changing chart displays with touch actions: swipe, rotate, and pinch

A short introduction to jQuery Mobile

This chapter is not a full tutorial for jQuery Mobile (jQM) by any means, but it is a quick-start guide for using it with Highcharts. JQuery Mobile is a web development framework for mobile devices built on top of jQuery. It is designed to be compatible across all mobile platforms, and the UI look and feel emulate native mobile applications. The benefit of this is low cost development in a single-source code, without the need for testing across all mobile platforms and browsers.

Before we drill down into how Highcharts can be integrated with jQM, a few important concepts need to be understood.

Understanding mobile page structure

The most important concept in jQM is to understand the structure of a mobile page, which is not the same as a normal HTML page. A mobile page is constructed inside an HTML `<div>` box with a jQM-specific attribute, `data-role='page'`, marked as a boundary. In fact, the `data-*` syntax is **Customer Data Attributes** defined in HTML5 standard. This allows web developers to store custom data specific to the page or application, which can easily access the data attribute values. For more information on APIs for HTML visit `http://dev.w3.org/html5/spec/single-page.html#custom-data-attribute`. Within a mobile page, normal HTML tags, such as input, hyperlinks, select, and so on, are used.

An HTML document can contain multiple mobile pages and links through an anchor and the `id` attribute. An anchor is the same as a normal HTML anchor (for example, `#chart`). The framework resolves the anchor reference and retrieves a mobile page with the matching `id` attribute, which has the following syntax:

```
<div data-role="page" id="chart">
```

The following is an example of a single mobile page in an HTML document:

```
<!DOCTYPE html>
<html>
<head>
  <title>My Page</title>
  <meta name="viewport"
        content="width=device-width, initial-scale=1,
                 maximum-scale=1, user-scalable=0">
  <!-- CDN loading of jQuery and jQM -->
  <link rel="stylesheet"
   href="https://code.jquery.com/mobile/1.4.3/jquery.mobile-1.4.3.min.
css" />
  <script
   src="http://code.jquery.com/jquery-2.1.1.min.js"></script>
  <script
   src="http://code.jquery.com/mobile/1.4.3/jquery.mobile-1.4.3.min.
js"></script>
</head>
<body>
  <div data-role="page">
     <div data-role="header">
        <h1>jQuery Mobile</h1>
     </div><!-- /header -->
     <div data-role="content">
        ....
```

```
      </div>
   </div><!-- /page -->
</body>
</html>
```

Depending on the purpose of the mobile application, all the pages can be built into a single HTML document or they can exist in separate documents. One important aspect is that, if multiple mobile pages are defined within a document, the first page in the `<body>` tag is always loaded on the screen. In jQM, a page is generally composed of the head and content, optionally a footer and navigation bar. Each component also has a `<div>` box with `data-role` to indicate the type of component within a mobile page. The following code shows how multiple mobile pages in a document are loaded:

```
<div data-role="page" >
   <div data-role="header">
      <h1>jQuery Mobile</h1>
      <a href="#config" data-rel='dialog'
         data-icon="gear">Options</a>
   </div><!-- /header -->
   <div data-role="content">
       ....
   </div>
</div><!-- /page -->

<!-- Page for the option dialog -->
<div data-role="page" id='config'>
   <div data-role="header">
       <h1>Config</h1>
   </div><!-- /header -->

   <div data-role="content">
       <a href="#" data-role="button"
          data-rel="back" >Cancel</a>
   </div>
</div><!-- /page -->
```

As we can see, there are two `<div>` boxes with the `data-role='page'` attribute. The first `<div>` box is the same as the previous example with an additional `Options` link button that redirects to the second mobile page, `id='config'`. The `data-icon="gear"` attribute decorates the button with a gear icon provided by the framework.

For the full list of icons visit `http://demos.jquerymobile.com/1.4.3/icons`. When the button is pressed, it will open the second page as a modal dialog box due to the `data-rel='dialog'` attribute. The following screenshot shows the view of the first mobile page as it appears on an iPhone:

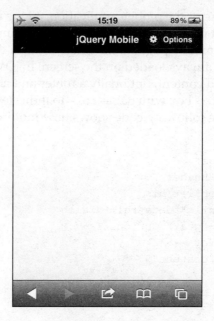

Understanding page initialization

In this section, we will look at why we don't use the traditional DOM-ready method to run initialization code for mobile pages. Suppose a page's content requires some sort of initialization, then using the traditional DOM-ready method `$.ready` can have a negative effect. This is because the `$.ready` method runs as soon as all the DOMs inside the document are loaded. In other words, we have no control over when to run the jQM page initialization code if it is inside the DOM ready handler.

However, jQM provides a specific event, `pageinit`, that caters for this scenario. All we need to do is to assign an `id` value inside the `<div data-role='page'>` markup, then define the `pageinit` event handler for that `id` value. Whenever a page is going to be initialized for display, this event is triggered. Note that the `$.ready` method is still going to be called, but we just don't use it in jQM. To demonstrate this concept, let us use the previous multipage example with an additional `$.ready` call:

```
<script type="text/javascript">
    $(document).on('pageinit', '#main_page', function() {
         alert('jQuery Mobile: pageinit for Main page');
    });

    $(document).on('pageinit', '#config', function() {
         alert('jQuery Mobile: pageinit for Config page');
    });

    $(document).ready(function() {
         alert('jQuery ready');
    });
</script>
</head>
<body>
  <!--  MAIN PAGE -->
  <div data-role="page" id='main_page'>
     <div data-role="header">
        <h1>jQuery Mobile</h1>
           <a href="#config" data-rel='dialog'
              data-icon="gear"
              class='ui-btn-right'>Options</a>
     </div><!-- /header -->

     <div data-role="content" id=''>
     </div>
  </div><!-- /page -->

  <!-- CONFIG PAGE -->
  <div data-role="page" id='config' >
     <div data-role="header">
        <h1>Config</h1>
     </div><!-- /header -->

     <div data-role="content">
        <a href="" data-role="button"
           data-rel="back" >Cancel</a>
     </div>
  </div><!-- /page -->
```

There are two mobile pages defined in this example: `main_page` and `config`. Each mobile page is tied to its `pageinit` event handler. With the `$.ready` method, we can observe the call sequence with other `pageinit` events. When we first load the document to the browser, we see the following screenshot:

Remember that jQM always displays the first page in the HTML body. That means the `pageinit` event for `main_page` is fired as soon as the DOM for `main_page` is fully loaded and initialized for the display. It is also important to understand that, at the point of execution, the DOM for the subsequent `config` page is not loaded yet. When we touch the **OK** button, the execution resumes and the DOM for the `config` page is then loaded. Hence all the DOMs in the document are loaded and the `$.ready` method is then called; it shows the second alert message as shown in the following screenshot:

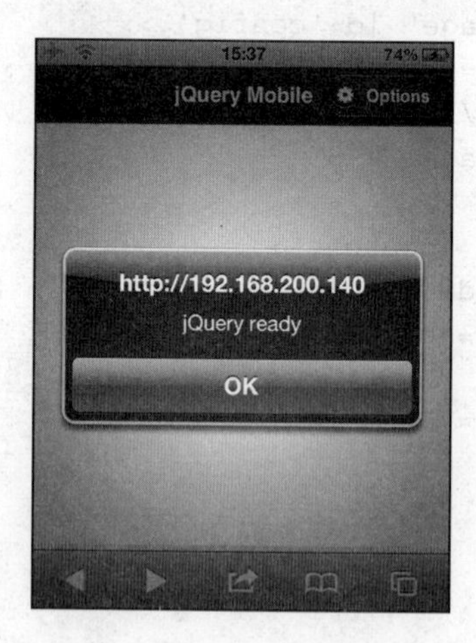

When we touch the **OK** button, the alert box disappears and the control returns back to the browser. Now, if we touch the **Options** button in the top right-hand corner, the `config` dialog page is initialized and displayed on the screen. Hence the `pageinit` handler for the `config` page is called:

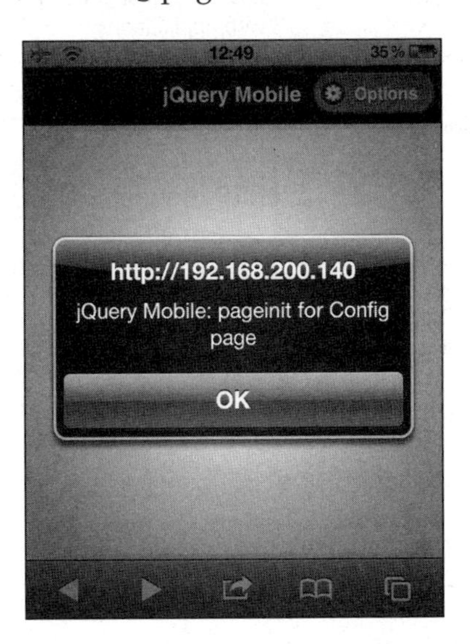

Linking between mobile pages

The second important concept in jQM is how the mobile pages are linked together. Understanding this concept can help us to design a web mobile application with a fluid user experience. In jQM, there are two ways to load an external mobile page: HTTP and Ajax. Depending on how we set the `data-` attribute, it interprets the `href` value and decides which way to load a mobile page. By default, apart from the first document load that is a normal HTTP transfer, the mobile page is loaded through Ajax.

The following block diagram explains how multiple mobile page blocks are managed within a document:

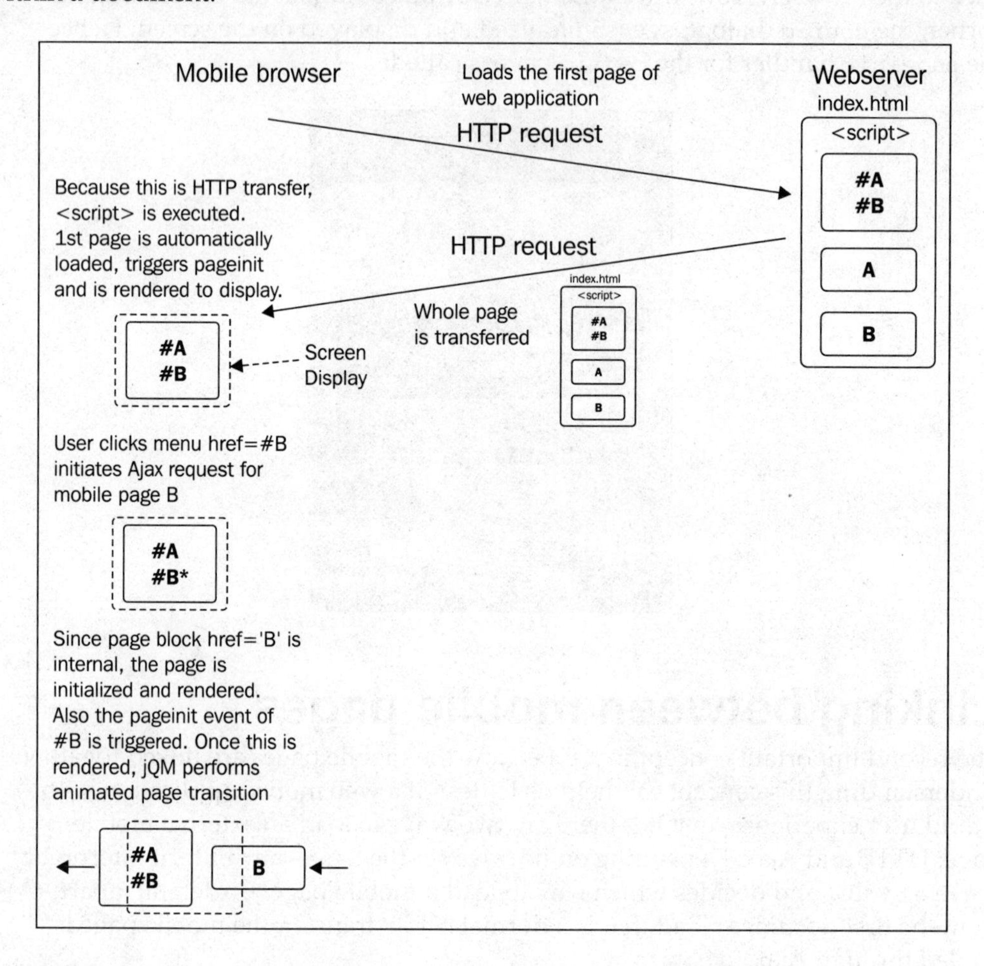

When a mobile page invokes another mobile page, the jQM framework basically parses the `href` value. Since this is an anchor reference, it indicates that this is an internal mobile page block. The framework locates the page block from the current DOM by matching the `id` value. It then initializes and renders the page, which also triggers the `pageinit` event for page B as shown in the previous block diagram.

Suppose we have two separate HTML documents in which a button on one page is referring to the other document. The following block diagram describes the scenario:

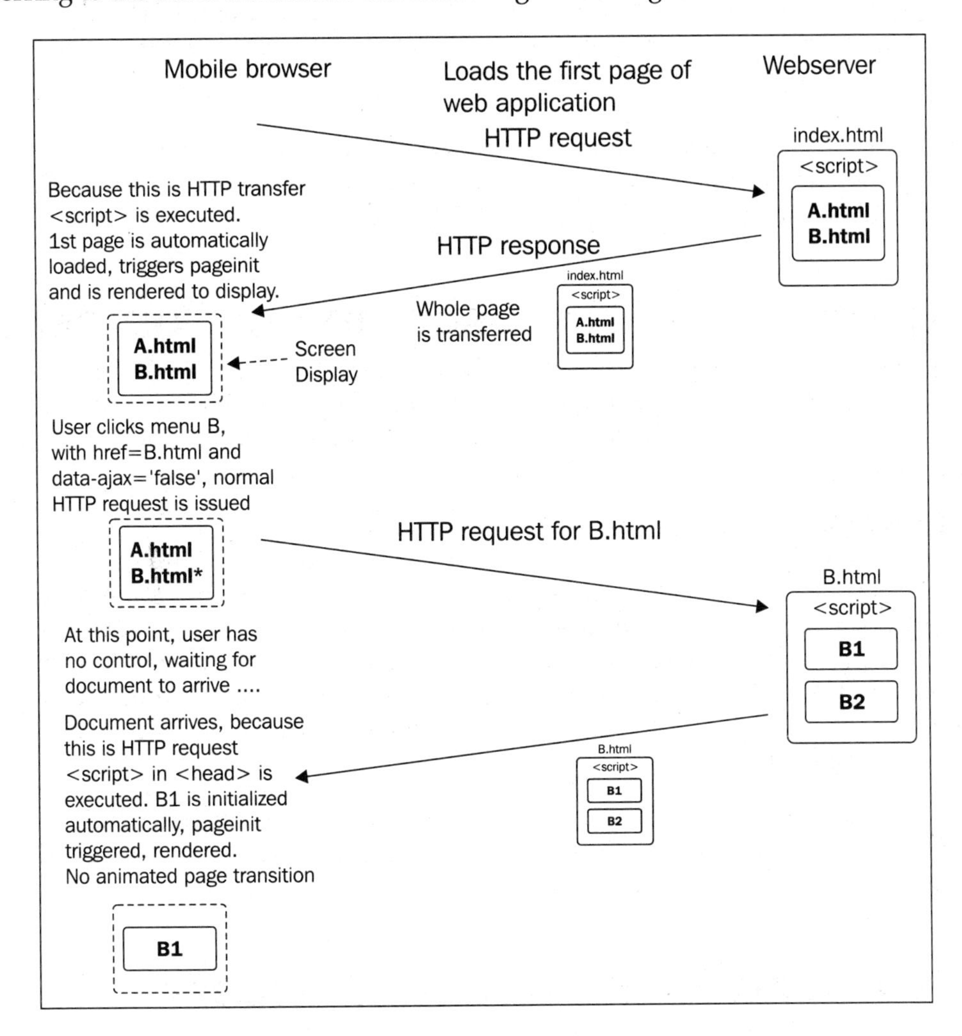

In this case, we add an attribute, `data-ajax="false"` (for the sake of a simpler approach to managing the JavaScript code), to tell jQM that this button requires a document load instead of a background Ajax load. This is important because otherwise the `pageinit` handler code (or any JavaScript file) inside the `<script>` tag will not be loaded for the new mobile page, `B.html`.

> JavaScript code can be embedded inside a `<script>` tag within a mobile page block and executed. The downside of this approach is that it requires more code management, as each page block has its own `pageinit` handler code.

There is an alternative way to load Ajax in multiple documents, but we will leave it here. This is more than sufficient to implement a simple mobile web application. Readers can learn more from the jQuery Mobile documentation.

Highcharts in touch-screen environments

The good thing about Highcharts is that it works perfectly well on both desktop browsers and web mobile environments without requiring any change of code. The only part that needs some consideration is events handling, because mobile devices are all touch-screen based and that means the mouse cursor is invisible.

In Highcharts, all the mouse hover events can still be triggered in touch devices, even though the mouse cursor is not shown. For instance, suppose we define a series with the `mouseOut`, `mouseOver`, and `click` events handling. If we touch the series, both the `mouseOver` and `click` events are triggered. However, if we touch another series causing the previous selected series to be unselected, a `mouseOut` event for the first series is fired. Needless to say, the sequence of events would be different with a real pointing device. In general, we should refrain from using any mouse hover events in touch-screen based devices.

In the next section, we will learn how to integrate jQM with Highcharts, including how to apply touch events to charts, how to use the chart `click` events to launch another chart and mobile page, and so on.

Integrating Highcharts and jQuery Mobile using an Olympic medals table application

In this section, we will build a mobile application for browsing the results of the Olympic 2012 medals table. This application is only tested on iPhone and iPad. The startup screen provides four menus for looking up the results, as shown in the following screenshot:

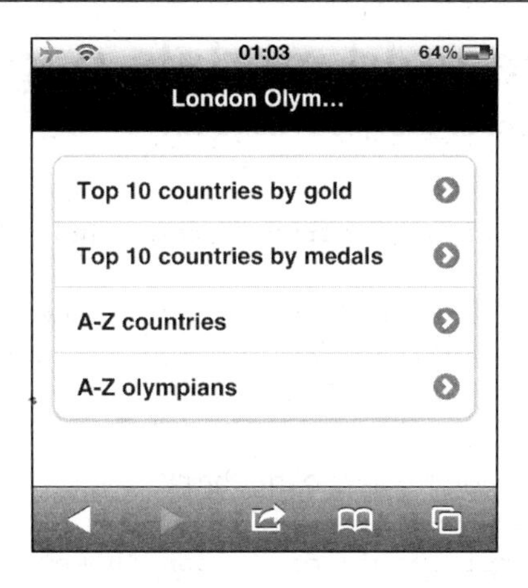

The initial page is made up of a list of four hyperlinks referring to other pages, as shown in the following code:

```
<head>
    <!-- CDN load of Highcharts, jQuery and jQM -->
    ....
</head>
<body>
<div data-role="page">
   <div data-role="header">
      <h1>London Olympic 2012 </h1>
   </div><!-- /header -->

   <div data-role="content">
       <ul data-role="listview" data-inset="true">
           <li><a href="./gold.html"
                data-ajax="false" >Top 10 countries by gold</a></li>
           <li><a href="./medals.html"
                data-ajax="false" >Top 10 countries by medals</a></
li>
           <li><a href="#">A-Z countries</a></li>
           <li><a href="#">A-Z olympians</a></li>
       </ul>
   </div>
</div><!-- /page -->
</body>
```

So, when the **Top 10 countries by gold** button is clicked, the `gold.html` file is HTTP-loaded (not Ajax) because we define `data-ajax="false"`. Since it is an HTTP load, the whole page, `gold.html`, is loaded onto the browser as well as everything within the `<script>` tags that are being executed.

Loading up the gold medals page

The following is the content of the `gold.html` file:

```
<head>
   <!-- CDN load of Highcharts, jQuery and jQM -->
   ....
   <script type="text/javascript" src="common.js"></script>
   <script type="text/javascript" src="gold.js"></script>
</head>
<body>
<div data-role="page" id='gold_chart'>
   <div data-role="header">
    <a href="olympic.html" data-icon="home"
       data-iconpos="notext">Home</a>
    <h1>London Olympic 2012 - Top 10 countries by gold</h1>
    <a href="#options" data-rel='dialog'
       data-icon="gear" id='options'>Options</a>
   </div><!-- /header -->

   <div data-role="content">
     <div id='chart_container'></div>
   </div>
  </div><!-- /page -->

  <!-- options dialog -->
  <div data-role="page" id='options' >
    ....
  </div>
</body>
```

Since this whole HTML document is HTTP-loaded onto the browser, the `common.js` and `gold.js` files are also loaded (See `http://joekuan.org/Learning_Highcharts/Chapter_12/`). The file `common.js` contains common routine code shared in the demo, such as device detection, orientation detection, chart creation, and so on. The `gold.js` file contains the `pageinit` handler code for all the mobile pages in the `gold.html` file. As the mobile page block, `gold_chart`, is the first defined block in the document, it is automatically loaded and rendered to the display; thus, the `pageinit` event for the `gold_chart` page block is triggered.

Detecting device properties

For detecting mobile devices, the techniques range from string matching of the `navigator.userAgent` option, `jQuery.support`, `jQuery.browser` (deprecated), CSS media queries, to a third-party plugin such as Modernizer (see `http://modernizr.com/` for details). However, there is no standard way of doing this. Perhaps it is due to the diverse requirements for compatibility checks. It is beyond the scope of this book to debate the merits of each technique. For this demo, all we are interested in is the difference in screen size; that is, if there is more space in the display (such as with tablet devices), then we display the full country name in the charts instead of the country code, which is used for smaller devices (touch phones). We assume the following technique is sufficient to differentiate between phone and tablet devices:

```
function getDevice() {
    return ($(window).width() > 320)  ? "tablet" : "phone";
}
```

The `$(window).width` property returns the width of the device in pixels, regardless of the device orientation. As for getting the current device orientation, we have the following method:

```
function getOrientation() {
     return (window.innerHeight/window.innerWidth) > 1 ?
             'portrait' : 'landscape';
}
```

Plotting a Highcharts chart on a mobile device

The following is the `pageinit` handler code for the `gold_chart` mobile page:

```
$(document).on('pageinit', '#gold_chart',
      function() {
        var device = getDevice();

        // current orientation
        var orientation = getOrientation();

        $.olympicApp = $.olympicApp || {};

        // Setup the point click events for all the
        // country medal charts - used by plotGoldChart method
        var pointEvt = {
            events: {
               click: function(evt) { ... }
            }
        };
```

```
        // Switch between column and pie chart
        $('#chart_container').on('swipeleft',
            function(evt) { ... } );

        $('#chart_container').on('swiperight',
            function(evt) { ... } );

        // Switch between column and bar chart on
        // smaller display
        $(document).on('orientationchange',
            function(evt) { ... } );

        // General method for plotting gold medal chart
        // Used by dialog box also to toggle chart options
        // such as column stacking, labels etc
        $.olympicApp.plotGoldChart = function(chart, options) {
             .....
        };
        // Create and display Highcharts for gold medal chart
        $.olympicApp.goldChart = createChart({
              device: device,
              orientation: orientation,
              load: $.olympicApp.plotGoldChart,
              type: (orientation == 'landscape') ?
                   'bar' : 'column',
              // legend and title settings specific
              ....

        });
    }
  );
```

The touch events such as `swipeleft`, `swiperight`, and `orientationchange` will be discussed later on. The event handler, `pointEvt`, drills down further to another chart when the user taps on a country bar in the gold medal chart. We will also explore this interaction later on. Let's first focus on the last part of the code, which creates the chart. The `createChart` method is a general routine to create a Highcharts graph that has the common options shared by all the chart mobile pages. For example, the `renderTo` option is always set to `chart_container`, which is inside the `data-role='content'` attribute. The following code shows the `createChart` implementation:

```
// Main routine for creating chart
function createChart(options) {

    // Based on the device display and current orientation
    // Work out the spacing options, labels orientation
    return new Highcharts.Chart({
        chart: {
```

```
                renderTo: 'chart_container',
                type: options.type,
                events: {
                    load: function(chart) {
                        // Execute the page general plot routine
                        options.load &&
                        options.load(chart, options);
                    }
                },
                spacingLeft: ....,
                ....
            },
            title: { text: options.title },
            xAxis: {
               labels: ....
            },
            ....
        });
    }
```

Note that there is no series defined in the `options` parameter and the `options.load` property is set up to call the `plotGoldChart` function once the chart is created and loaded into the browser. The following code snippet is part of the `plotGoldChart` function:

```
// chart is the target chart object to apply new settings,
// options is an object containing the new settings
$.olympicApp.plotGoldChart =
 function(chart, options) {

    // Get the top 10 countries with the
    // most gold medals
    $.getJSON('./gold_10.json',

        function(result) {

            var device = getDevice();

            // Remove any series in the chart if exists
            ....

            // If display pie chart,
            // then we only plot the gold medals series
            if (options && options.type == 'pie') {
                var goldOpt = {
```

```
            data: [],
            point: pointEvt,
            type: 'pie',
            dataLabels: { ... }
        };
        $.each(result.rows,
            function(idx, data) {
                goldOpt.data.push({

                    // If device is phone,
                    // set short country labels
                    // otherwise full names
                    name: (device === 'phone') ?
                        data.code : data.country,
                    y: data.gold,
                    color: pieGoldColors[idx]
                });
        });
        chart.addSeries(goldOpt, false);

        // Disable option button for pie chart
        $('#options').addClass('ui-disabled');

    } else {
        // Sorting out chart option - stacking,
        // dataLabels if specified in the option
        // parameters
        var dataLabel = (options &&
                         options.dataLabel) ? {
            enabled: true,
            rotation:
              (getOrientation() == 'landscape') ?
              0 : -45,
            color: '#404040'
            } : {
            enabled: false
        } ;
        var stacking = (options &&
            options.stacking) || null;

        var bronzeOpt = {
            data: [], name: 'Bronze',
            color: '#BE9275',
            stacking: stacking,
```

```
                dataLabels: dataLabel,
                point: pointEvt
              };
            var silverOpt = {
                data: [], name: 'Silver',
                color: '#B5B5B5',
                stacking: stacking,
                dataLabels: dataLabel,
                point: pointEvt
            };
            var goldOpt = {
                data: [],
                name: 'Gold',
                color: '#FFB400',
                point: pointEvt,
                stacking: stacking,
                dataLabels: dataLabel
            };
            var category = [];

            $.each(result.rows,
                function(idx, data) {
                   // Display country code on phone
                   // otherwise name
                      category.push((device === 'phone') ?
                            data.code : data.country);
                   goldOpt.data.push(data.gold);
                   silverOpt.data.push(data.silver);
                   bronzeOpt.data.push(data.bronze);
            });

            chart.xAxis[0].setCategories(category);
            chart.addSeries(bronzeOpt, false);
            chart.addSeries(silverOpt, false);
            chart.addSeries(goldOpt, false);
            // Enable the option button for the
            // column chart
            $('#options').removeClass('ui-disabled');
        }
        chart.redraw();
    });   // function(result)
};  // function(chart, …
```

The `plotGoldChart` method is a general routine to plot a series into an existing chart. The `options` parameter is a configuration object with new settings to be applied to the chart. First, the function invokes an Ajax call, `gold_10.json`, to get the top 10 countries with the most gold medals. Here is what the result looks like in JSON format:

```
{"rows": [{
  "country":"United States","gold":46,"silver":29,"bronze":29,
  "total":104,"code":"USA"
},{
  "country":"China","gold":38,"silver":27,"bronze":23,
  "total":88,"code":"CHN"
},{
  "country":"Great Britain & N.Ireland",
  "gold":29,"silver":17,"bronze":19, "total":65,"code":"GBR"
},{
  "country":"Russia Federation",
  "gold":24,"silver":26,"bronze":32,"total":82,"code":"RUS"
}, {
  ....
}]
```

Upon the results being returned, the handler function examines the `options` parameter for series type, device orientation, and other fields (stacking and data labels from the `config` dialog, which we will discuss later). Then, it creates the chart based on the settings. If the `type` property is `column`, then we create three column series called `Gold`, `Silver`, and `Bronze` with the point `click` event configured. If the `type` value is `pie`, then it creates a single pie series of gold medals with a gradual change of colors and data labels.

So, when the `gold_chart` page is first loaded, a column chart is created and displayed. The following screenshot shows the initial column chart in portrait mode:

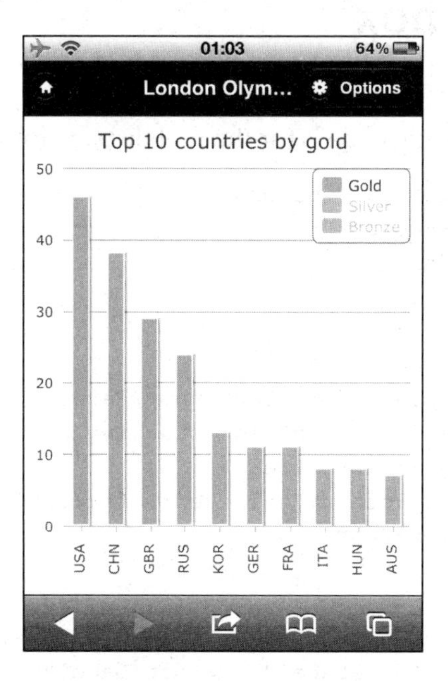

If we touch the legend items to display the number of silver and bronze medals, the chart looks like the following screenshot:

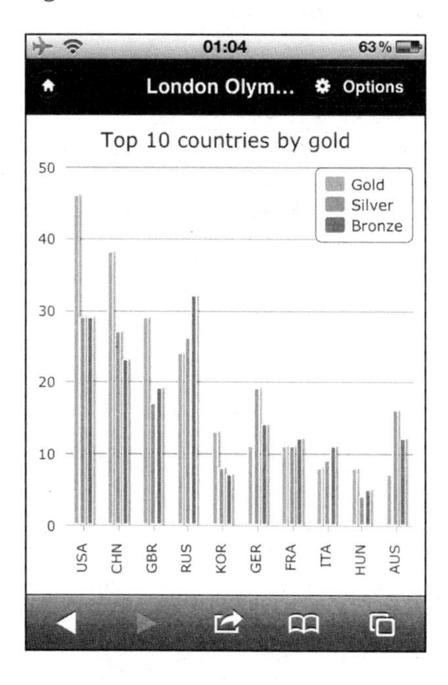

Switching graph options with the jQuery Mobile dialog box

The **Options** button in the top right-hand corner is only enabled if the current display chart is a column chart. It launches an option dialog box for switching stack columns and data labels. The following code is for the mobile page for the dialog box:

```
<div data-role="page" id='options' >
     <div data-role="header">
        <h1>Config</h1>
     </div><!-- /header -->

     <div data-role="content">
        <label for="stacking">Stacking:</label>
        <select name="stacking" id="stacking"
                data-role="slider">
            <option selected="selected">Off</option>
            <option>On</option>
        </select>

        <label for="dataLabel">Show Values:</label>
        <select name="dataLabel" id="dataLabel"
                data-role="slider">
            <option selected="selected">Off</option>
            <option>On</option>
        </select>

        <a href="#" data-role="button"
           data-rel="back" id='updateChart' >Update</a>
        <a href="#" data-role="button"
           data-rel="back" >Cancel</a>
     </div>
</div><!-- /page -->
```

The `<select>` markups in jQM are rendered as slider switches with the `data-role='slider'` attribute and the hyperlinks are rendered as dialog buttons with the `data-role='button'` attribute. The following screenshot shows the dialog page:

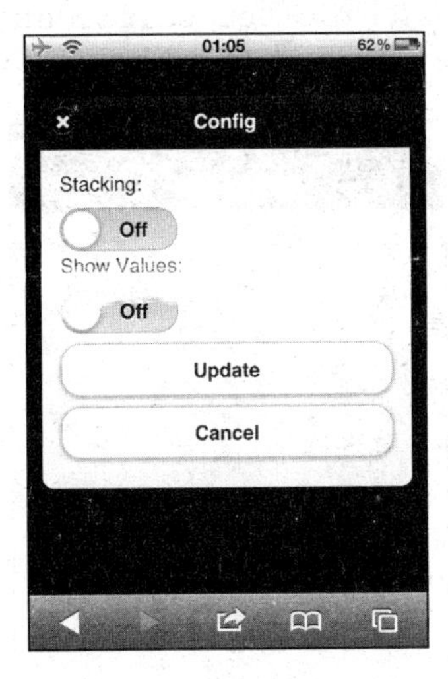

Likewise, we program the `pageinit` handler for the dialog page to initialize the **Update** button action:

```
$(document).on('pageinit', '#options',
        function() {
            var myApp = $.olympicApp;

         $('#updateChart').click(function() {

             var stacking =
                 ($('#stacking').val() === 'Off') ?
                    null: 'normal';
             var dataLabel =
                 !($('#dataLabel').val() == 'off');

             myApp.plotGoldChart(myApp.goldChart, {
                 stacking: stacking,
                 dataLabel: dataLabel
             });
         });
});
```

Actually, the action code for the button is very simple. Since we define the **Update** button with the `data-rel='back'` attribute, as soon as we tap the button, the dialog box is closed and we go back to the previous page. The option values from the `<select>` inputs are passed to the `plotGoldChart` routine to redraw the current chart. The following is a screenshot with only **Show Values** switched on:

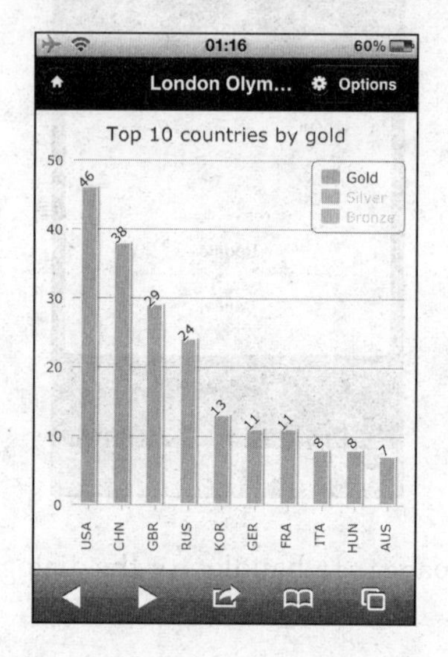

The following screenshot shows a column chart with both stacking and data labeling switched on:

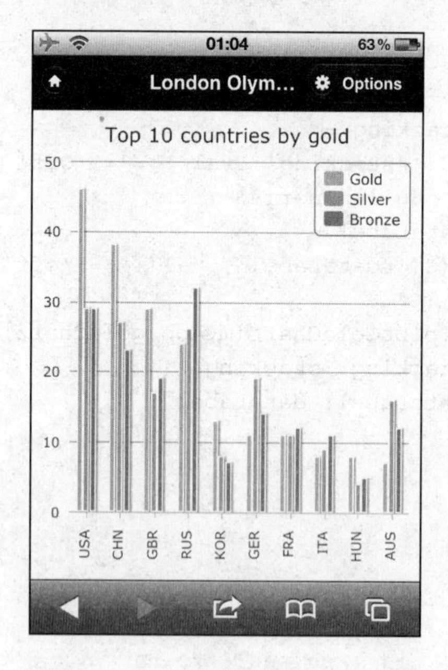

Changing the graph presentation with a swipeleft motion event

Here, we enhance the experience by adding a `swipeleft` event to a chart. What we try to achieve is to apply a swipe motion from the right-hand side to the left-hand side on an existing column chart. This action switches the column chart to a pie chart with the same dataset, and vice versa with the swiperight motion:

```
// Switch to pie chart
$('#chart_container').on('swipeleft',
    function(evt) {
        var myApp = $.olympicApp;
        if (myApp.goldChart.series[0].type == 'column') {
            myApp.plotGoldChart(myApp.goldChart, {
                type: 'pie'
            });
        }
    });
// Switch back to default column chart
$('#chart_container').on('swiperight',
    function(evt) {
        var myApp = $.olympicApp;
        if (myApp.goldChart.series[0].type == 'pie') {
            myApp.plotGoldChart(myApp.goldChart);
        }
});
```

The guard condition inside the handler is to stop redrawing the chart with the same presentation. The following is the view after the `swipeleft` action:

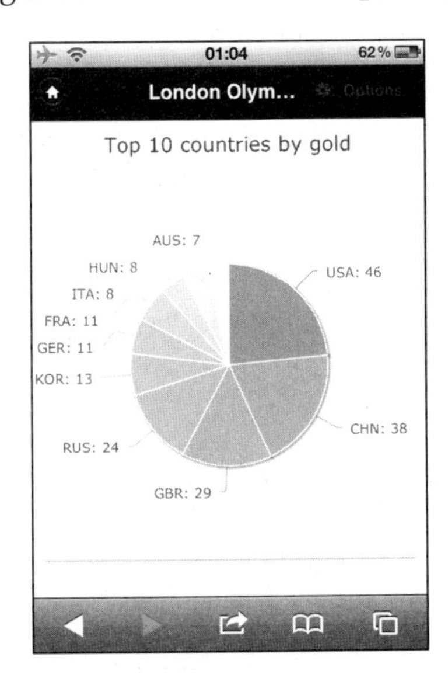

Switching the graph orientation with the orientationchange event

Assume that we are viewing the column chart on a touch-phone device in the portrait position. If we rotate the device, the chart will resize itself but the scale along the *y* axis is squashed. As a result, there is less clarity when comparing how well each country did. To overcome that, we use another jQuery Mobile event, `orientationchange`, that triggers when the mobile device is rotated. Here is the implementation for the handler:

```
var device = ($(window).width() > 750) ? 'tablet' : 'phone';

// Switch between vertical and horizontal bar
$(document).on('orientationchange',
      function(evt) {

        // We only do this for phone device because
        // the display is narrow
        // Tablet device have enough space, the
        // chart will look fine in both orientations
        if (device == 'phone') {

          var myApp = $.olympicApp;
          var orientation = getOrientation();

          // I have to destroy the chart and recreate
          // to get the inverted axes and legend box
          // relocated
          myApp.goldChart.destroy();

          // create the chart optimized for horizontal
          // view and invert the axis.
          myApp.goldChart = createChart({
               device: device,
               orientation: orientation,
               inverted: (orientation === 'landscape'),
               load: myApp.plotGoldChart,
               legend: ....,
          });

          // Hide the address bar
          window.scrollTo(0,1);
        }
      }
   );
```

We recreate the chart with the `inverted` option set to `true` to swap both x and y axes, as well as positioning the legend in the lower-right corner instead. A method for the chart `load` event is also set up in the configuration. In the end, an inverted chart is produced, as shown in the following screenshot:

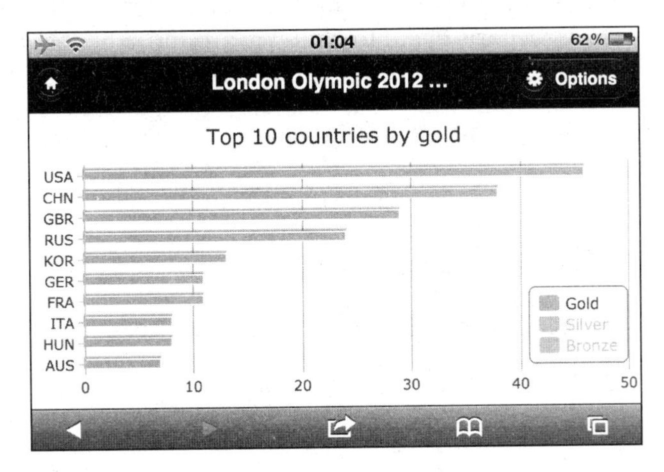

The following is a screenshot from a tablet device showing the gold and silver medals' chart:

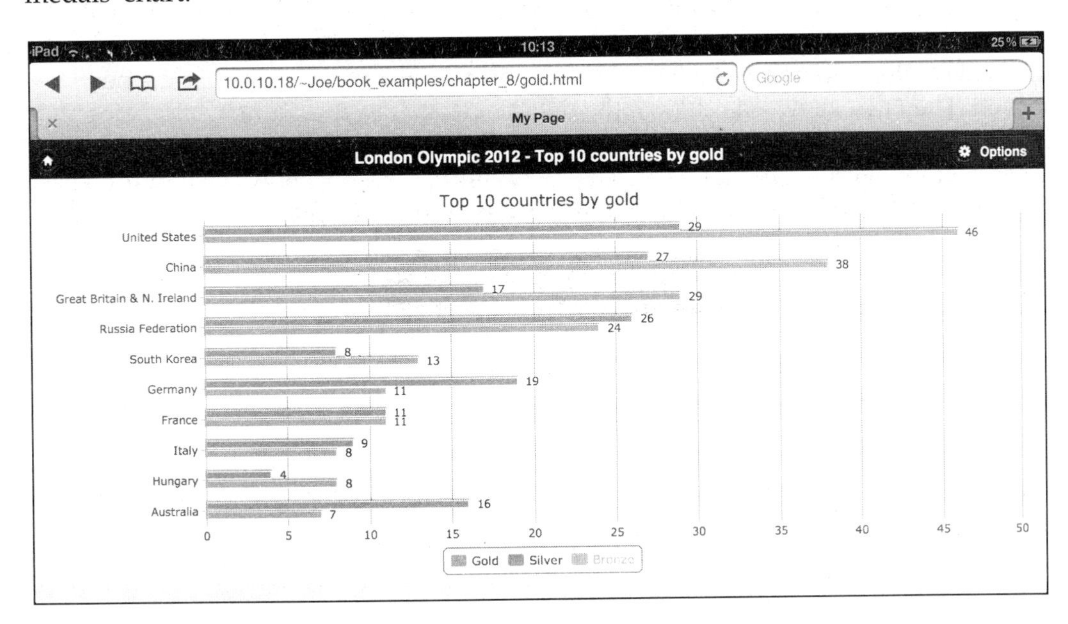

The `plotGoldChart` method detects a larger display and renders the chart (using the `setCategories` method) with full country names instead of the country code from the JSON result.

Drilling down for data with the point click event

So far we have only fiddled with the top countries ordered by gold medals. Let's see how we can use a Highcharts event to navigate around other charts in jQuery Mobile. Back with the `pageinit` handler code for `chart_page`, we declared the `pointEvt` variable, which is a `click` event handler shared by all the series in gold medal charts. The following code is for the event:

```
var pointEvt = {
    events: {
        click: function(evt) {
            document.location.href = './sport.html?country='
            // Country code or name
            + encodeURIComponent(this.category) +
            // Medal color
            '&color=' + this.series.name;
        }
    }
};
```

This event is triggered by touching a bar in a column chart or a slice in a pie chart. As a result, it loads a new document page (`sport.html`) with a bar chart. The URL for the document page is built inside the handler with the selected country code and the medal color as parameters. The HTML content of the page is listed in the next section. The `this` keyword refers to the data point (that is, the country bar) being clicked. The bar chart displays the list of sports in which the selected country won medals, along with the medal color. The following screenshot shows a chart for a list of sports in which Great Britain and Northern Ireland won gold medals:

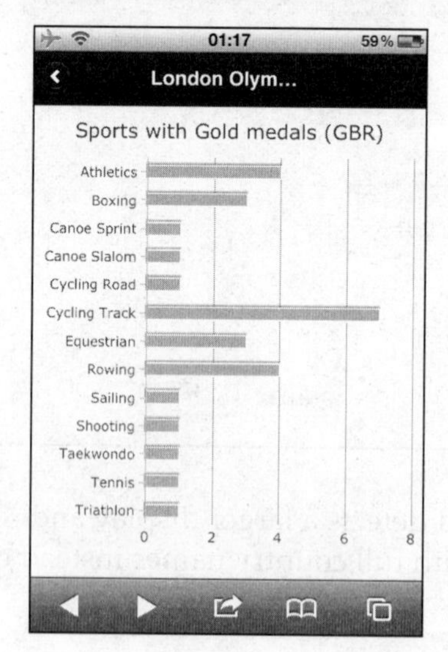

Inside the new page, it uses similar code to the gold medal countries chart to produce the graph shown in the preceding screenshot. The only difference is that it is embedded with the point `click` callbacks. We will see that in the next section.

Building a dynamic content dialog with the point click event

Now we know which sports have achieved gold medals in the Olympics, but we want to further find out who the medalists are. Let's touch the **Athletics** bar in the chart. A dialog box appears and presents a list of athletes in thumbnails along with their names, photos, and their event information, as shown in the following screenshot:

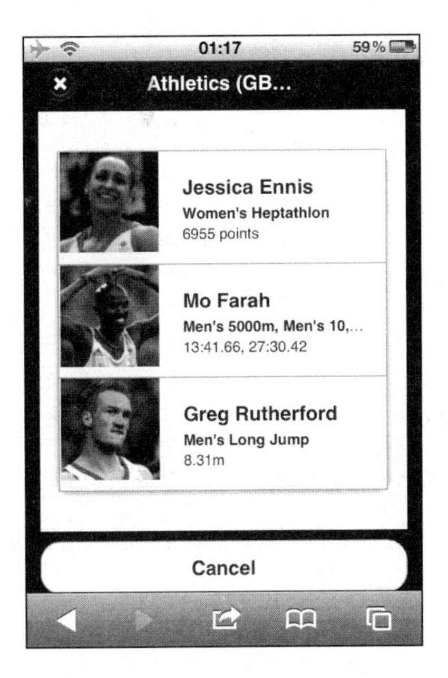

Notice that the dialog box shown in the preceding screenshot is not a static HTML page. This is constructed via a point `click` event and builds the dialog box content dynamically from the result. The problem is that, in order to launch a dialog page in jQM, we need to have a button somewhere on the page to start from. The trick is to create a jQM hidden button and a program to invoke the click action from inside the event handler. The following code is the HTML (`sport.html`) for both the hidden button and dialog page:

```
<!-- hidden dialog -->
<a id='hiddenDialog' href="#atheletes_dialog"
   data-rel="dialog" data-transition="pop"
```

```
    style='display:none;'></a>
<!-- medalists page -->
<div data-role="page" id='atheletes_dialog' >
    <div data-role="header">
        <h1></h1>
    </div><!-- /header -->

    <div data-role="content">
        <ul data-role="listview" data-inset="true"
            id='atheletes_list' >
        </ul>
    </div>

    <a href="" data-role="button" data-rel="back">Cancel</a>
</div><!-- /page -->
```

The following is the implementation of the `click` handler for the sports chart:

```
point: {
    events: {
        click: function(evt) {
            var params = {
                country : urlParams.country,
                color : urlParams.color.toLowerCase(),
                sport : this.category
            };

            // Set the title for dialog
            $('#atheletes_dialog h1').text(this.category +
                " (" + urlParams.country + ") - " +
                urlParams.color + " medalists");

            // Simulate a button click to launch the
            // list view dialog
            $('#hiddenDialog').click();

            // Launch ajax query, append the result into
            // a list and launch the dialog
            $.getJSON('./olympic.php', params,
                function(result) {

                    $("#atheletes_list").empty();
                    $.each(result.rows,
                        function(idx, athelete) {
                            // Formatting the image, name, and
                            // the sport event
                            var content = "<li><img src='" +
                                athelete.image + "' />" + "<h3>" +
                                athelete.name + "</h3><p><strong>"
                                + athelete.event +
```

```
                        "</strong></p><p>" +
                        athelete.desc + "</p></li>";

                    $("#atheletes_list").append(content);
                });
                // Need this to apply the format to
                // the new entry
                $('#atheletes_list').listview('refresh');
            });  // getJSON
        }
    }
```

First, we assemble the title for the dialog page ready to launch. Then, we trigger an action click to the hidden button with the call as follows:

```
$('#hiddenDialog').click();
```

This in turn generates a click event to launch the dialog page. Then, we issue an Ajax query for the list of medalists with the current selected country, medal color, and sport as the filters. The server page, `olympic.php`, contains the Olympic result of each nation. It sorts the result according to the URL parameters and formats the ordered list in JSON format. We then convert each item from the JSON result and insert them into the `<ul>` list, `atheletes_list`.

Applying the gesturechange (pinch actions) event to a pie chart

So far, we have only explored actions involving a single touch point. Our next goal is to learn how to apply more advanced action events with multi-touch. One of the most common actions is the pinch-in/out for zooming out/in respectively. The Safari browser for iOS supports this motion with the `gesturestart`, `gesturechange`, and `gestureend` events. Whenever there are two or more fingers touching the screen, the `gesturestart` event is fired. Then, the `gesturechange` event is triggered when the fingers are moved on the screen. When the fingers leave the screen, the `gestureend` event is generated. In returning control to the event handler, if the action is recognized, a certain property in the event object is updated. For instance, the `scale` property in the event object is set to larger than 1.0 for pinch-out and less than 1.0 for pinch-in. For the `GestureEvent` class reference, please see `https://developer.apple.com/library/mac/documentation/AppleApplications/Reference/SafariWebContent/HandlingEvents/HandlingEvents.html`.

In this section, we will apply the pinch motions to a pie chart. For the pinch-out action, we turn the pie chart into a doughnut chart with extra information on the outer ring—and vice versa for pinch-in, turning the doughnut chart back to a pie chart. First of all, let's launch a new chart, **Top 10 countries by medals**, the second item from the front menu. The following screenshot is the output of the chart:

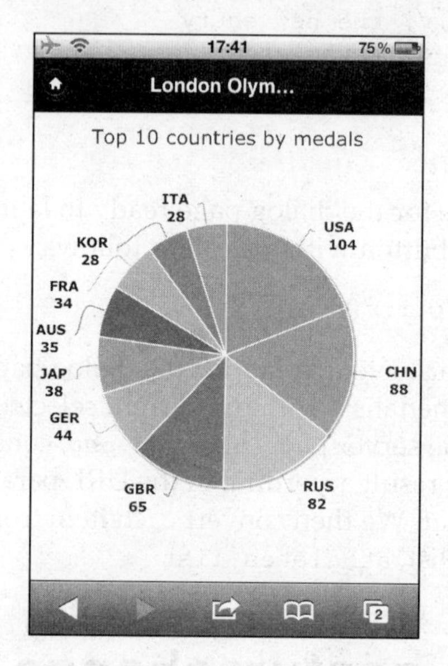

When we perform a pinch-out action, the chart is redrawn as shown in the following screenshot:

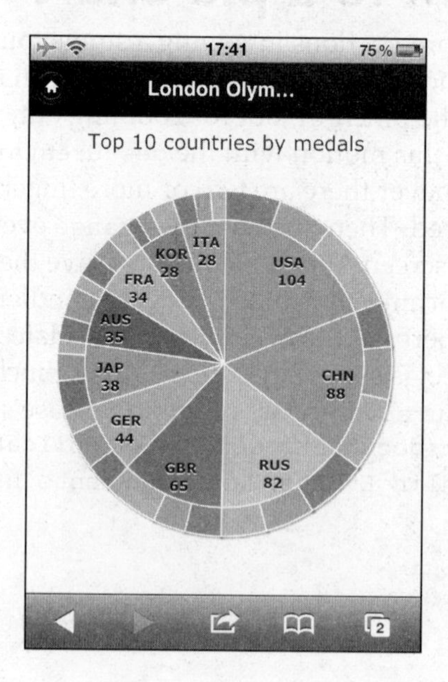

The outer ring shows the ratio of each color medal for each country. Moreover, the original pie chart data labels move inwards to make space for the outer ring. Let's see how the `gesturechange` event is implemented. The following is the code inside the `pageinit` handler:

```
$('#chart_container').on('gesturechange',
    function(evt) {
        evt = evt.originalEvent || evt;
        var myApp = $.olympicApp;

        if (evt.scale > 1) {
            // Pinch open - from pie chart to
            // donut chart
            if (!myApp.medalChart.series[1].visible) {
                myApp.medalChart.destroy();
                myApp.medalChart = createChart({
                    orientation: getOrientation(),
                    device: device,
                    outerRing: true,
                    load: myApp.plotMedalChart
                });
            }
        } else if (myApp.medalChart.series[1].visible) {
            // Pinch close
            myApp.medalChart.destroy();
            myApp.medalChart = createChart({
                orientation: getOrientation(),
                device: device,
                load: myApp.plotMedalChart
            });
        }
});
```

We bind the gesture event to the chart container. This event handler is called whenever there is a multi-touch gesture made on the screen, such as a pinch or rotate motion. In order to make sure this is a pinch action, we need to look into the original event generated by the browser that is wrapped by the jQuery layer. We will examine whether the `scale` property has been set and decide whether it is pinch-in or pinch-out, then we will recreate the pie or doughnut chart if necessary.

Summary

The goal of this chapter was to deploy Highcharts graphs on mobile touch devices. To do that, we used a mobile web development framework, jQuery Mobile. We worked through a very brief introduction and the concepts of the framework. We then examined how to integrate Highcharts with jQuery Mobile.

Then, we demonstrated a mobile application to show the results of the Olympic 2012 medals table. A menu of charts was built using the jQuery Mobile dialog page, and then we showed how to use the single-touch, multi-touch, and orientation events to navigate between charts. We also showed how to use the Highcharts click event to build a dialog page dynamically.

In the next chapter, we will learn how to apply Highcharts with ExtJs, a very powerful and popular **Rich Internet Application** (**RIA**) framework for building a desktop-style application.

13
Highcharts and Ext JS

This chapter starts with an introduction of Sencha's Ext JS. Since the Ext JS framework covers a broad range of features, it comprises a large collection of classes. Therefore, a quick guide on a small set will be given, especially for the user interface components likely to be used with Highcharts. Then, we will learn which Highcharts extension we have for Ext JS and how to create a Highcharts graph within Ext JS. We will also learn about a small set of APIs provided by the extension. After that, we will use network data to build a simple application to demonstrate how the Ext JS components can interact with Highcharts. Finally, we will have a brief look at a commercial Ext JS application working together with Highcharts. In this chapter, we will cover the following topics:

- Introducing and giving a quick tutorial on Sencha Ext JS classes
- Introducing the Highcharts extension for Ext JS
- Demonstrating how to convert a working Highcharts configuration for the extension
- Preparing the Ext JS JsonStore object for the extension
- Describing APIs provided by the extension module
- Illustrating how to create an Ext JS application with the Highcharts extension

A short introduction to Sencha Ext JS

Sencha's Ext JS is one of the most comprehensive **Rich Internet Application** (**RIA**) frameworks on the market. An RIA framework can produce a web frontend that behaves like a desktop application. Ext JS supports many features such as proxy storage, charting, managing SVG, tabs, toolbars, a myriad of different form inputs, and many, many others. There are other popular RIA frameworks, such as the Java-based **Google Web Toolkit** (**GWT**) and Python-based Dojo. Both frameworks can be integrated with Highcharts via third-party contributed software.

See `http://www.highcharts.com/download` under the section *Third Party Implementation* for the full list of software contributed by other developers.

The Highcharts extension was originally written by Daniel Kloosterman for Ext JS 2+ as an adapter, as it didn't support any charts. In Ext JS 3, it started adopting the YUI charting library as the charting solution. However, the charts lacked features and style, and the main drawback was that they required Flash to run. Since Ext JS 3.1, I have been maintaining the Highcharts extension and added features, such as support for donut charts and enhancements for some of the APIs.

Although Ext JS 4 comes with its own chart library, some users still prefer Highcharts over Ext JS 4 charts for style and flexibility. Moreover, Ext JS 4 can run alongside version 3 codes, so it is desirable to enhance the extension to natively support Ext JS 4, which I have done. The extension implementation has always been to follow the original approach, which is to preserve the use of Highcharts configurations as much as possible.

At the time of writing, Ext JS 5 has just been released and the changes from Ext JS 4 to Ext JS 5 are not as drastic as from Ext JS 3 to Ext JS 4. The Highcharts extension has been updated to be fully compatible with both Ext JS 4 and 5. In this chapter, we will focus on working with Ext JS 5. All the examples are simply from the previous edition, which is based on Ext JS 4, and they are updated to work with Ext JS 5.

There are demos online at `http://joekuan.org/demos/Highcharts_Sencha/desktop.extjs5/` and the extension can be downloaded from `http://github.com/JoeKuan/Highcharts_Sencha/`.

Unlike jQuery UI, an Ext JS application is programmed in pure JavaScript, without the need to collaborate with HTML markup or fiddle with particular CSS classes (strictly speaking, there are times when it is necessary to interface with HTML and CSS, but it is not common and is only in small doses). This empowers programmers to focus on developing the entire web application in a single language and to concentrate on application logic. That also pushes the server-side development to reside in data operations only, unlike some approaches that use server-side language with HTML and CSS to serve client pages.

Technically, JavaScript does not have classes: function itself is an object. The Ext JS framework provides access to its components through the class approach, organized in a hierarchical manner. In this chapter, we will use the word "class" to refer to Ext JS classes.

A quick tour of Ext JS components

There are a myriad of classes in Ext JS, and it is beyond the scope of this book to introduce them. Sencha provides three types of online documentation in terms of both quality and quantity: a reference manual, tutorials (written and video), and working demos. You are strongly advised to spend ample time reviewing these materials. In this section, a very brief introduction is given about some components, especially those that are likely to interface with Highcharts. This chapter is by no means enough to get you started with programming in Ext JS, but should be enough to give you an idea.

Implementing and loading Ext JS code

An Ext JS application can always be divided into multiple JavaScript files, but they should always start from one HTML file. The following code snippet demonstrates how to start up Ext JS from an HTML file:

```
<html>
  <head>
    <meta http-equiv="Content-Type"
          content="text/html; charset=UTF-8">
    <title>Highcharts for Ext JS 5</title>
    // At the time of writing, there is no CDN for ExtJs 5
    // Download http://www.sencha.com/products/extjs/download/ext-
js-5.0.0/3164
    <link rel="stylesheet" type="text/css"
          href="/extjs5/packages/ext-theme-classic/build/resources/
ext-theme-classic-all.css" />
  </head>
  <body></body>
  <script type="text/javascript" src="/extjs5/ext-all.js"></script>
  <script type='text/javascript'>
          Ext.require('Ext.window.Window');
          Ext.onReady(function() {
               // application startup code goes here
               ....
          });
  </script>
</html>
```

Ext JS 5 is packaged with various themes. The preceding example demonstrates how to load one of the available themes. We will apply different themes in the examples to show the look and feel of Ext JS 5. The script file, `ext-all.js`, contains all the Ext JS classes in a compressed format.

[Ext JS has the facility to build a custom class file to cut down loading for production deployments. We are leaving that for you to explore.]

`Ext.require` is to load specific classes used in the application. `Ext.onReady` is the DOM-ready method, the same as the `$.ready` jQuery method that the application startup code starts running inside this function.

Creating and accessing Ext JS components

Out of all the classes in Ext JS, we should start by discussing `Ext.Component`, which is the base class for Ext JS user interface components. Depending on the characteristics of the component, some of them such as `Panel`, `Window`, `FieldSet`, and `RadioGroup` can contain multiple components, because they are inherited through another class: `Container`. We will look at `Container` in more detail later.

To create an Ext JS object, we use the `Ext.create` method, which takes two parameters. The first parameter is the string presentation of a class path, for example `'Ext.window.Window'`, or an alias name such as `'widget.window'`. The second parameter is the object specifier, containing the initial values to instantiate a class:

```
var win = Ext.create('Ext.window.Window', {
        title: 'Ext JS Window',
        layout: 'fit',
        items: [{
              xtype: 'textarea',
              id: 'textbox',
              value: 'Lorem ipsum dolor sit amet, ... '
        }]
});

win.show();
```

The preceding code snippet is used to create a window widget and its content is defined through the `items` option. `Window` is a class derived from the `Container` class, which inherits the `items` option for containing other components. When the window is finally created and ready to render, it goes through each object specifier in the items array and creates each component.

The `xtype` option is the Ext-specific type, which has a short unique name to symbolize the component's class path. In Ext JS, all interface components have their own `xtype` names (this refers to the `Ext.Component` manual). The `xtype` option is commonly used for convenience to create components within the container, as opposed to `Ext.create` with a full pathname.

The id field is to give a unique ID name to a component. The purpose is to gain direct access to a component at any point inside a program. To retrieve the component with an ID value, we can execute the following line of code:

```
var tb = Ext.getCmp('textbox');
```

Alternatively, we can use the itemId option to assign a unique name. The difference is that the ID has to be globally unique to the application, whereas itemId only has to be unique within the parent container, to avoid name conflict elsewhere in the application. To access a component with the itemId value, we need to call getComponent from the immediate parent container, as follows:

```
var tb = win.getComponent('textbox');
```

Moreover, we can chain the call all the way from the top level to the desired component, as follows:

```
var val =
win.getComponent('panel').getComponent('textbox').getValue();
```

The 'textbox' (with itemId defined) component is constructed inside the parent container, 'panel', which resides inside the window object. Although the getCmp method provides direct, easy access to a component, it should generally be avoided as part of best practices, due to slower performance and undesired effects if a duplicate ID is accidentally used.

For the sake of avoiding long sample code, we use the getCmp call in some of the demos.

Note that Sencha also provides convenient component navigation, up and down methods which search for target component with CSS style selector. Here is an example:

```
var val = win.down('#textbox').getValue();
```

As we can see the preceding expression is much more simplified and direct. The down method basically traverses down to its children components and so on until come across the first component with the matching criteria. In this case, the matching expression '#textbox' means a component with itemId specified as a textbox. Many different search expression can be used, another example is down('textarea') which means searching for the first child component with xtype value of textarea.

Using layout and viewport

As we mentioned earlier, some types of components have the ability to contain other components, because they are extended from the `Container` class. Another feature of the `Container` class is to arrange the layout between the contained components; the layout policy is specified via the `layout` option. There are about a dozen layout policies: among them `'anchor'`, `'border'`, and `'fit'` are most commonly used (the `card` layout is also used often, but through the tab panel). The `border` layout is widely used in GUI programming. The layout is finely divided into the `'north'`, `'east'`, `'south'`, `'west'`, and `'center'` regions.

When developing an application that requires utilizing the whole browser space, we generally use a `Viewport` class coupled with a `border` layout. `Viewport` is a special type of container whose size automatically binds to the browser. The following is a simple example of using a viewport:

```
var viewport = Ext.create('Ext.container.Viewport', {
   layout: 'border',
   defaults: {
      frame: true
   },
   items: [{
      region: 'north',
      html: '<h1>North</h1>'
   }, {
      region: 'east',
      html: '<h1>East</h1>',
      width: '15%'
   }, {
      region: 'south',
      html: '<h1>South</h1>'
   }, {
      region: 'west',
      html: '<h1>West</h1>',
      width: '20%'
   }, {
      region: 'center',
      html: '<h1>Center</h1>'
   }]
});
```

The following screenshot shows the border layout in a gray theme:

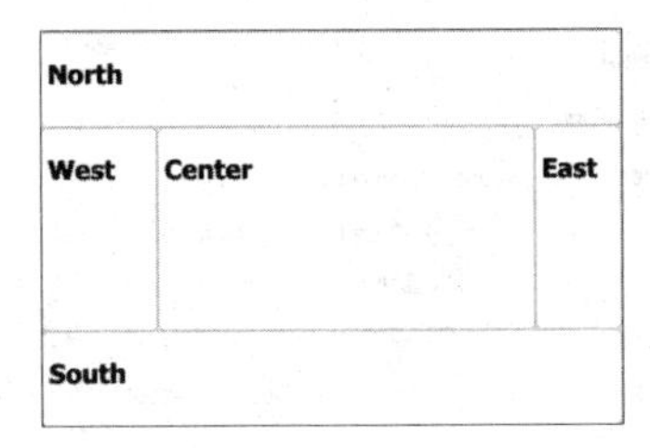

Panel

Panel is a basic container component, and is generally used as a building block with the layout format and then combined with more panels or components. Another general use is to extend the Panel class to a special purpose type of panel, for example PortalPanel in the online portal demo. The most widely used panel types are GridPanel, FormPanel, and TabPanel.

GridPanel

GridPanel is used for displaying data in table format and it comes with lots of useful features, such as drag-and-drop column ordering, column sorting, flexible data rendering, enable or disable column display functions, and many others. GridPanel can also be used with different plugins such as row editor, allowing a user to edit field values on the fly. The class comes with a large set of events settings that can establish smooth coordination with other components. Nonetheless, the most tightly coupled component is the store object, which we will demonstrate in a later section.

FormPanel

FormPanel is a panel for accommodating field input components in form style, that is, labels on the left-hand side, inputs on the right-hand side, and the buttons array. Ext JS provides a great selection of form inputs, such as date time fields, comboboxes, number fields, sliders, and many others. Underneath the FormPanel layer, there is a BasicForm component, which contributes to field validations, form submission, and loading services with the store's Record class for adding and editing entries.

The following is a screenshot of `FormPanel` with various inputs:

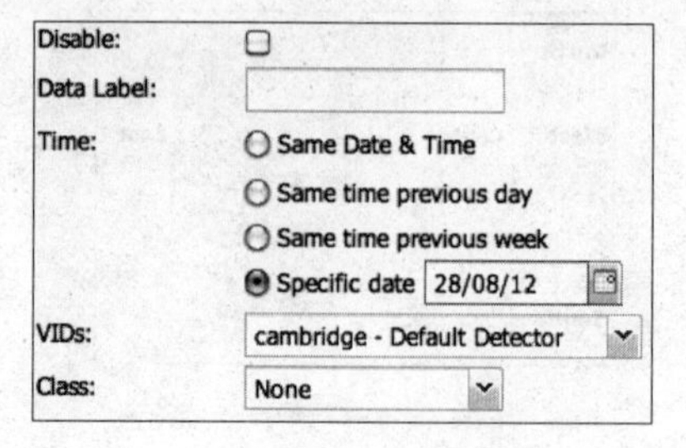

TabPanel

`TabPanel`, as its name implies, is a panel associated with tabs. It supports creating and removing tabs on the fly and scrolling between tabs. The following code snippet shows how to create a tab panel:

```
items:[{
     xtype: 'tabpanel',
     items: [{
          title: 'Tab 1',
          xtype: 'form',
          items: [{
                .....
          }]
     }, {
        title: 'Tab 2',
        ....
     }]
}]
```

The following is a screenshot of tabs within the tab panel, with a scrolling feature:

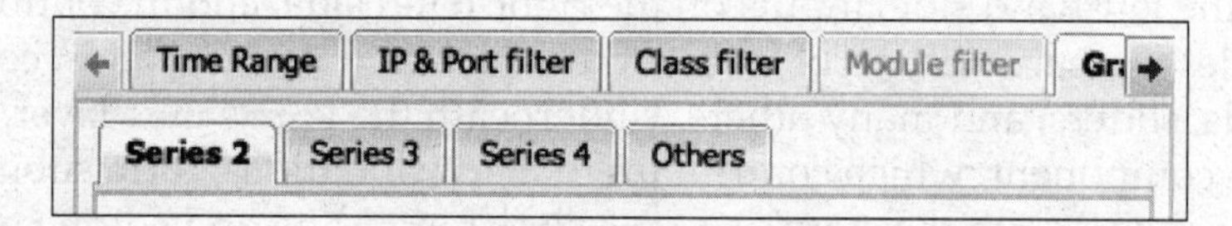

Window

`Window` is a special type of panel that is not bound to any parent container and is free-floating within the application. It offers many features found in normal desktop windows, such as resize and maximize/minimize, and also comes with options for adding a toolbar, footer bar, and buttons. Later, we will see the `Window` panel in action in an example.

Ajax

The Ext JS framework provides its own method, `Ajax.request`, for issuing Ajax queries. This is used when the returned JSON data is not required to be converted into table rows and field columns. The method is generally coupled with `Ext.decode` to convert the returned JSON format string into a JavaScript object and directly access individual fields inside the object.

The following code snippet shows a sample of issuing an Ajax query:

```
Ext.Ajax.request({
      url: 'getData.php ',
      params: { id: 1 },
      success: function(response) {
            // Decode JSON response from the server
            var result = Ext.decode(response.responseText);
            if (result && result.success) {
                  ....
            } else {
                  ....
            }
      }
});
```

Store and JsonStore

`Store` is a general purpose storage class for modeled data. There are several classes derived from `Store`, but the most important one for Highcharts is `JsonStore`. It is a proxy-cached storage class responsible for issuing an Ajax query and unpacks the returned JSON data into modeled data. The `JsonStore` class is often used for accessing database data that resides on the server side. A store object can bind with more than one component, for example a `JsonStore` object can bind to a grid panel and a column chart. Clicking on a column order direction in the grid panel can change the row sequence in `JsonStore`, affecting the order of the columns displayed in the chart. In other words, the `Store` class acts as a skeleton to hold several components working together effortlessly and systematically.

It is important to note that the load method in the `Store` class is asynchronous. An event handler should be assigned to the load event if we want to tie an action after the data is loaded. The action can be specified through `listeners.click` via either the object specifier or `store.on` method.

Example of using JsonStore and GridPanel

So far, a number of Ext JS components have been introduced; we should see how they work together. Let's build a simple window application that contains a table (`GridPanel`) showing a list of hosts with their download usage that are returned from the server. Let's assume that we have rows of data returned in JSON format from the server:

```
{ "data": [
        { "host" : "192.168.200.145", "download" : 126633683 },
        { "host" : "192.168.200.99" , "download" : 55840235 },
        { "host" : "192.168.200.148", "download" : 54382673 },
        ...
] }
```

First we define the data model to correspond with the JSON data. For the sake of simplicity, we can put all our demo code inside `Ext.onReady` rather than in a separate JavaScript file:

```
Ext.onReady(function() {
    Ext.define('NetworkData', {
        extend: 'Ext.data.Model',
        fields: [
            {name: 'host',  type: 'string'},
            {name: 'download', type: 'int'}
        ]
    });
});
```

> It is not mandatory to accept field names returned by the server. `Ext.data.Model` offers the `mapping` option to map an alternative field name to use on the client side.

The next step is to define a `JsonStore` object with the URL, connection type, and the data format type. We will bind the `JsonStore` object with the `NetworkData` data model defined in the preceding code snippet:

```
var netStore = Ext.create('Ext.data.JsonStore', {
    autoLoad: true,
    model: 'NetworkData',
    proxy: {
        type: 'ajax',
        url: './getNetTraffic.php',
        reader: {
            type: 'json',
```

```
                idProperty: 'host',
                rootProperty: 'data'
            }
        }
    });
```

`idProperty` is used to define which field is regarded as an ID if the default `'id'` fieldname is not provided, so that methods such as `Store.getById` can function properly. The `root` option tells the reader (`JsonReader`) which property name holds the array of row data in the JSON response from the server. The next task is to build a `Window` panel with a `GridPanel` class, as follows:

```
var win = Ext.create('Ext.window.Window', {
    title: 'Network Traffic',
    layout: 'fit',
    items: [{
        xtype: 'grid',
        height: 170,
        width: 270,
        store: netStore,
        columns: [{
            header: 'IP Address',
            dataIndex: 'host',
            width: 150
        }, {
            header: 'Download',
            dataIndex: 'download'
        }]
    }]
}).show();
```

We instruct the grid panel to bind with the `netStore` object and define a list of columns to display. We then match each column to the store's data field through the `dataIndex` option. The following is a screenshot showing part of a window (crisp theme) with a grid panel inside it:

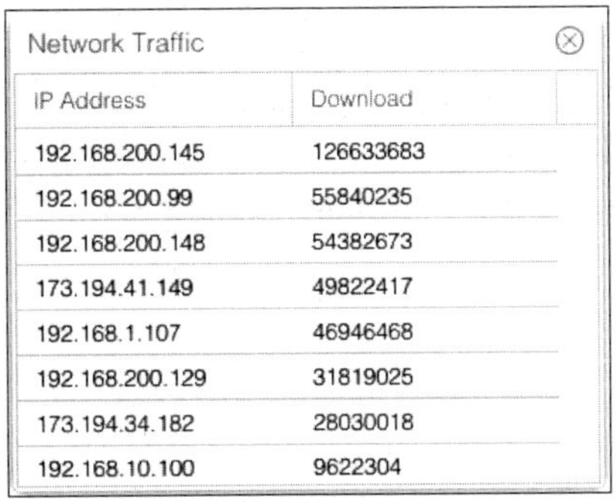

Network Traffic

IP Address	Download
192.168.200.145	126633683
192.168.200.99	55840235
192.168.200.148	54382673
173.194.41.149	49822417
192.168.1.107	46946468
192.168.200.129	31819025
173.194.34.182	28030018
192.168.10.100	9622304

The Highcharts extension

In this section, we will examine how simple it is to create a Highcharts component in Ext JS. We do this by importing from an existing Highcharts configuration. Let's continue from the previous JsonStore example and incorporate it within the extension.

Step 1 – removing some of the Highcharts options

Let's assume that we already have a working independent Highcharts configuration, as follows:

```
var myConfig = {
        chart: {
                renderTo: 'container',
                width: 350,
                height: 300,
                ....
        },
        series: [{
            type: 'column',
            data: [ 126633683, 55840235, .... ]
        }],
        xAxis: {
            categories: [ "192.168.200.145",
                          "192.168.200.99", ... ],
            ....
        },
        yAxis: { .... },
        title: { .... },
        ....
};
```

The first step is to remove all the fields that the extension will handle internally and pass them to Highcharts. For this reason, we need to remove `chart.renderTo` and the dimension options. We also need to remove the `chart.series` array, because eventually `JsonStore` will be the source of graph data. We also want to remove `chart.xAxis.categories` as it contains graph data.

Step 2 – converting to a Highcharts extension configuration

The next step is to construct a new configuration for the extension derived from the old Highcharts configuration. Let's start a new configuration object, `myNewConfig`, with the size properties:

```
var myNewConfig = {
        width: 350,
        height: 300
};
```

The next step is to create a new option, `chartConfig`, which is required by the extension. We put the rest of the properties left in the `myConfig` object towards `chartConfig`. The following code snippet shows what the new config should look like:

```
// ExtJS component config for Highcharts extension
var myNewConfig = {
        width: 450,
        height: 350,
        chartConfig: {
              // Trimmed Highcharts configuration here
              chart: { .... },
              xAxis: { .... },
              yAxis: { .... },
              title: { .... },
              ....
        }
};
```

Step 3 – constructing a series option by mapping the JsonStore data model

Recalling the data model of the store object, we have the following code snippet:

```
        fields: [
            { name: 'host', type: 'string' },
            { name: 'download', type: 'int' }
        ]
```

The next task is to build a series array with options matching the data model of `JsonStore`. The new series array has a similar structure to the one in Highcharts options. We also need to link the store object inside the object configuration. Eventually, the options object should become like the following code snippet:

```
var myNewConfig = {
        width: 450,
        height: 350,
        store: netStore,
        series: [{
              name: 'Network Traffic',
              type: 'column',
              // construct the series data out of the
              // 'download' field from the return Json data
              dataIndex: 'download'
        }],
        // construct the x-axis categories data from
        // 'host' field from the return Json data
        xField: 'host',
        chartConfig: {
              ....
        }
};
```

The `dataIndex` option is used for mapping the *y* value from `JsonStore` into the series data array. As the `'host'` field is string-type data, it is used as categories. Therefore, we specify the `xField` option outside the series array shared by the series.

Step 4 – creating the Highcharts extension

The final step is to put everything together to display a chart in Ext JS. We can create a Highcharts component first and put it inside an Ext JS container object, as follows:

```
var hcChart = Ext.create('Chart.ux.Highcharts', myNewConfig);
var win = Ext.create('widget.window', {
      title: 'Network Traffic',
      layout: 'fit',
      items: [ hcChart ]
}).show();
```

Or alternatively, we can create the whole thing through one configuration using `xtype`, as follows:

```
var win = Ext.create('widget.window', {
      title: 'Network Traffic',
      layout: 'fit',
```

```
        items: [{
                xtype: 'highchart',
                itemId: 'highchart',
                height: 350,
                width: 450,
                store: netStore,
                series: [{ .... }],
                xField: 'host',
                chartConfig: {
                    chart: { .... },
                    xAxis: { .... },
                    yAxis: { .... },
                    ....
        }]
    }).show();
```

The following screenshot shows a Highcharts graph inside an Ext JS window (classic theme):

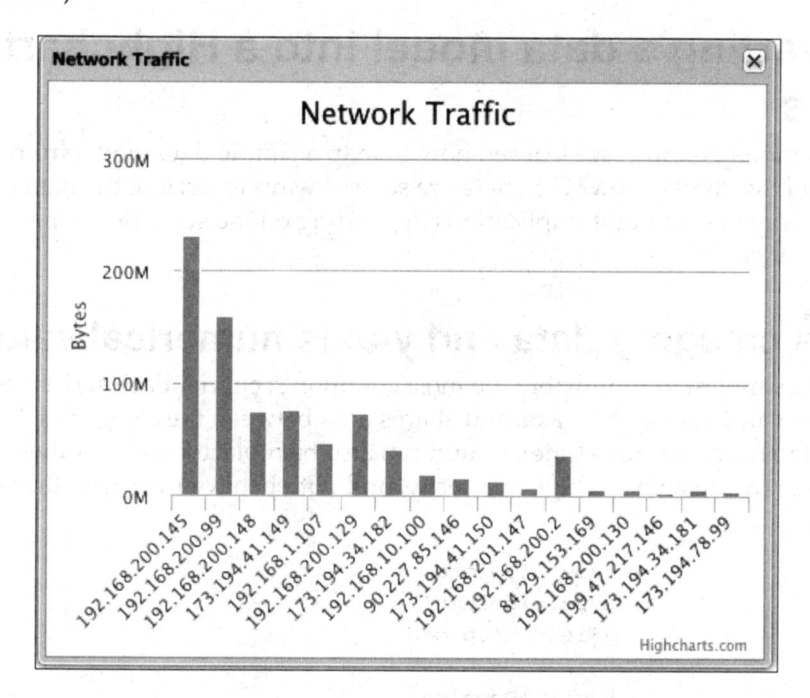

In order to display data at startup, the JsonStore must be instantiated by setting the `autoLoad` option to `true` or calling the `Store.load` method manually at the start of the program.

Passing series-specific options in the Highcharts extension

If we need to pass series-specific options, for example color, data point decorations, and so on, then we simply put them into the series configuration in the same way we normally do in Highcharts:

```
.....
store: netStore,
    series: [{
            name: 'Network Traffic',
            type: 'column',
            dataIndex: 'download',
            color: '#A47D7C'
    }],
```

The extension will copy these options across at the same time as creating the series.

Converting a data model into a Highcharts series

In the previous example, we learned how to map a simple data model from the Ext JS store into Highcharts. However, there are several ways to declare the data mapping, and each way has different implications depending on the scenarios, especially in multiple series.

X-axis category data and y-axis numerical values

This is the simplest and probably the most common scenario. Each series has numerical values along the *y* axis and shares data between the categories. For historical reasons, the `dataIndex` option can also be replaced with another option name, `yField`, which has a higher priority, and both behave in exactly the same way:

```
series: [{
        name: 'Upload',
        type: 'column',
        yField: 'upload'
}, {
        name: 'Download',
        type: 'column',
        yField: 'download'
}],
// 'Monday', 'Tuesday', 'Wednesday' ....
xField: 'day'
```

Numerical values for both x and y axes

Another scenario is where both the *x* and *y* axes are made up of numerical values. There are two different ways to specify the data mapping. First, each series holds the *y* axis values and shares common *x* axis values. In this case, the series are specified in the same way as the previous example:

```
series: [{
        name: 'Upload',
        type: 'line',
        yField: 'upload'
 }, {
        name: 'Download',
        type: 'line',
        yField: 'download'
}],
// Time in UTC
xField: 'time'
```

Another situation is that each series holds its own pairs of *x* and *y* values, as follows:

```
series: [{
        name: 'Upload',
        type: 'line',
        yField: 'upload',
        xField: 'upload_time'
 }, {
        name: 'Download',
        type: 'line',
        yField: 'download',
        xField: 'download_time'
}]
```

The difference between the two settings is that the first configuration ends up with two line series in the graph with data points aligning along the *x* axis, whereas the latter one doesn't, and the store data model is different as well.

Performing preprocessing from store data

Suppose that we need to perform a preprocessing task on the server data before we can plot the chart. We can do this by overriding a template method in the series configuration.

Inside the extension code, each series is actually instantiated from a `Serie` class. This class has a standard method defined, `getData`, which is for retrieving data from the store. Let's visit the original implementation of `getData`:

```
getData : function(record, index) {
    var yField = this.yField || this.dataIndex,
        xField = this.xField,
        point = {
            data : record.data,
            y : record.data[yField]
        };
    if (xField)
        point.x = record.data[xField];
    return point;
},
```

The classes and methods in this extension are named that way with the word "`Serie`" by the original author.

Basically, `getData` is called for every row returned from `JsonStore`. The method is passed with two parameters. The first one is an Ext JS `Record` object, which is an object representation of a row of data. The second parameter is the index value of the record inside the store. Inside the `Record` object, the `data` option holds the values according to the model definition when the store object is created.

As we can see, the simple implementation of `getData` is to access `record.data` based on the values of `xField`, `yField`, and `dataIndex` and format it into a Highcharts `Point` configuration. We can override this method as we declare a series to suit our need for data conversion. Let's continue the example: suppose the server is returning the data in a JSON string:

```
{"data":[
    {"host":"192.168.200.145","download":126633683,
     "upload":104069233},
    {"host":"192.168.200.99","download":55840235,
     "upload":104069233},
    {"host":"192.168.200.148","download":54382673,
     "upload":19565468},
     ....
```

`JsonStore` interprets the preceding data as rows with the following model definition:

```
fields: [
    {name: 'host',  type: 'string'},
    {name: 'download', type: 'int'},
    {name: 'upload', type: 'int'}
]
```

We need to plot a column chart with each bar as the total of the upload and download fields, so we define the `getData` method for the series as shown next. Note that we don't need to declare `yField` or `dataIndex` anymore, because the `getData` method for this particular series has already taken care of the field mappings:

```
series: [{
    name: 'Total Usage',
    type: 'column',
    getData: function(record, index) {
        return {
            data: record.data,
            y: record.data.upload +
               record.data.download
        };
    }
}],
xField: 'host',
....
```

Plotting pie charts

Plotting pie charts is slightly different to line, column, and scatter charts. A pie series is composed of data values where each value is from a category. Therefore, the module has two specific option names, `categorieField` and `dataField`, for category and data, respectively. To plot a pie chart, the series is needed to specify the following:

```
series: [{
    type: 'pie',
    categorieField: 'host',
    dataField: 'upload'
}]
```

The `getData` method of the `PieSeries` class subsequently converts the mapped data from the store into the `Point` object, with values assigned to the `name` and `y` fields.

Plotting donut charts

Let's remind ourselves that a donut chart is actually a two-series pie chart in which the data in the inner pie is a subcategory of the outer pie. In other words, each slice in the inner series is always the total of its outer portions. Therefore, the data returned from `JsonStore` has to be designed in such a way that these can be grouped into subcategories by field name. In this case, the JSON data should be returned, as follows:

```
{ "data": [
      { "host" : "192.168.200.145", "bytes" : 126633683,
        "direction" : "download"},
      { "host" : "192.168.200.145", "bytes" : 104069233,
        "direction" : "upload"},
      { "host" : "192.168.200.99", "bytes" : 55840235,
        "direction" : "download"},
      { "host" : "192.168.200.99", "bytes" : 104069233,
        "direction" : "upload"},
      ....
] }
```

Then, we use an extra Boolean option, `totalDataField`, for the inner pie series to indicate that we want to use `dataField` to scan for the total value for each `"host"` category. As for the outer series, we just define it as a normal pie series, but with `"direction"` and `"bytes"` as `categorieField` and `dataField`, respectively. The following is the series definition for the donut chart:

```
series: [{
        // Inner pie
        type: 'pie',
        categorieField: 'host',
        dataField: 'bytes',
        totalDataField: true,
        size: '60%',
        ....
}, {
        // Outer pie
        type: 'pie',
        categorieField: 'direction',
        dataField: 'bytes',
        innerSize: '60%',
        ....
}]
```

The following screenshot shows what a donut chart looks like in Ext JS (aria theme):

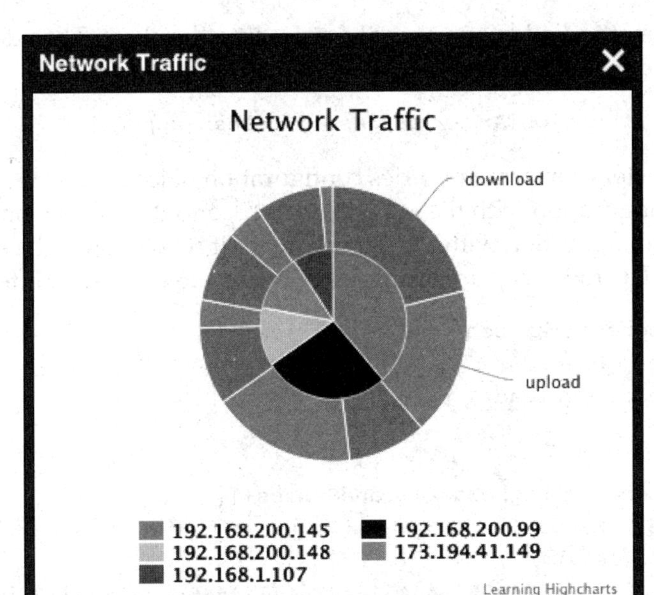

Inside the extension, the implementation of the `getData` method for the `PieSeries` class is significantly different from other series types, in order to handle both pie and donut series data. Therefore, it is not advisable to overwrite this method. Later on, we will see how pie and donut charts are plotted with this module.

Module APIs

The Highcharts extension comes with a small set of APIs. Most of them are helper functions to modify series in the Ext JS layer. As for the Highcharts native APIs, they can be invoked through the `chart` property inside the extension component, for example:

```
win.down('#highchart').chart.getSVG({ ... });
```

In the preceding line of code, `'highchart'` is the `itemId` value when the chart component is created. The down method is Ext JS's convenient way of using the CSS selection style to navigate through the hierarchical components.

As mentioned earlier, the `chartConfig` option contains all the Highcharts configurations. Once the chart component is created, it saves `chartConfig` inside the component. Hence, the `chartConfig` property possesses all the initial configurations that have created the chart. Later, we will see how this `chartConfig` property plays a role with regards to API calls.

addSeries

The `addSeries` method adds one or more series into the chart. The added series is/are also stored inside the `chartConfig.series` array, as follows:

```
addSeries : function(Array series, [Boolean append])
```

The series parameter is an array of series configuration objects. `addSeries` not only allows series configuration with the `xField`, `yField`, and `dataIndex` options, but also supports series configuration with the data array, so it won't go via the store object to extract the data. The following are examples of using `addSeries` in different ways:

```
Ext.getComponent('highchart').addSeries([{
        name: 'Upload',
        yField: 'upload'
}], true);

Ext.getComponent('highchart').addSeries([{
        name: 'Random',
        type: 'column',
        data: [ 524524435, 434324423, 43436454, 47376432 ]
}], true);
```

The optional append parameter sets the series parameter to either replace the currently displayed series or append the series to the chart. The default is `false`.

removeSerie and removeAllSeries

The `removeSerie` method removes a single series in the chart and the `removeAllSeries` method removes all the series defined for the chart. Both methods also remove the series configuration in `chartConfig.series`, as follows:

```
removeSerie : function(Number idx, [Boolean redraw])
removeAllSeries : function()
```

The `idx` parameter is the index value in the series array. The optional `redraw` parameter sets whether to redraw the chart after the series is removed. The default is `true`.

setTitle and setSubTitle

Both `setTitle` and `setSubTitle` change the current chart title as well as the title settings in `chartConfig`, as follows:

```
setSubTitle : function(String title)
setTitle: function(String title)
```

draw

So far, we have mentioned `chartConfig` but haven't really explained what it does in the module. The `draw` method actually destroys the internal Highcharts object and recreates the chart based on the settings inside the current `chartConfig`. Suppose we have already created a chart component but we want to change some of the display properties. We modify properties inside `chartConfig` (Highcharts configurations) and call this method to recreate the internal Highcharts object:

```
draw: function()
```

Although we can call Highcharts' native APIs via the internal `chart` option without destroying and recreating the chart, not all Highcharts elements can be changed with API calls, for example series color, legend layout, the column stacking option, invert chart axes, and so on.

As a result, this method enables the extension component to refresh the internal chart with any configuration change, without the need to recreate the component itself. Hence, this empowers the Ext JS application by not removing it from the parent container and reinserting a new one. Also, the layout in the parent container is not disrupted.

Event handling and export modules

Specifying chart event handlers for the extension is exactly the same as how we normally declare event handlers in Highcharts. Since this is now under both the Ext JS and jQuery environments, the implementation can use both Ext JS and jQuery methods.

The Highcharts exporting chart module is unaffected by the extension. The export settings simply bypass this extension and work straightaway.

Extending the example with Highcharts

In this section, we will build a larger example that includes other types of panels and charts. The application is built with a viewport showing two regions—the `'center'` region is a tab panel containing three tabs for each different type of network data graph, and the `'west'` region shows the table data of the current graph on display. The graph in the first tab is **Bandwidth Utilisation**, which indicates the data rate passing through the network.

The following screenshot shows the front screen of the application (neptune theme):

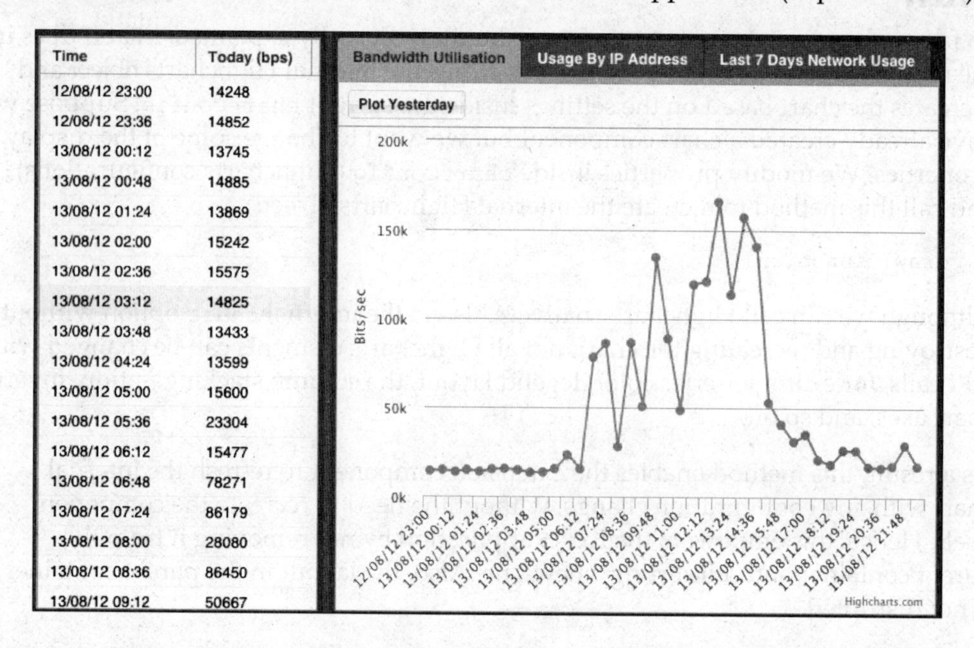

Plot Yesterday in the toolbar is a toggle button that triggers an additional series, **Yesterday**, to be plotted on the same chart. An extra column of data called **Yesterday** is also displayed in the left-hand side table, as shown in the following screenshot:

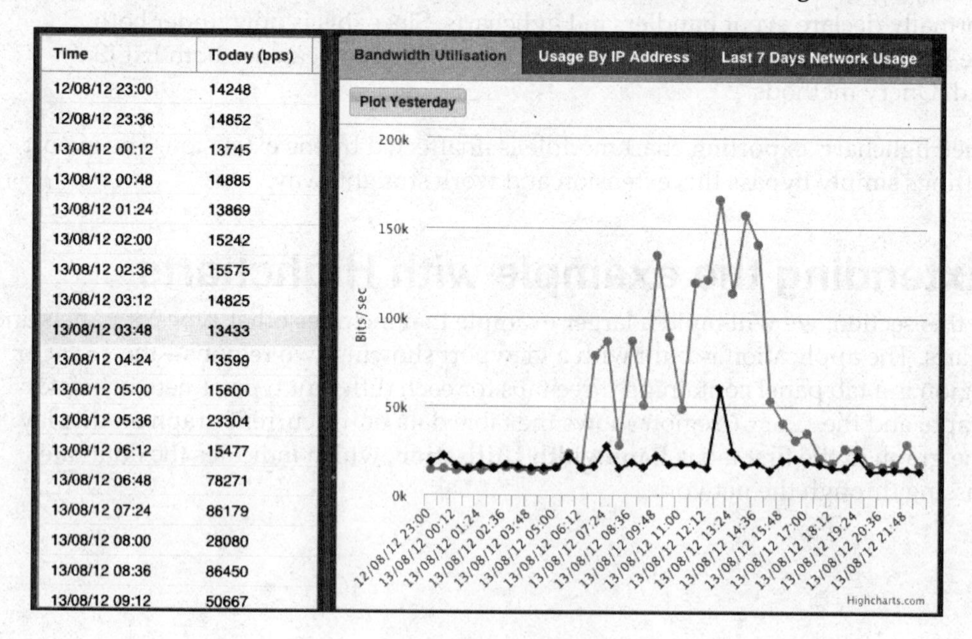

The **Plot Yesterday** button handler uses the `addSeries` and `removeSeries` methods internally to toggle the **Yesterday** series. The following is the implementation:

```
toggleHandler: function(item, pressed) {
    // Retrieve the chart extension component
    var chart = Ext.getCmp('chart1').chart;
    if (pressed && chart.series.length == 1) {
        Ext.getCmp('chart1').addSeries([{
            name: 'Yesterday',
            yField: 'yesterday'
        }], true);
        // Display yesterday column in the grid panel
        ....
    } else if (!pressed && chart.series.length == 2) {
        Ext.getCmp('chart1').removeSerie(1);
        // Hide yesterday column in the grid panel
        ....
    }
}
```

Let's move on to the second tab, which is a column chart showing a list of hosts with their network usage in uplink and downlink directions, as follows:

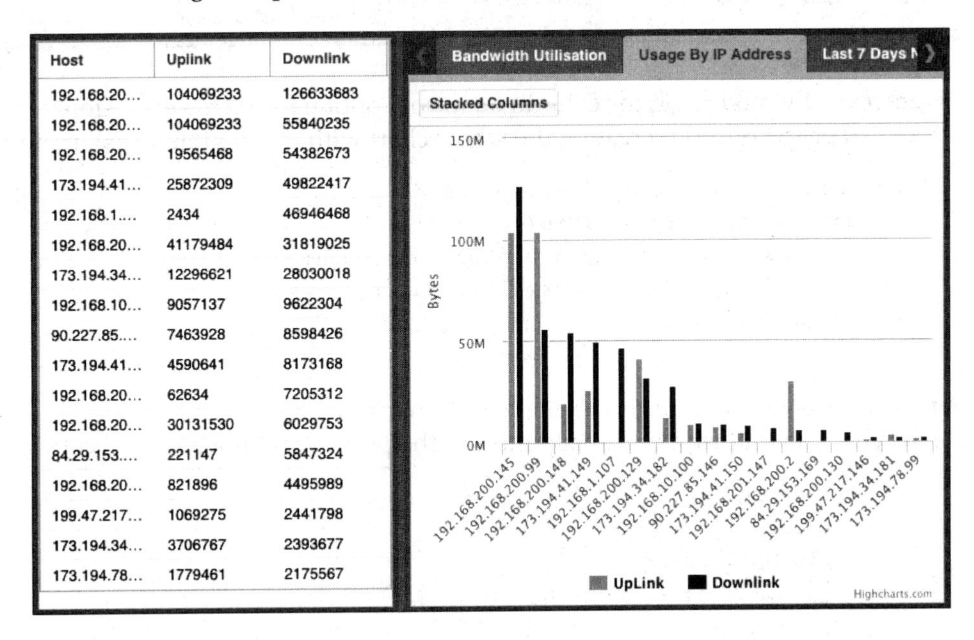

When we click on the **Stacked Columns** button, the bars of both series are stacked together instead of aligned adjacent to each other, as follows:

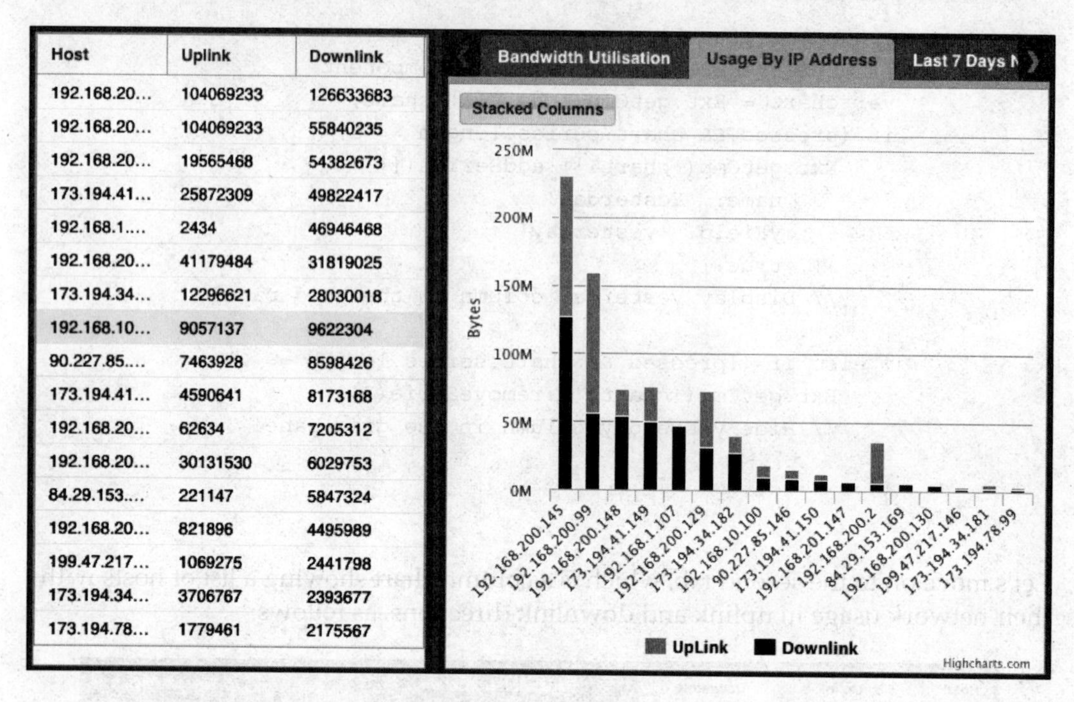

This is achieved by modifying the column `stacking` option inside the extension `chartConfig` property and recreating the whole chart with the module's `draw` method:

```
toggleHandler: function(item, pressed) {
    var chart2 = Ext.getCmp('chart2');
        chart2.chartConfig.plotOptions.column.stacking =
            (pressed) ? 'normal' : null;
        chart2.draw();
}
```

Note that we declare the default `stacking` option inside `chartConfig` when we create the chart, so that we can directly modify the property in the handler code later:

```
chartConfig: {
    .... ,
    plotOptions: {
        column: { stacking: null }
    },
    ......
```

The final tab is **Last 7 Days Network Usage**, which has a pie chart showing the network usage for each of the last seven days, as shown in the following screenshot:

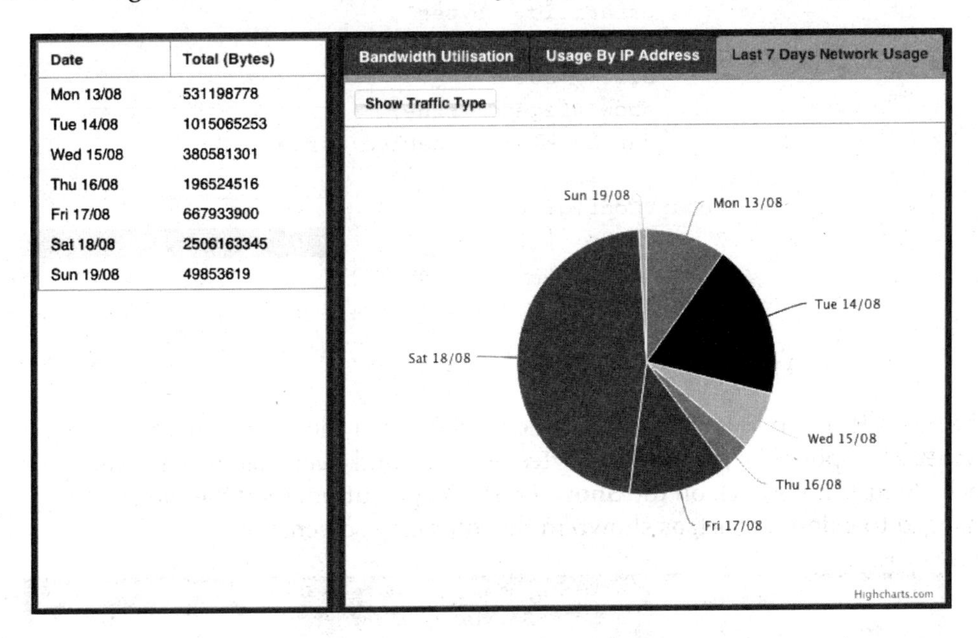

Let's see how this pie chart is implemented in detail. `JsonStore` is adjusted to return data in the following format:

```
{"data": [
     {"date": "Mon 13/08", "type": "wan",
      "bytes": 92959786, "color": "#8187ff" },
     {"date": "Mon 13/08", "type": "lan",
      "bytes": 438238992, "color": "#E066A3" },
     {"date": "Tue 14/08", "type": "wan",
      "bytes": 241585530, "color": "#8187ff" },
     {"date":"Tue 14/08", "type": "lan",
      "bytes": 773479723, "color": "#E066A3" },
     .....
```

Then, we define the tab panel content, as follows:

```
items:[{
      xtype: 'highchart',
      id: 'chart3',
      store: summStore,
      series: [{
          type: 'pie',
```

```
                name: 'Total',
                categorieField: 'date',
                dataField: 'bytes',
                totalDataField: true,
                size: '60%',
                showInLegend: true,
                dataLabels: { enabled: true }
            }],
            chartConfig: {
                chart: {...},
                title: {  text: null },
                legend: { enabled: false }
            }
        }]
```

The series is set up as an inner series, hence the use of the `totalDataField` and `dataField` options to get the total bytes of `"lan"` and `"wan"` as the slice value for each `'host'`. If we click on the **Show Traffic Type** button, then the pie chart is changed to a donut chart, as shown in the following screenshot:

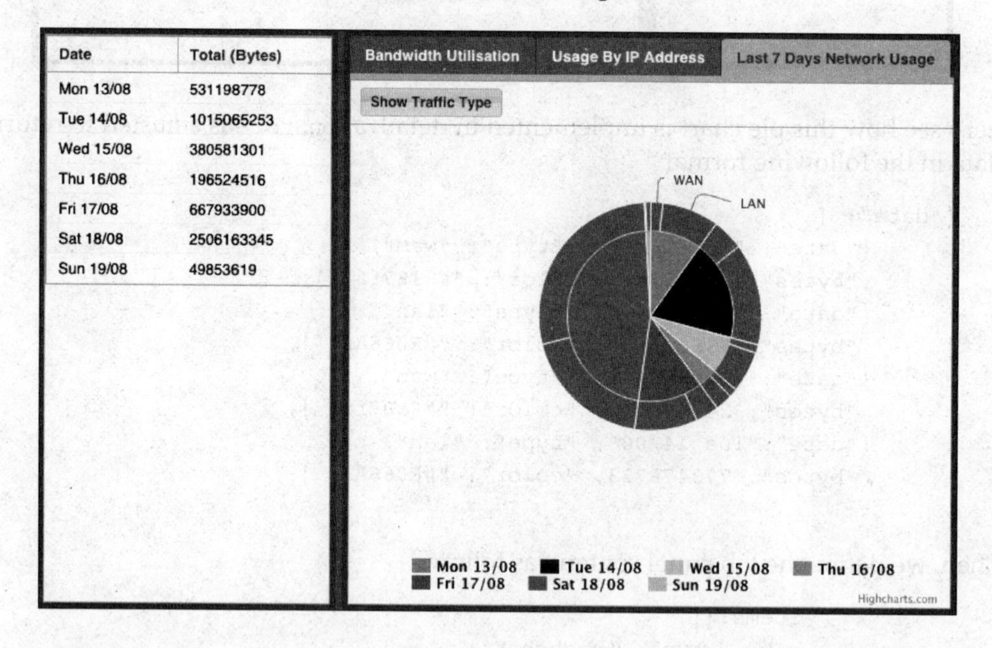

The original data labels in the first pie chart are replaced with items inside the legend box. An outer series is displayed with a fixed color scheme to show the **LAN** and **WAN** portions of traffic. The following is the **Show Traffic Type** button's button handler code:

```
toggleHandler: function(item, pressed) {
    var config = Ext.getCmp('chart3').chartConfig;
    if (pressed) {
        Ext.getCmp('chart3').addSeries([{
            type: 'pie',
            center: [ '50%', '45%' ],
            categorieField: 'type',
            dataField: 'bytes',
            colorField: 'color',
            innerSize: '50%',
            dataLabels: {
                distance: 20,
                formatter: function() {
                  if (this.point.x <= 1) {
                    return this.point.name.toUpperCase();
                  }
                  return null;
                }
            },
            size: '60%'
        }], true);

        config.legend.enabled = true;
        config.series[0].dataLabels.enabled = false;
        config.series[0].size = '50%';
        config.series[0].center = [ '50%', '45%' ];
    } else {
        Ext.getCmp('chart3').removeSerie(1);
        config.legend.enabled = false;
        config.series[0].dataLabels.enabled = true;
        config.series[0].size = '60%';
        config.series[0].center = [ '50%', '50%' ];
    }
    Ext.getCmp('chart3').draw();
}
```

If the toggle button is enabled, then we add an outer pie series (with the `innerSize` option) via the `addSeries` method. Moreover, we align the outer series accordingly with the traffic `'type'`, and so `categorieField` and `dataField` are assigned to `'type'` and `'bytes'`. Since more information is needed to display the second series, we set the inner series to a smaller size for more space. In order to only show the first two data labels in the outer series, we implement `dataLabels.formatter` to print the label when `this.point.x` is 0 and 1. After that, we disable the data labels by returning null in the `formatter` function. Finally, the `draw` method is used to reflect all the changes.

Displaying a context menu by clicking on a data point

For interactive applications, it would be handy to allow users to launch specific actions by clicking on a data point. To do that, we need to handle Highcharts' click events. Here, we create a simple menu for showing the difference between the selected point and the average value of the series. The following is the sample code:

```
point: {
  events: {
    click: function(evt) {
      var menu =
        Ext.create('Ext.menu.Menu', {
          items: [{
            text: 'Compare to Average Usage',
            scope: this,
            handler: function() {
              var series = this.series,
                  yVal = this, avg = 0, msg = '';

              Ext.each(this.series.data, function(point) {
                avg += point.y;
              });
              avg /= this.series.data.length;

              if (yVal > avg) {
                msg =
                  Highcharts.numberFormat(yVal - avg) +
                  " above average (" +
                  Highcharts.numberFormat(avg) + ")";
              } else {
                msg =
                  Highcharts.numberFormat(avg - yVal) +
                  " below average (" +
                  Highcharts.numberFormat(avg) + ")";
```

```
                    }

                    Ext.Msg.alert('Info', msg);
                }
            }]  // items:
        });

        menu.showAt(evt.point.pageX, evt.point.pageY);
    }
  }
}
```

First we create a simple Ext JS `Menu` object with the menu item **Compare to Average Usage**. The `click` handler is called with the mouse event parameter, `evt`, and then we obtain the mouse pointer location, `pageX` and `pageY`, and pass it to the menu object. As a result, the Ext JS menu appears next to the pointer after clicking on a data point.

The `'this'` keyword in the `click` event handler refers to the selected point object. We then use the `scope` option to pass the Highcharts point object to the menu handler layer. Inside the handler, the `'this'` keyword becomes the data point object instead of the Ext JS menu item. We extract the series data to calculate the average and compute the difference with the selected point value. Then, we display the message with the value. The following is the screenshot of the menu:

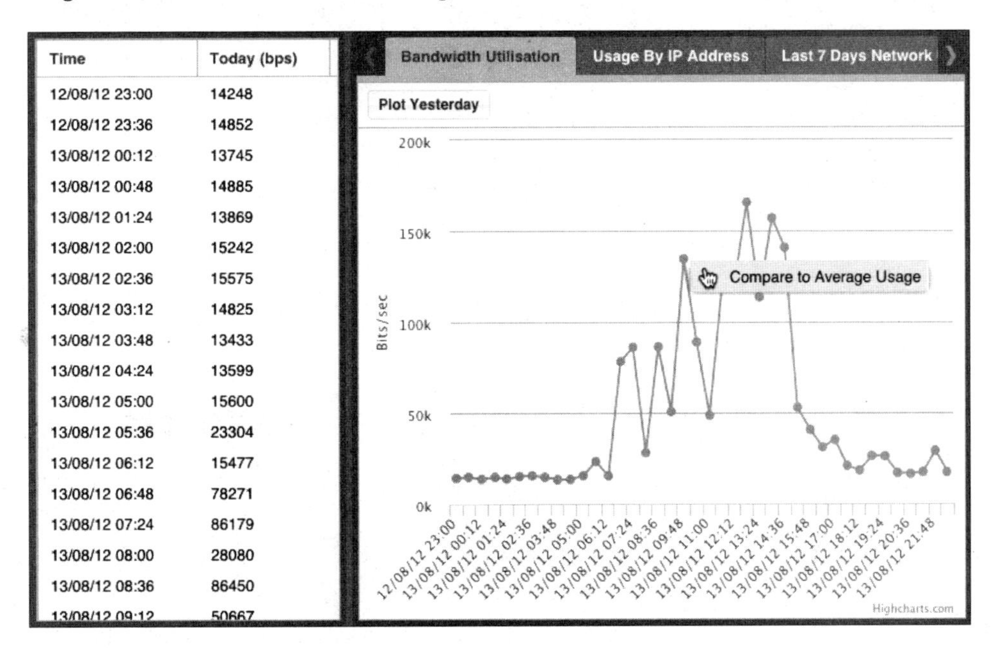

A commercial RIA with Highcharts – Profiler

So far, we have demonstrated how Highcharts can be applied within the Ext JS framework. However, the demo itself seems rather shrink-wrapped for an RIA product. In this section, we will have a quick glance at a commercial application, **Profiler,** a tool for profiling companies network scenario developed by iTrinegy. Due to the nature of its business, a stack of diagnostic graphs is required for this type of application. The whole application is designed as a collection of portals for monitoring network traffic from multiple sites. Users can drill down from utilization graph to top downlink usage by IP address graph, modify filter properties to display relative data in multiple series, and so on.

In order to fine-tune the profiling parameters and provide a portal interface, a framework offering dynamic and calibrated user interfaces is needed. For this reason, Ext JS is a suitable candidate, as it offers a rich set of professional looking widget components, and its cross-browser support makes building complicated RIA software manageable. The following is the interface for launching a bandwidth utilization report graph with specific parameters:

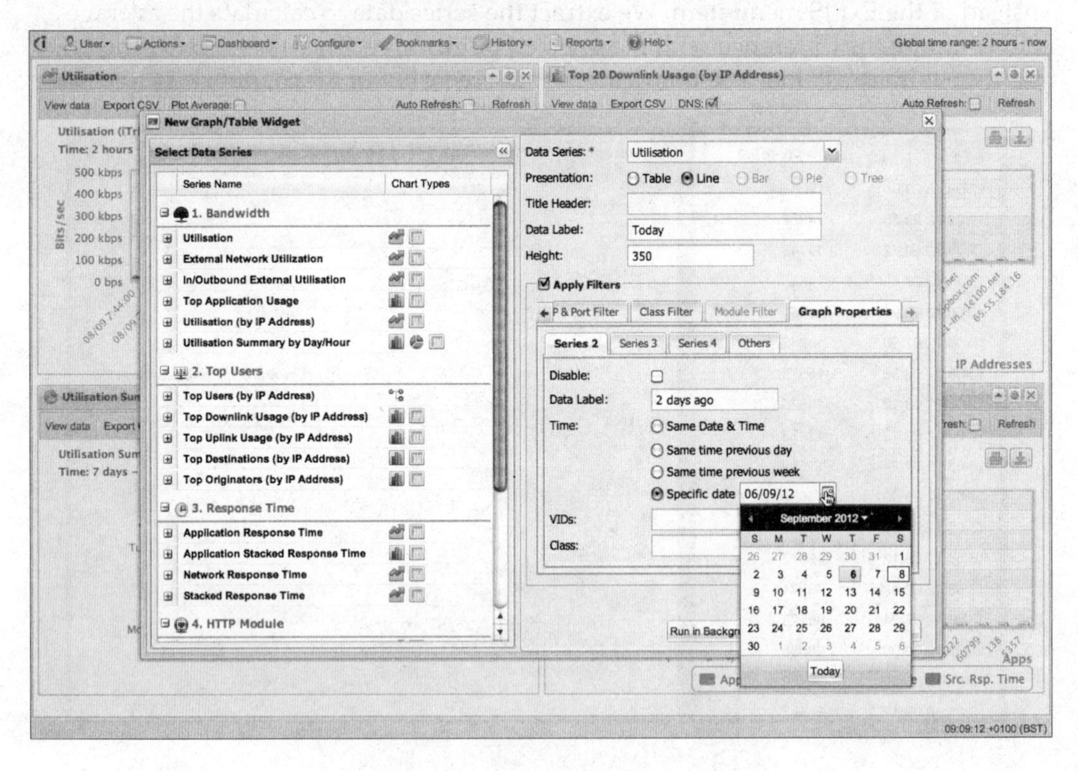

The Highcharts events are easily bound with Ext JS components so that a fully interactive navigation style becomes possible. For instance, if a peak appears on the **Utilisation** graph, the users can either click on the peak data point or highlight a region for a specific time range, then a context menu with a selection of network graphs pops up. This action means that we can append the selected time region to be part of the accumulated filters and navigate towards a specific graph. The following is a screenshot of the context menu, which shows up in one of the graphs:

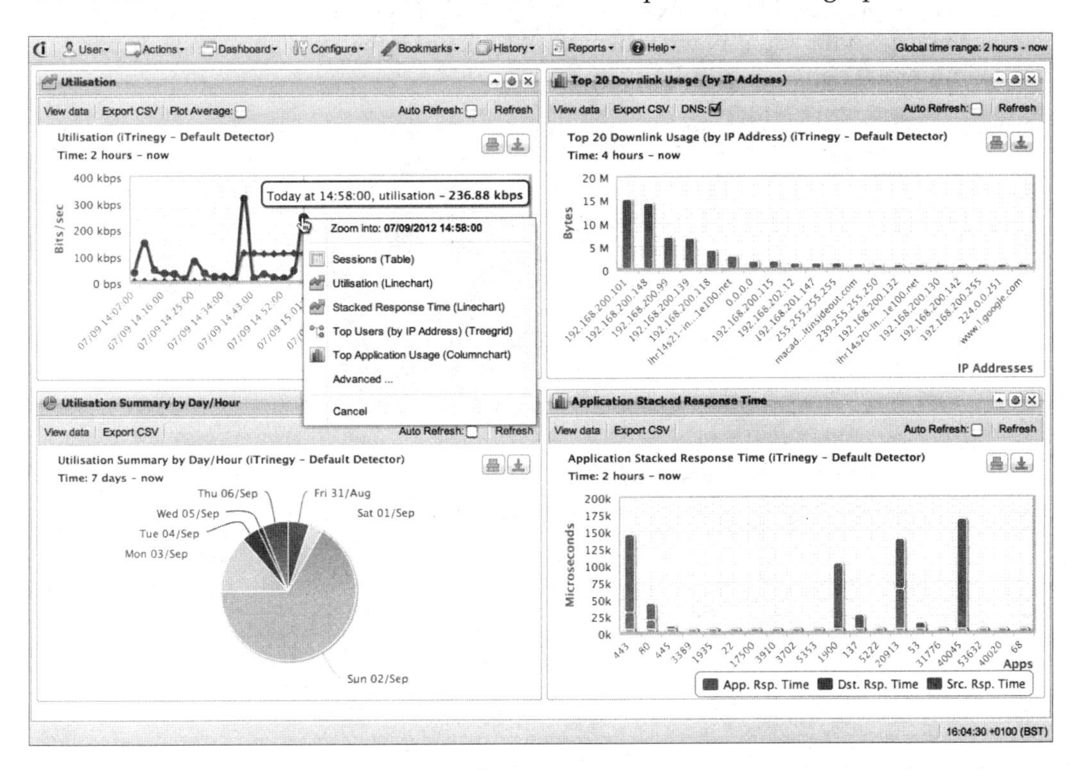

If we proceed by selecting the same graph again, **Utilisation**, it means we want to zoom into greater detail within the selected time region. This doesn't use the Highcharts default zoom action, which just stretches the graph series and redraws the axes. In fact, it launches another Ajax query with the selected time and returns graph data in finer granularity, so the peak in the graph can be diagnosed further. In other words, the application enables the user to visually filter through a sequence of different graphs. At the same time, the user gradually refines the filters in different dimensions. This process dissects the problem into the root cause in a prompt, intuitive, and effective fashion.

Summary

In this chapter, we learned the very basics of Ext JS, which is a framework for building **Rich Internet Applications (RIAs)**. We looked at a quick introduction of a dozen Ext JS components that are likely to be used with the Highcharts extension for Ext JS. Then, we explored how to create a Highcharts component from an existing Highcharts configuration in a step-by-step approach. We looked into the small set of APIs that are provided by the extension module and built a simple application with network usage data. Finally, we took a brief look at Highcharts and Ext JS applied to a commercial network profiling application.

In the next chapter, we will explore how to run Highcharts on the server side.

14
Server-side Highcharts

The first edition of *Learning Highcharts* covered a number of approaches to run Highcharts on the server side. Since then, there has been significant development in this area. It turns out that Highcharts adopts PhantomJS (the headless webkit) for the server solution and PhantomJS/Batik for the server implementation in Java. We will also explore how we can create our own Highcharts server process using PhantomJS and how to use the official server-side script released by Highcharts.

In this chapter, we will cover the following topics:

- Why we want to run Highcharts on the server side
- Why PhantomJS and Batik are adopted by Highcharts
- The basics of PhantomJS and Batik
- Creating our own PhantomJS script to export charts
- How to use the Highcharts server-side script in both command-line and server modes

Running Highcharts on the server side

The main reason for running Highcharts on the server side is to allow the client-based graphing application to be automated and accessible on the server side. In some cases, it is desirable to produce graphs at the frontend as well as delivering automated reports with graphs at the backend. For the sake of consistency and development costs, we would like to produce the same style of graphs at both ends. Here are other scenarios where we may want to generate graphs on the server side:

- The application is required to run a scheduled task on the server side. It generates a regular summary report with graphs (for example, the Service Level Agreement report) and automatically e-mails the report to clients or users with a managerial role.

- The nature of the data means it requires a long time to compute for a graph. Instead, users send the parameters over to the server to generate a graph. Once it is finished, the chart setup is saved, then the users are notified to see a live Highcharts chart from the precomputed JSON setup.
- The application involves a vast amount of recurring data that is only kept for a certain period, such as data trend graphs that are automatically produced and stored in an image format for your records.

Highcharts on the server side

In the first edition of this book, we mentioned a number of technologies that can be used to produce chart images purely on the server side. Within those technologies, PhantomJS is the most prominent. In a nutshell, it is a standalone program that is capable of running JavaScript on the server. Besides this, it is easy to use, has minimum setup, and is programmable and robust.

The alternative approach was to use Rhino, a Java implementation of the JavaScript engine, to run JavaScript on the server side so that Highcharts can be run on the server side to export a chart into an SVG file. Then, the SVG file is forwarded to Batik, a generic Java-based SVG toolkit, to produce an image file from SVG.

Since then, Highcharts have extensively experimented with different approaches and concluded that incorporating PhantomJS is the solution moving forward. There are a number of reasons for this decision. First, Rhino has rendering problems compared to PhantomJS, which makes PhantomJS a better choice. Moreover, PhantomJS can also export images, although it has scalability issues in rendering charts when the number of data points increases to around 1,500. ImageMagick, the image converter, was also considered, but it also has specific performance and reliability issues. For details of the findings, please see `http://www.highcharts.com/component/content/article/2-articles/news/52-serverside-generated-charts#phantom_usage`.

For a server-side solution required to implement in Java, Batik is a more natural choice for formatting SVG, whereas PhantomJS is launched to run Highcharts for SVG content. As for a non-Java approach, PhantomJS itself is good enough to drive the whole server-side solution.

Batik – an SVG toolkit

Batik is part of the Apache foundation projects, `http://xmlgraphics.apache.org/batik/`. Its purpose is to provide a web service to view, generate, and transform SVG data. For instance, Highcharts uses this third-party software to convert SVG data into an image format. When the user clicks on the export button, Highcharts internally forwards the chart's SVG data and the user-selected image format request to Batik.

Then, Batik receives the SVG data and transforms the data into the desired image format. The following diagram summarizes how a normal Highcharts chart uses the export service with Batik:

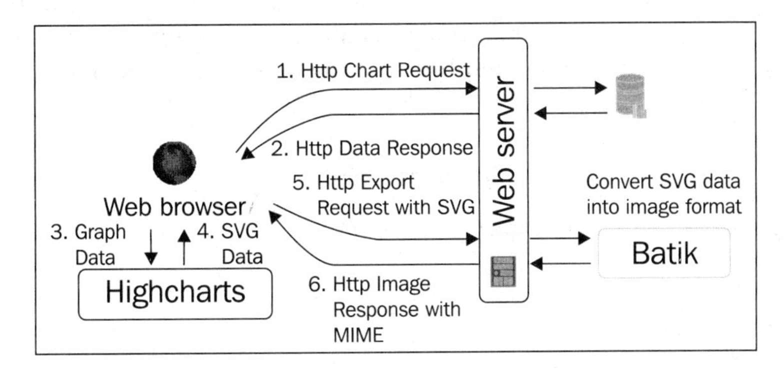

To install Batik, download the latest distribution from `http://xmlgraphics.apache.org/batik/download.html#binary` and follow the installation instructions. As for Ubuntu, simply do:

```
apt-get install libbatik-java
```

Out of the entire Batik package, we only need the image converter component, which is the `batik-rasterizer.jar` file. To transcode from an SVG to a PNG file, we can use the following command:

```
java -jar batik-rasterizer.jar chart.svg
```

The preceding command will convert `chart.svg` and create `chart.png` in the same directory.

PhantomJS (headless webkit)

A webkit is basically the backend engine that drives browsers such as Safari and Google Chrome. It implements almost everything in HTML5 except the browser's user interface. PhantomJS (found at `http://phantomjs.org/`, created and maintained by Ariya Hidayat) is a headless webkit, which means that the webkit engine can be run as a standalone program. It is useful in a number of ways, and one of them is server-side SVG rendering.

Creating a simple PhantomJS script

Although Highcharts released a PhantomJS script to export charts on the server side, it is worth understanding the concept of PhantomJS and how it works. Suppose we already have a web server and PhantomJS installed and running. To run an HTML page on PhantomJS from a command line, run the following command:

```
phantomjs loadPage.js
```

The `loadPage.js` page can be as simple as this:

```
var page = require('webpage').create();

page.onError = function(msg, trace) {
    console.error(msg);
    phantom.exit(1);
};

page.onConsoleMessage = function(msg) {
    console.log(msg);
};

page.open('http://localhost/mychart.html', function(status) {
    if (status === 'success') {
        console.log('page loaded');
        phantom.exit();
    }
});
```

Inside the PhantomJS process, it first loads the `webpage` module and creates a `page` object.

This is only a short example for illustration. For a proper way of handling error messages, please refer to the PhantomJS API documentation.

The `page.onError` and `page.onConsoleMessage` methods redirect the page's error and output messages to the terminal output via `console.log`. Note that `console.log` in this instance is referring to our terminal console. If `console.log` is called inside a page, it will only stay within the page object life cycle and we will never see those messages unless `page.onConsoleMessage` is defined to redirect them.

The preceding script only opens the HTML page into a `webpage` object and then terminates, which is not particularly useful.

Creating our own server-side Highcharts script

Let's use PhantomJS in a slightly more advanced way. In PhantomJS, we don't need to rely on a web server to serve a page. Instead, we load a Highcharts page file locally and include the series data from another JSON file. Then, we render the result into an image file. So here is how we will run the server-side script on a command line:

```
phantomjs renderChart.js chart.html data.json chart.png
```

The `chart.html` page is just a simple Highcharts page that we would normally create. In this exercise, we will leave the series data as a variable, `seriesData`. The following shows how we structure the `chart.html` page:

```
<html>
  <head>
     <meta> ....
     <script src='..../jquery.min.js'></script>
     <script src='..../Highcharts.js'></script>
     <script type='text/javascript'>
     $(function () {
     $(document).ready(function() {
        chart = new Highcharts.Chart({
            chart: {
                ....
            },
            plotOptions: {
                ....
            },
            ....,
            series: [{
                  name: 'Nasdaq',
                  data: seriesData
            }]
        });
     });
  });
  </script>
  </head>
  <body>
   <div id="container" ></div>
  </body>
</html>
```

Then, `data.json` is just a simple JSON file containing the array of *x* and *y* series data. Here is some of the content:

```
[[1336728600000,2606.01],[1336730400000,2622.08],
 [1336732200000,2636.03],[1336734000000,2637.78],
 [1336735800000,2639.15],[1336737600000,2637.09],
 ....
```

For the PhantomJS file, `renderChart.js`, it is surprising how little extra code (highlighted in bold) we need to add to achieve the result:

```
var page = require('webpage').create(),
    system = require('system'),
    fs = require('fs');

// Convert temporary file with series data - data.json
var jsonData = fs.read(system.args[2]);
fs.write('/tmp/data.js', 'var seriesData = ' + jsonData + ';');

page.onError = function(msg, trace) {
    console.error(msg);
    phantom.exit(1);
}

page.onConsoleMessage = function(msg) {
    console.log(msg);
};

// initializes the seriesData variable before loading the script
page.onInitialized = function() {
     page.injectJs('/tmp/data.js');
};

// load chart.html
page.open(system.args[1], function(status) {
    if (status === 'success') {
        // output to chart.png
        page.render(system.args[3]);
        phantom.exit();
    }
});
```

We first load the `system` and `fs` modules, which are used in this example to select command-line arguments and process file I/O on the JSON file. The script basically reads (`fs.read`) the content of the JSON file and converts the content into a JavaScript expression and saves (`fs.write`) it in a file. Then, we define the `onInitialized` event handler for the page object that is triggered before the URL is loaded. So, we insert (`injectJs`) the JavaScript expression of `seriesData` before the page object loads the `chart.html` page. Once the page is loaded, we export (`page.render`) the page content into an image file.

Notice that the resulting image file is not quite correct in that the line series is actually missing. However, if we observe the image more carefully, actually the line has just started being drawn (see the following screenshot):

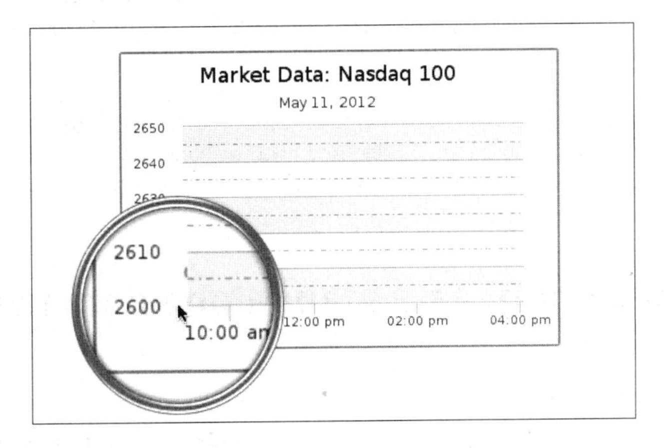

This is because of the chart default animation settings. After we turn the initial animation off by setting the `plotOptions.series.animation` option to `false`, the line series appears:

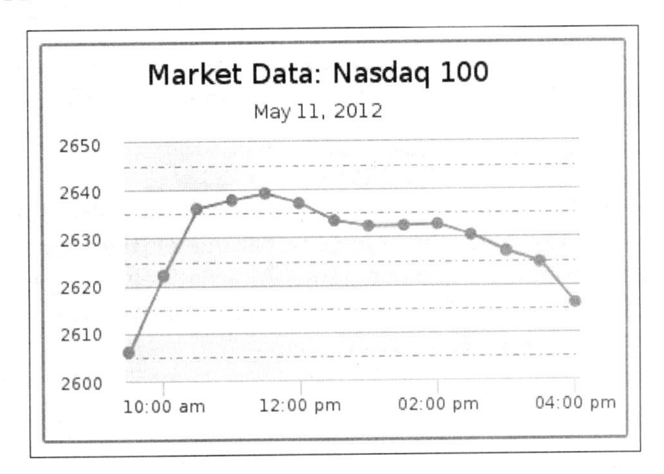

Running the Highcharts server script

So far, our script example is lacking in features and error checking functionality, and is far from perfect. Nonetheless, we can see how easy it is to create our own server-side Highcharts process to produce images. In this section, you will learn how to use the official server-side script by Highcharts, which has more features and can be used in different scenarios.

Server script usage

Since version 3, Highcharts is packaged with server-side script, `highcharts_convert.js`, which is located in the `exporting-server/phantomjs` directory. The script can be run as either a command line or as a listening server.

For a full description of usage and parameters, refer to `https://github.com/highslide-software/highcharts.com/tree/master/exporting-server/phantomjs`.

Running the script as a standalone command

Here is a typical command-line format for `highcharts_convert.js`:

```
phantomjs highcharts-convert.js -infile file
  -outfile chart1.png | -type ( png | jpg | pdf | svg ) -tmpdir dir
  [-scale num | -width pixels ] [ -constr (Highcharts | Highstocks) ]
  [-callback script.js ]
```

The following is a list of parameters used in the preceding command:

- `-infile`: This is the input source for the script, which can be either a chart configuration in the JSON format (general usage) or an SVG file. The server script automatically detects the content type, and processes and exports the chart/content in the desired format.
- `-outfile`, `-type`, `-tmpdir`: The two ways to specify the output format are by `-type` or `-outfile`. With the `-outfile` parameter, the script will derive the image format from the extension name. Alternatively, `-type`, for example, type `png`, formats into a PNG image file and combines with `-tmpdir` to save the output file in a specific location.

- `-scale`, `-width`: There are two optional parameters to adjust the output image size, by `-scale` or by `-width`. As the name suggests, one is to adjust the size by scaling and the other is by the absolute size.
- `-constr`: The `-constr` parameter is to instruct the script whether to export the chart as a Highcharts or Highstock chart (another product for financial charts).
- `-callback`: The `-callback` parameter is to execute additional JavaScript code on the chart once it is loaded and before the chart is exported.

Let's apply the previous chart configuration file into this command line. Furthermore, we are going to superimpose a watermark, `SAMPLE`, on top of the chart with the callback argument.

First, we save the whole chart configuration object into a file including the series data:

```
{ chart: {
    renderTo: 'container',
    height: 250,
    spacingRight: 30,
    animation: false
  },
  . . . .
}
```

Then, we create a callback script with the following code to add the watermark, `watermark.js`:

```
function(chart) {
    chart.renderer.text('SAMPLE', 220, 200).
          attr({
              rotation: -30
          }).
          css({
              color: '#D0D0D0',
              fontSize: '50px',
              fontWeight: 'bold',
              opacity: 0.8
          }).
          add();
}
```

Finally, we run the following command:

`phantomjs highcharts-convert.js -infile options.json -outfile chart.png -width 550 -callback watermark.js`

The command generates the output as it runs:

```
Highcharts.options.parsed
Highcharts.cb.parsed
Highcharts.customCode.parsed
/tmp/chart.png
```

It also produces the following screenshot:

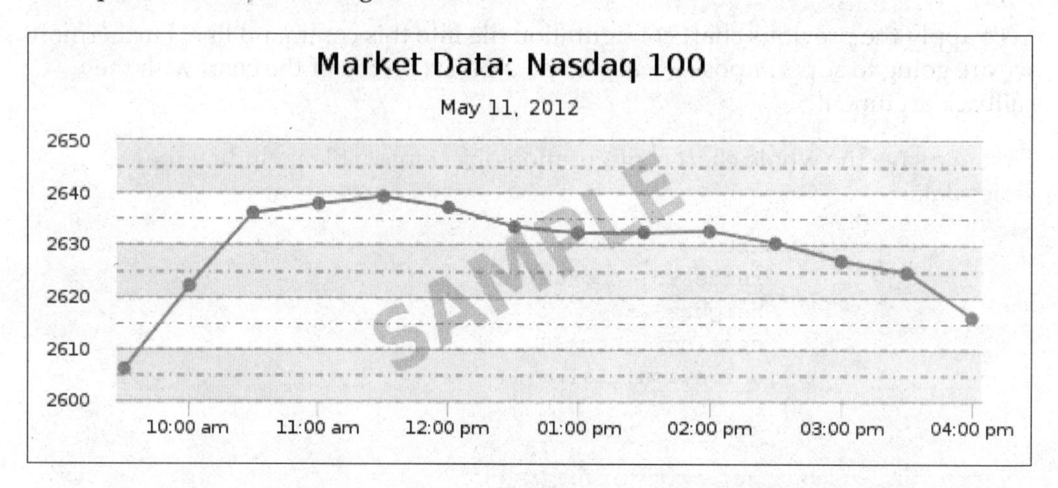

Running the script as a listening server

To run the script as a server listening for HTTP requests, we launch the script with the following command:

```
phantomjs highcharts-convert.js -host address -port num
            -type ( png | svg | jpg | pdf )
```

Let's start a Highcharts export server with the following command:

```
phantomjs highcharts-convert.js -host 127.0.0.1 -port 9413 -type png
```

This starts a server listening only to local incoming connections on port 9413, and the following message outputs to the screen:

```
OK, PhantomJS is ready.
```

Passing options to the listening server

Once the server process is ready, we can start sending POST requests embedded within the Highcharts configuration data. The Highcharts options used inside the request are the same ones we used in the command line. Let's reuse the configuration from the last exercise and pack them into a POST request.

First, we need to "stringify" the whole chart configuration as a value for the `infile` option. Next, we treat the callback method in the same manner. Then, we put the rest of the options into one JSON format and save it in a file called `post.json`:

```
{ "infile" : " { chart: { .... }, series { .... } } " ,
  "callback" : "function(chart) { .... } ",
  "scale" : 1.2
}
```

The next task is to package this data into a POST query. Since the purpose of this chapter is the server-side process, we should operate in a command-line style. Hence, we use the `curl` utility to create a POST request. The following command can do the job:

```
curl -X POST -H "Content-Type: application/json" -d @post.json http://
localhost:9413/ | base64 -d > /tmp/chart.png
```

The preceding `curl` command is to create a POST request with the JSON content type. The `-d @` argument notifies the `curl` command about which file contains the POST data. Since HTTP is an ASCII protocol, the response of the result binary image data is returned in base-64 encoding. Therefore, we need to pipe the POST response data to another utility, base64, to decode the data and write it to a file.

Summary

In this chapter, we described the purpose of running Highcharts on the server side and you learned which technology Highcharts has opted to use on a server. You learned the basics of PhantomJS and the role of Batik. You extended your understanding of PhantomJS to create your own server-side script for Highcharts. Besides that, we experimented with how to run the official PhantomJS script released by Highcharts in both single command-line and server mode.

In the next chapter, we will take a glimpse at which online services Highcharts offer and explore some of the Highcharts plugins.

15
Highcharts Online Services and Plugins

In the previous chapter, you learned how to run Highcharts on the server side. This enables Highcharts to expand its reach to online services. We will visit these services in this chapter and explore what benefits we can gain from them. As well as that, we examine how we can extend Highcharts with plugins. In this chapter, we will cover the following topics:

- What service `export.highcharts.com` provides
- A step-by-step exercise to create an online chart from the new cloud service—`cloud.highcharts.com`
- What a Highcharts plugin is
- Two plugin examples - regression and draggable points
- Creating a new user experience by interoperating both plugins
- How to write a plugin - extend existing methods, export a new method, and handle events

Highcharts export server – export.highcharts.com

In the last chapter, we looked into running Highcharts on the server side. However, some users may not want to set up their own server operations. This is where `export.highcharts.com` comes in. Originally, it was only set up for the exporting module so that users running Highcharts on the Internet could export their charts freely. Later, the URL was expanded to support online services. This let users enter their own Highcharts configuration and download the resulting chart images.

The following is part of the `export.highcharts.com` web page:

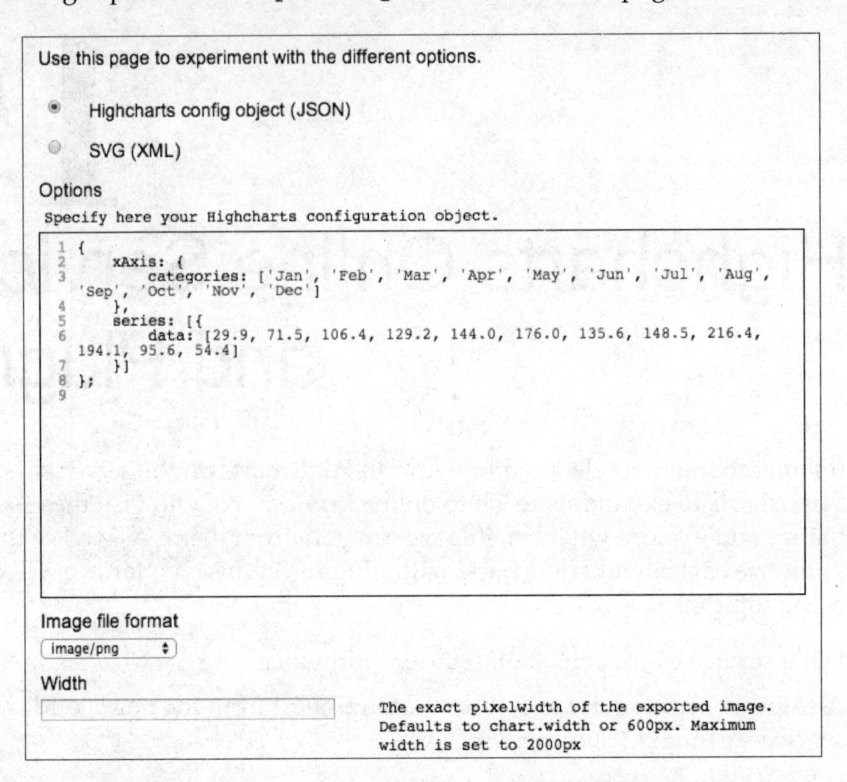

As we can see, the user input indeed corresponds to the parameters of the server-side script, `highcharts_convert.js`, which we covered in a previous chapter. Both the web interface and server process are implemented in Java, which deliver the user's options to the PhantomJS/`highcharts_convert.js` process and exports it into SVG. Once the Java server receives the SVG result, it launches Batik to format into image files. The source for the whole web service solution is available in the `exporting-server/java/highcharts-export` directory.

The downside to the online export service is that it is not WYSIWYG, and so can be unintuitive to use. For this reason, a new web service with much richer user experience was born—**Highcharts Cloud Service**. We will take a ride in the next section and see what difference it brings.

Highcharts Cloud Service

In this section, we will review a brand new online chart service developed by the Highcharts team, Highcharts Cloud Service (`http://cloud.highcharts.com`). The following screenshot shows the initial welcome screen:

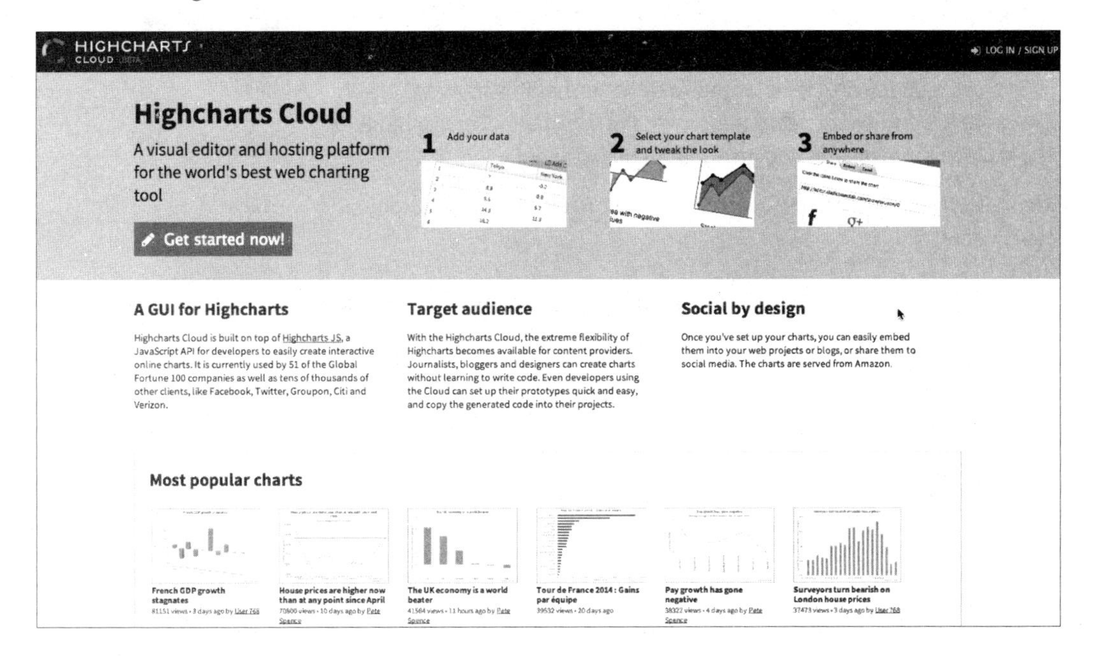

Highcharts Cloud Service is a major milestone in terms of expanding the product line. It is designed for users to:

- Create HTML5 charts even without any JavaScript or Highcharts knowledge (in SIMPLE mode)
- Prototype their charts interactively without any installation and setting up on the web server and Highcharts
- Embed charts in online articles, applications, or web pages with a simple hyperlink
- Store their charts in the cloud rather than locally
- Share their charts easily with other people

The following is a screenshot of a news website linking a chart created from the cloud service:

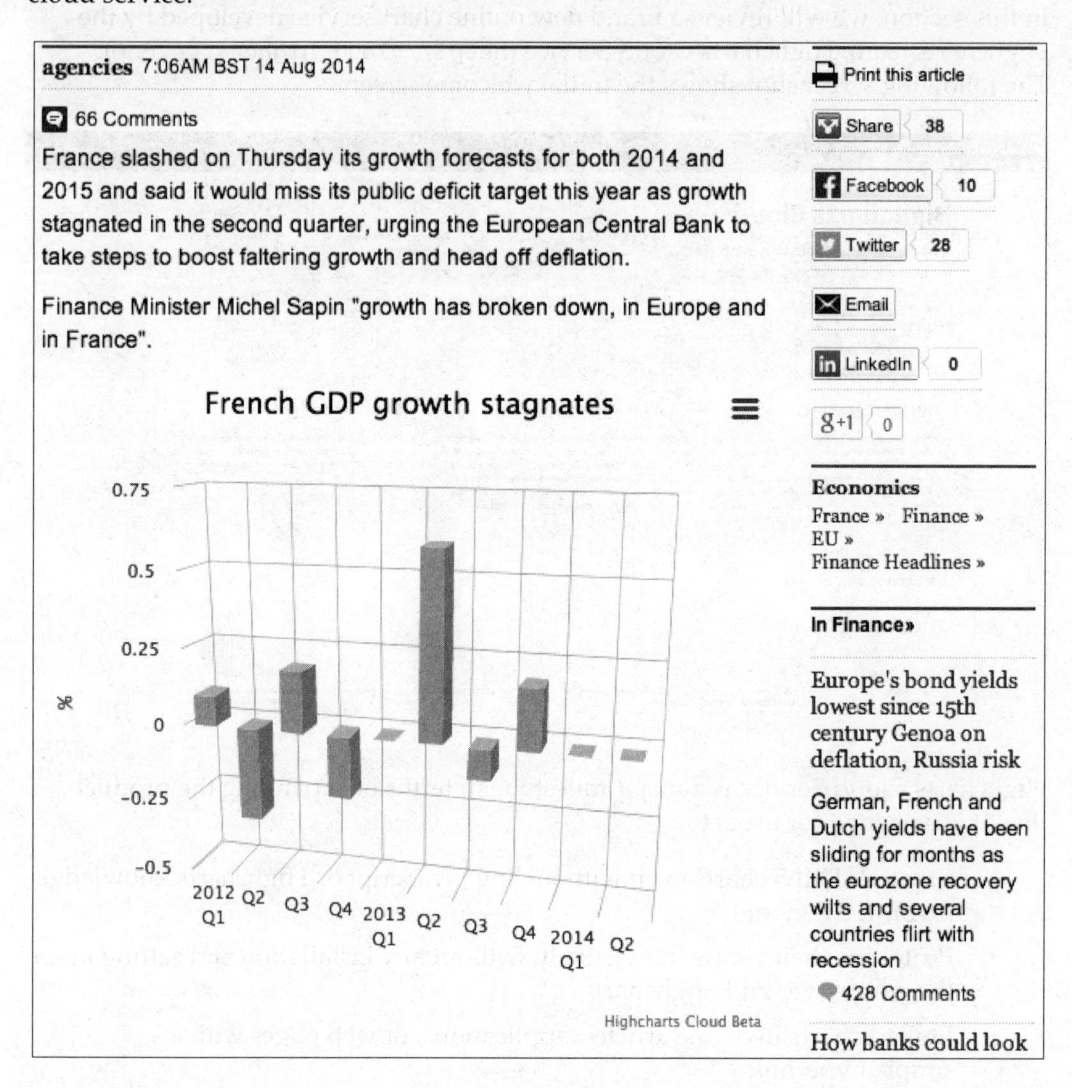

agencies 7:06AM BST 14 Aug 2014

66 Comments

France slashed on Thursday its growth forecasts for both 2014 and 2015 and said it would miss its public deficit target this year as growth stagnated in the second quarter, urging the European Central Bank to take steps to boost faltering growth and head off deflation.

Finance Minister Michel Sapin "growth has broken down, in Europe and in France".

Print this article

Share 38

Facebook 10

Twitter 28

Email

LinkedIn 0

g+1 0

Economics

France » Finance » EU » Finance Headlines »

In Finance»

Europe's bond yields lowest since 15th century Genoa on deflation, Russia risk

German, French and Dutch yields have been sliding for months as the eurozone recovery wilts and several countries flirt with recession

428 Comments

How banks could look

Let's try to create our first chart using the cloud service. The web interface is wizard-based and intuitive for any non-technical users. The following is the initial screen of the cloud service:

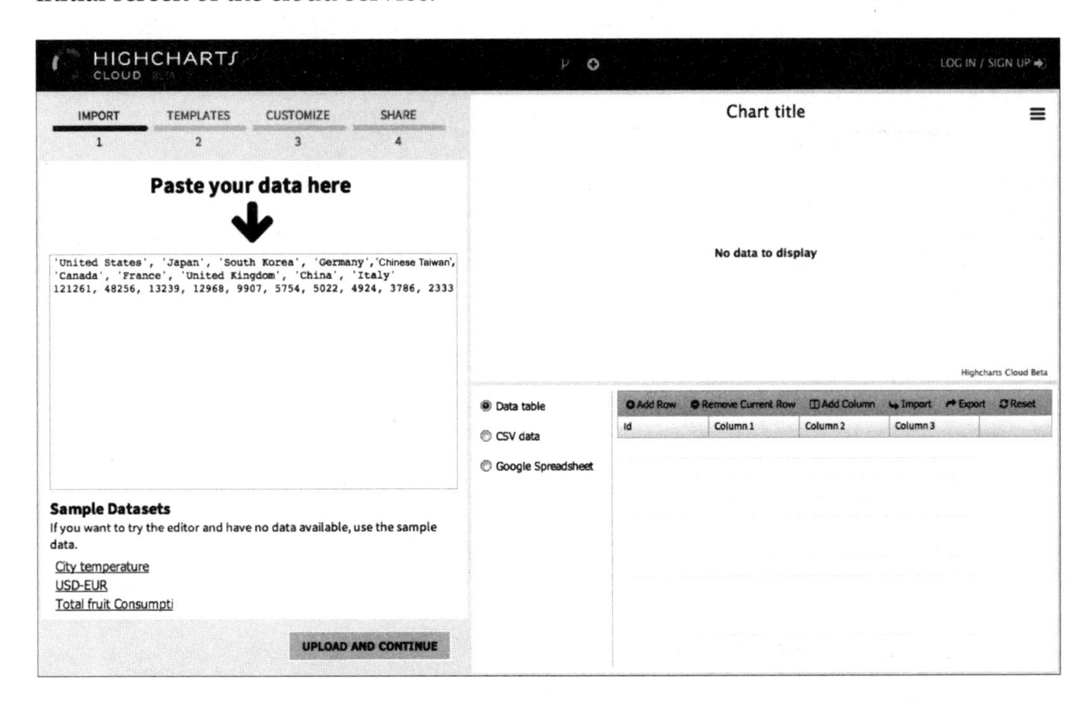

There are three major sections in the interface: the left wizard panel, the top-right result chart panel, and the bottom-right series data editor. At the top of the left wizard panel, it shows which stage we are currently in. At the first stage (**IMPORT**), we can either paste our CSV data into the text area and click on upload or manually enter series data through the bottom-right editor.

In the preceding screenshot, we have already pasted some data in the left panel. When we click on the **Upload and Continue** button, the application progresses to stage 2 (**TEMPLATES**). Here is the screenshot:

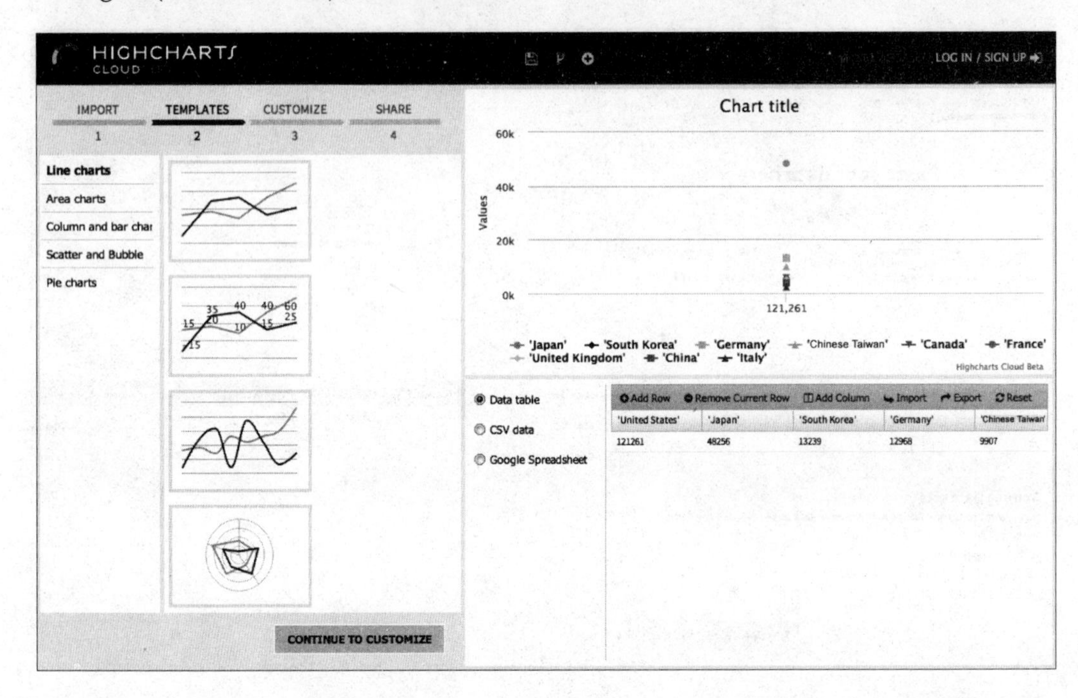

First, we can see the top-right panel updated with the default line series and the bottom-right editor panel is populated with the series data. Although the top-right chart doesn't show anything meaningful, it will become clearer as we configure the chart in a later stage. At this point, we can further edit the series data in the editor panel if we need to. Let's select a series in the left panel, **3D column chart**, which immediately updates the top-right chart:

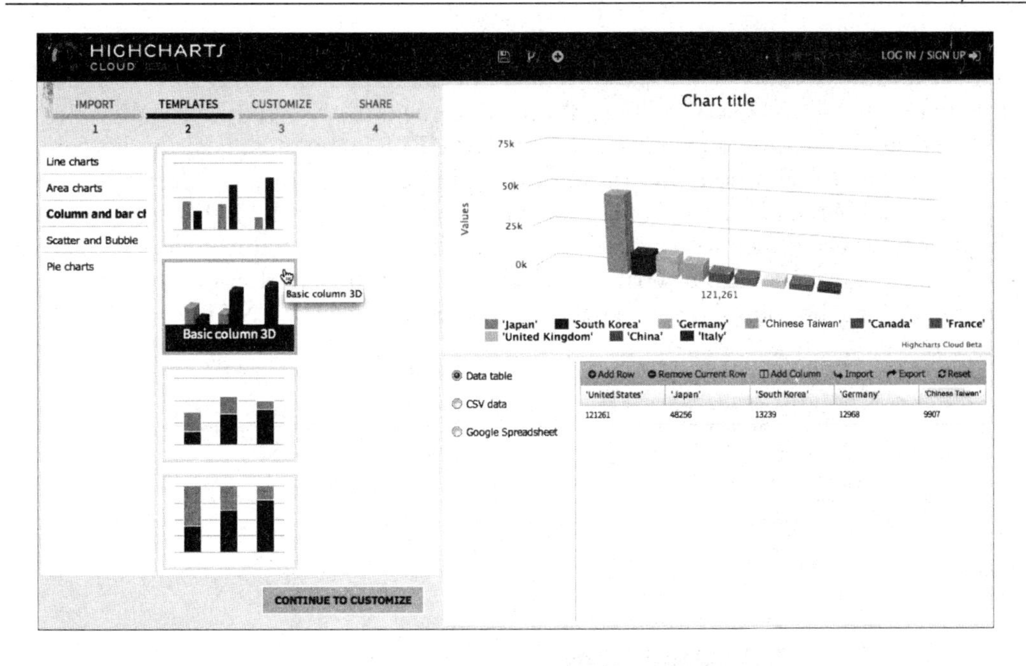

A 3D column chart is displayed in the top-right panel, but the axis and chart title are still incorrect. We can either click on the **CONTINUE TO CUSTOMIZE** button (shown in the preceding screenshot) or on **CUSTOMIZE** to go to the next stage and tune each component in the chart:

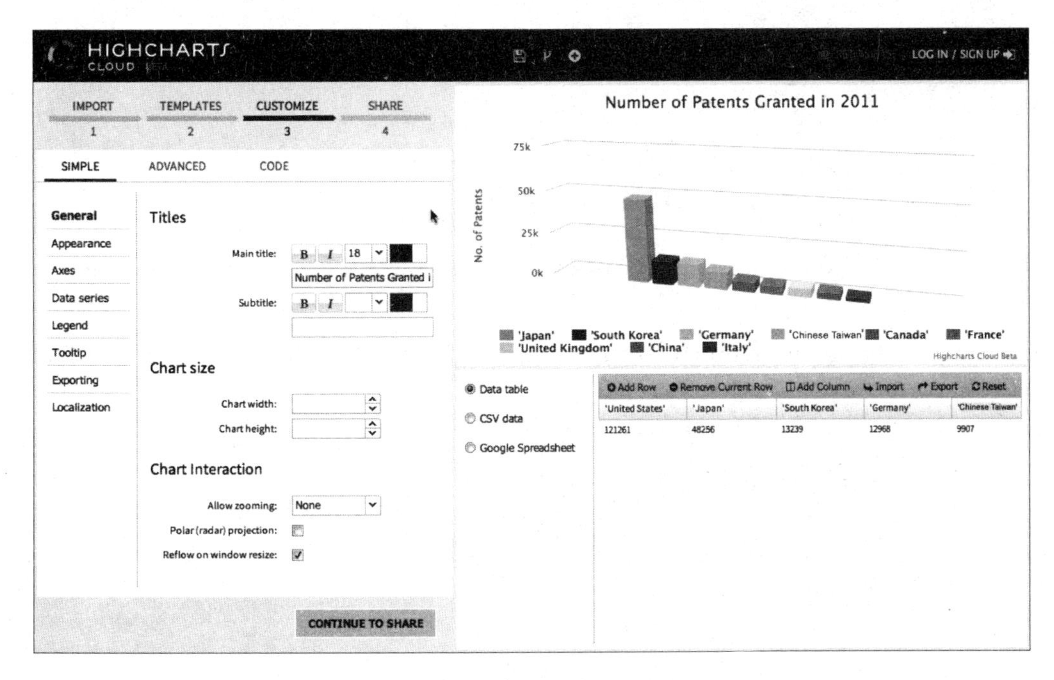

As we can see, there are different areas in the chart that we can choose to configure. In this example, we have changed the chart title, axis type, title, and label format. Note that at the **CUSTOMIZE** stage, there are three tabs shown underneath. This lets the user choose how to update the chart. **SIMPLE** is the most basic and is for non-technical users without any programming experience, or for quick simple changes. **ADVANCED** mode is for users who are familiar with Highcharts' options. The user interface is a simple properties update in name and value style. The **CODE** level is for users who wish to write JavaScript code for the chart, for example, the event handler. The following screenshot shows both the **ADVANCED** and **CODE** user interfaces:

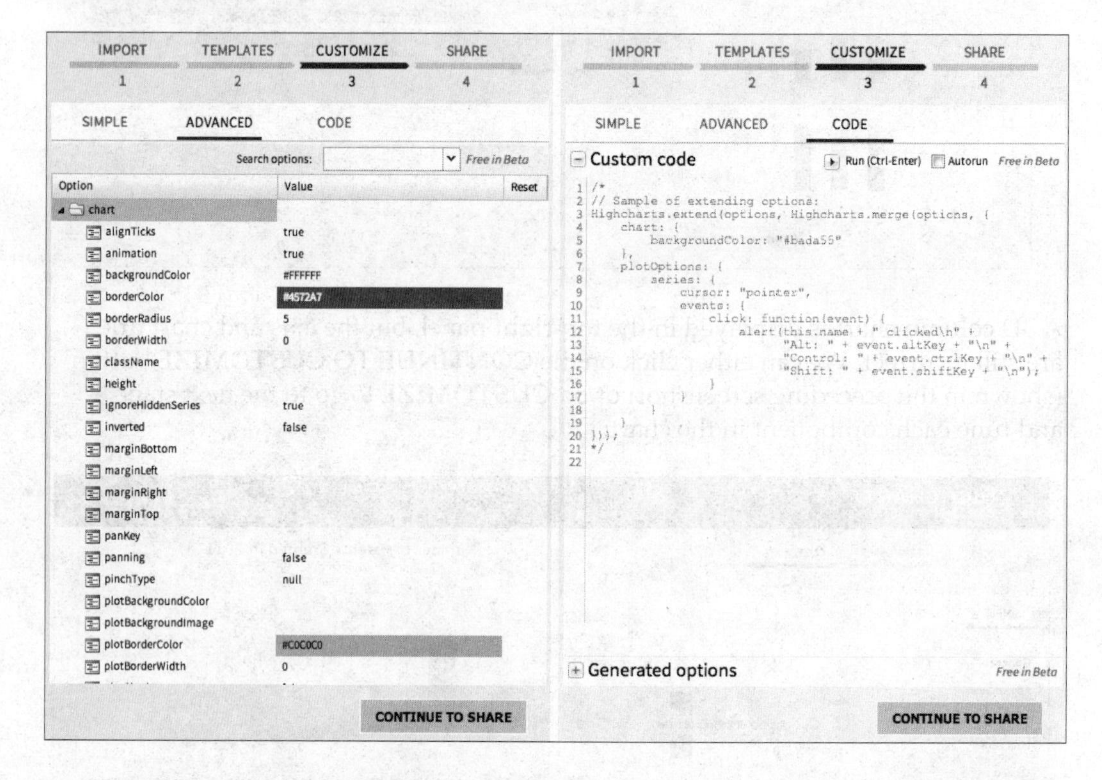

Once we are happy with our final chart, we can click on **CONTINUE TO SHARE** to generate a hyperlink for the chart.

Highcharts plugins

Highcharts can be extended through plugins that allow us to add functionality without disturbing the core layer of code and are easy to share. There is a library of plugins available online contributed by Highcharts staff and other users at `http://www.highcharts.com/plugin-registry`. One distinct advantage of developing features out of plugins is that we can pick and choose the plugin features and build a compressed JavaScript library from them. In fact, we can already do something similar with the Highcharts library on the download page.

In this section, we will take a tour of a couple of plugins that you may find handy.

The regression plot plugin

When we create a scatter plot with lots of data points, it is often worthwhile to overlay them with a regression line. Of course, we can always achieve this by adding a line series manually. However, we still need to write the code for regression analysis. It is much more convenient to include a plugin. The Highcharts regression plugin created by Ignacio Vazquez does the job nicely. First, we include the plugin:

```
<script src="http://rawgithub.com/phpepe/highcharts-regression/master/
highcharts-regression.js"> </script>
```

Then, we create our scatter chart as usual. Since we include the regression plugin, it provides additional regression options:

```
series: [{
    regression: true ,
    regressionSettings: {
        type: 'linear',
        color:  'rgba(223, 83, 83, .9)'
    },
    name: 'Female',
    color: 'rgba(223, 83, 83, .5)',
    data: [ [161.2, 51.6], [167.5, 59.0],
          ....
```

Here is the chart from the demo (`http://www.highcharts.com/plugin-registry/single/22/Highcharts%20regression`):

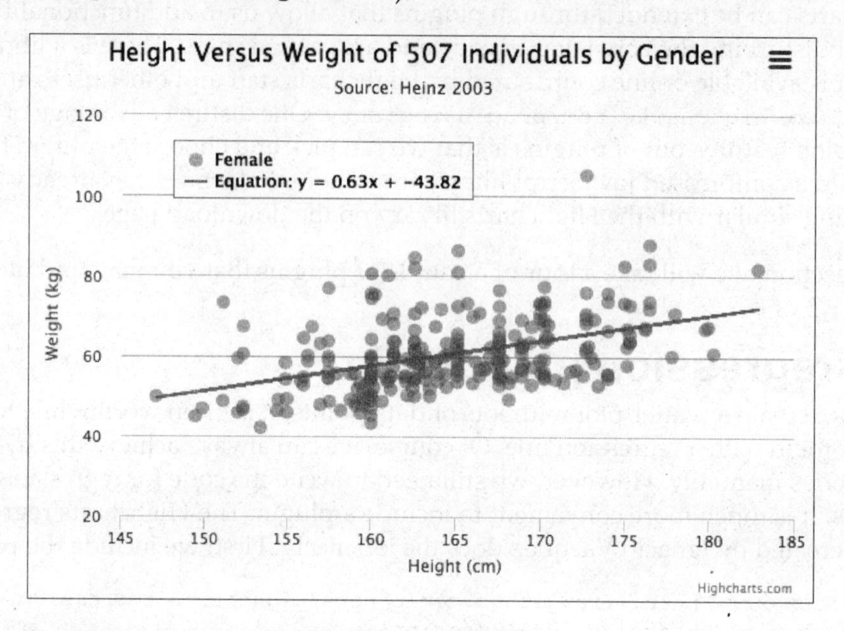

The draggable points plugin

Here is another remarkable plugin by Torstein that enables chart viewers to drag any series data points. We import the plugin with the following line:

```
<script src="http://rawgithub.com/highslide-software/draggable-points/
master/draggable-points.js"></script>
```

This plugin brings two new point events, `drag` and `drop`, which we can define the handlers for via the `plotOptions.series.point.events` option (or the `events` option in a data point object). Here is the example code from the demo:

```
events: {
    drag: function(e) {
        // Update new data value in 'drag' div box
        $('#drag').html(
            'Dragging <b>' + this.series.name + '</b>, <b>' +
            this.category + '</b> to <b>' +
            Highcharts.numberFormat(e.newY, 2) + '</b>'
        );
    },
    drop: function() {
        $('#drop').html(
```

```
                'In <b>' + this.series.name + '</b>, <b>' +
                 this.category + '</b> was set to <b>' +
                 Highcharts.numberFormat(this.y, 2) + '</b>'
            );
        }
    }
```

When we select a data point and move the mouse, a drag event is triggered and the demo code will update the textbox below the chart, as seen in the following screenshot. The plugin provides several new options to control how we can drag and drop the data points. The following is a usage example:

```
series: [{
    data: [0, 71.5, 106.4, .... ],
    draggableY: true,
    dragMinY: 0,
    type: 'column',
    minPointLength: 2
}, {
    data: [0, 71.5, 106.4, ....],
    draggableY: true,
    dragMinY: 0,
    type: 'column',
    minPointLength: 2
}, {
```

The Boolean option `draggableX/Y` notifies which direction the data points can be dragged in. Furthermore, the drag range can be limited by the `dragMinX/Y` and `dragMaxX/Y` options. The following screenshot shows a column being dragged:

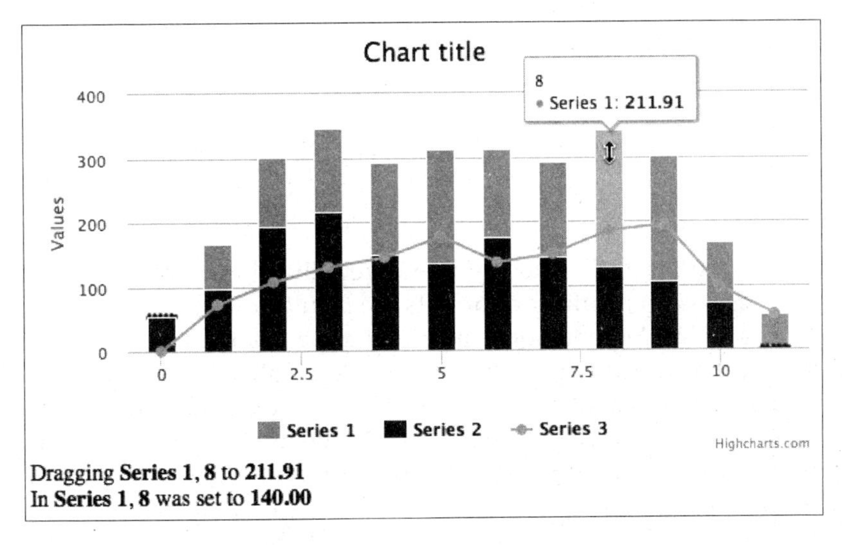

Creating a new effect by combining plugins

So far, we have seen the effect of two individual plugins. It's time for us to create a new user experience by loading these two plugins and combining their effects. The idea is to create a regression chart with movable data points, so that the regression line automatically adjusts in real time as we drag a data point. When doing so, we need to slightly modify the regression plugin code. Here is part of the original code:

```
(function (H) {

    H.wrap(H.Chart.prototype, 'init', function (proceed) {
        ....
        for (i = 0 ; i < series.length ; i++){
            var s = series[i];
            if ( s.regression && !s.rendered ) {
                // Create regression series option
                var extraSerie = {
                    ....
                };
                // Compute regression based on the type
                if (regressionType == "linear") {
                    regression = _linear(s.data) ;
                    extraSerie.type = "line";
                }
            }
        }
        // Append to series configuration array
        ....
    });

    function _linear(data) {
    ....
    }


    ....
})(Highcharts);
```

Basically, before the chart is created and rendered, the plugin scans the series data, computes the regression result, and formats the result into a line series option. To do that, the regression implementation is included as part of the `init` method for the `Chart` class, which is called when a `Chart` object is created. To extend an existing function in Highcharts, we call the `wrap` function on a method inside the object's prototype. In other words, when a `Chart` object is created, it will call the `init` function, which executes each function stacked internally (closure). We will further investigate this subject later.

For the purpose of updating the regression line at runtime, we need the ability to call `_linear` from outside the plugin. Here is a pseudo code of the new modification to add a new method, `updateRegression`:

```
(function (H) {
    H.wrap(H.Chart.prototype, 'init', function (proceed) {
     ....
    });

    H.Chart.prototype.updateRegression = function(point) {
          // Get the series from the dragged data point
          var series = point.series;
          var chart = series.chart;
          // Get the regression series associated
          // with this data series
          var regressSeries = chart.series[series.regressIdx];

          // Recompute based on the regression type and
          // update the series
          if (series.regressionType == "linear") {
            regression = _linear(series.data) ;
          }
          regressSeries.update(regression, true);
    };

    function _linear(data) {
    ....
    }

    ....
})(Highcharts);
```

Now we have a regression plugin with an accessible method, `updateRegression`, to call the inner scope function `_linear`. With this new plugin function, we can link the functionality with the `drag` event exported by the draggable plugin:

```
plotOptions: {
  series: {
    point: {
      events: {
          drag: function(e) {
             // Pass the dragged data point to the
             // regression plugin and update the
             // regression line
             Highcharts.charts[0]
```

```
                        .updateRegression(e.target);
            }
          }
        }
      }
    }
```

In order to observe the new effect more clearly, we use a smaller set of scatter plots. Here is the series configuration with both plugin options:

```
series: [{
    regression: true ,
    regressionSettings: {
        type: 'linear',
        color:  'rgba(223, 83, 83, .9)'
    },
    draggableX: true,
    draggableY: true,
    name: 'Female',
    color: 'rgba(223, 83, 83, .5)',
    data: [ [161.2, 51.6], [167.5, 59.0], [159.5, 49.2],
            [157.0, 63.0], [155.8, 53.6], [170.0, 59.0],
            [159.1, 47.6], [166.0, 69.8], [176.2, 66.8],
            ....
```

In the configuration, we have the scatter points draggable in both x and y directions and the regression type is linear. Let's load our new improved chart. The following is the initial screen:

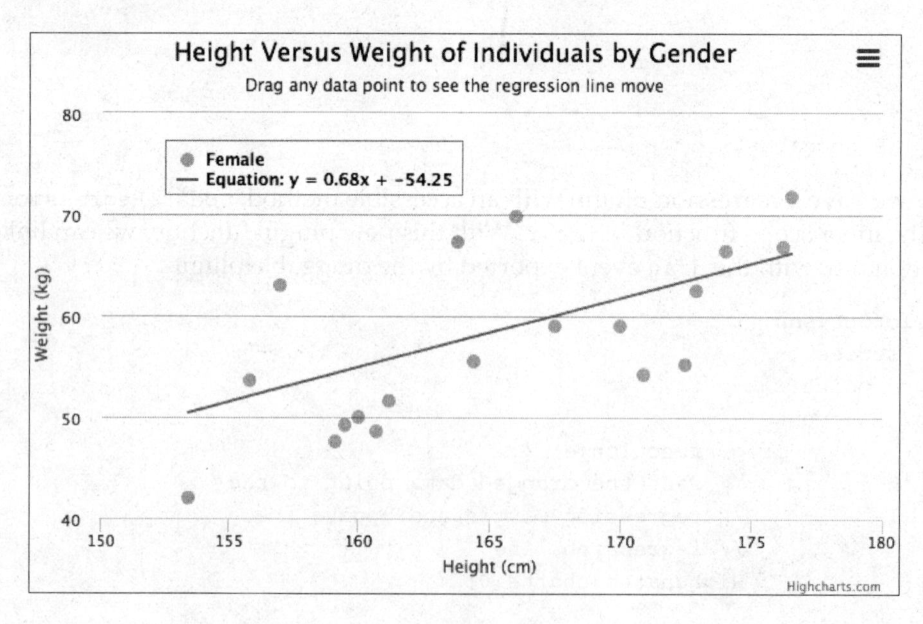

Let's hypothetically assume an overactive, unapproved slim-fast drug, "mouse down", has slipped onto the market, which has some unreported side effects. The unfortunate ones will shoot up and the really unfortunate ones have their heights gravitate. Here is the outcome of the new result:

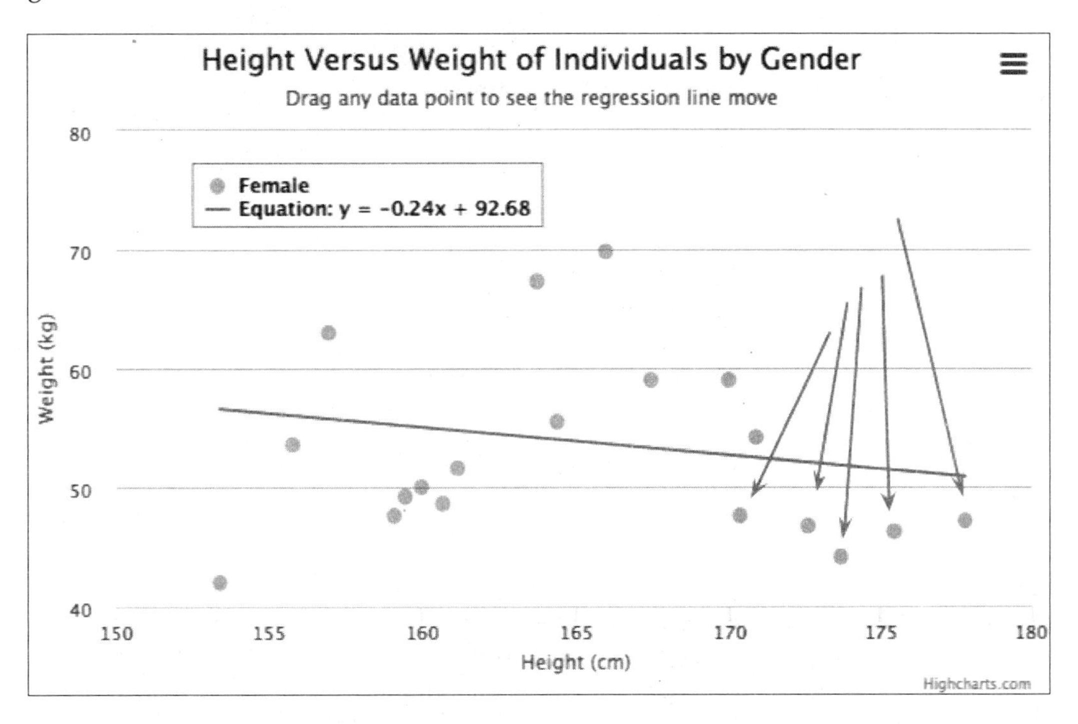

The regression line responds in real time as well as updating the top-left legend box as we mouse down those weights at the far-right data points.

Guidelines for creating a plugin

Some users create a plugin because certain tasks cannot be fulfilled by the API and the task is generic enough that it will be beneficial to other chart users. However, there is no standard API to create a plugin; developers have to be hands-on with their knowledge of Highcharts code. Nonetheless, there are a few guidelines that we can generalize from existing plugins.

Implementing the plugin within a self-invoking anonymous function

Always implement the plugin within a self-invoking anonymous function with Highcharts as the parameter. A self-invoking anonymous function is a pretty common technique in JavaScript. All the Highcharts plugins are implemented in this style. The following code shows an example:

```
(function (Highcharts) {
        ....
        // what happens in anonymous function,
        // stays in anonymous function
        function hangoverIn(place) {
            if (place === 'home') {
                return 'phew';
            } else if (place === 'hospital') {
                return 'ouch';
            } else if (place === 'vegas') {
                return 'aaahhhhh!!';
            }
        }
})(Highcharts);
```

None of the named functions and variables declared in the plugin are accessible externally because they are declared within the scope of a self-invoking anonymous function (closure and module pattern). Hence, the implementation is private to the outside world unless we assign properties in the Highcharts namespace.

Using Highcharts.wrap to extend existing functions

Depending on the plugin task, some plugins need to extend an existing function. For instance, the regression plugin calls `H.wrap` to extend the `init` function, which is called from the `Chart` constructor. See the following code:

```
(function (H) {
    H.wrap(H.Chart.prototype, 'init',
        function (proceed) {
            // Plugin specific code for processing
            // chart configuration
            ....
            // Must include this instruction for wrap
            // method to work
            proceed.apply(this,
```

```
                    Array.prototype.slice.call(arguments, 1));
            }
        );
    })(Highcharts);
```

`Highcharts.wrap` is a commonly used function within plugins. The way `wrap` functions work is to overwrite the `init` function with a new function body that includes the previous implementation `proceed`. When we extend the method with our new plugin code in an anonymous function, we have to accept the `proceed` argument, which represents the parent function body. Before or after our plugin code, we must call `proceed.apply` on the same arguments in order to complete the chain of executions.

For reference, we can always extend methods for a particular series, for example, `Highcharts.seriesTypes.column.prototype`, where `seriesTypes` is an object containing all the series classes. Alternatively, if the plugin needs to be set up for all the series, we can invoke the `wrap` method on `Highcharts.Series.prototype` instead (all series classes are extended from `Highcharts.Series`).

Using a prototype to expose a plugin method

Sometimes we may need to export specific methods for a plugin. To do so, we should always define new methods inside the `prototype` property, such as:

```
(function (H) {
     H.Chart.prototype.pluginMethod = function(x, y) {
         ....
     };
})(Highcharts);
```

This is because any code declared within the anonymous function is not accessible from the outside. Therefore, the only way to create a callable method is bound to the object passed to the anonymous function, which is the top level Highcharts object in this case. The `prototype` property is the standard way in JavaScript to inherit properties and methods from objects. The reason for attaching the method within the `prototype` property is because we don't know how developers will use the plugin. For instance, they may decide to create a new `Chart` object and call the plugin method. In such cases, the plugin code will still work.

Defining a new event handler

Another type of action for plugins is to define new events, as we saw in the draggable plugin. Here is the problem: we need access to a chart element to bind the event handler, that is, after the chart is rendered. However, the class `init` method is executed prior to the chart being rendered. This is where the `Highcharts.Chart.prototype.callbacks` array comes in. It is a place designed to store external functions that require access to elements. For example, the exporting module uses this array to insert buttons into the chart. Here is some pseudo code to set up events:

```
(function (H) {
    Highcharts.Chart.prototype.callbacks.push(function (chart) {
        // A DOM element, e.g. chart, container, document,
        var element = chart.container;
        // DOM event type: mousedown, touchstart, etc
        var eventName = 'mousedown';

        // We can call fireEvent to fire a new event
        // defined in this plugin
        fireEvent(element, newEvent, evtArgs,
                  function(args) {
                  ....
        });

        // Plugin event
        function handler(e) {
                // Plugin code
                ...
        }
        // Bind specific event to the defined handler
        Highcharts.addEvent(element, eventName, handler);
    });
})(Highcharts);
```

Highcharts has two event-related methods: `addEvent` and `fireEvent`. The former method is to bind an event to a chart element with a handler, whereas `fireEvent` triggers an event on an element.

The preceding code basically creates an anonymous function that organizes all the event(s) setup, such as defining the handler(s) and binding them to elements. The function must accept `chart` as the only parameter. Finally, we append the function into the `callbacks` array. The function will be automatically executed once the chart is fully rendered.

Summary

In this chapter, we visited the Highcharts online export server and the new cloud service. We took a short tour of the cloud service and demonstrated how we can create a chart online without any prior knowledge of JavaScript and Highcharts. Another topic covered was the Highcharts plugin in which we experimented with two plugins: regression and draggable data points. We then demonstrated how to modify one plugin to make both plugins work together to provide a new user experience.

So far, all my knowledge and experience of Highcharts is enclosed in this journal. To achieve two editions in 3 years has been much tougher than I expected. Hence, my journey with Highcharts has come to an end, and I can put my time back to where it should always be, family. My utmost gratitude to you for purchasing and reading this book. My goal was to illustrate to you how dynamic and impressive Highcharts can be, and I hope I have achieved this and ended this book on a high note.

Index

C

T

V

W

Z

Thank you for buying
Learning Highcharts 4

About Packt Publishing

Packt, pronounced 'packed', published its first book, *Mastering phpMyAdmin for Effective MySQL Management*, in April 2004, and subsequently continued to specialize in publishing highly focused books on specific technologies and solutions.

Our books and publications share the experiences of your fellow IT professionals in adapting and customizing today's systems, applications, and frameworks. Our solution-based books give you the knowledge and power to customize the software and technologies you're using to get the job done. Packt books are more specific and less general than the IT books you have seen in the past. Our unique business model allows us to bring you more focused information, giving you more of what you need to know, and less of what you don't.

Packt is a modern yet unique publishing company that focuses on producing quality, cutting-edge books for communities of developers, administrators, and newbies alike. For more information, please visit our website at `www.packtpub.com`.

About Packt Open Source

In 2010, Packt launched two new brands, Packt Open Source and Packt Enterprise, in order to continue its focus on specialization. This book is part of the Packt Open Source brand, home to books published on software built around open source licenses, and offering information to anybody from advanced developers to budding web designers. The Open Source brand also runs Packt's Open Source Royalty Scheme, by which Packt gives a royalty to each open source project about whose software a book is sold.

Writing for Packt

We welcome all inquiries from people who are interested in authoring. Book proposals should be sent to `author@packtpub.com`. If your book idea is still at an early stage and you would like to discuss it first before writing a formal book proposal, then please contact us; one of our commissioning editors will get in touch with you.

We're not just looking for published authors; if you have strong technical skills but no writing experience, our experienced editors can help you develop a writing career, or simply get some additional reward for your expertise.